Annual Reports on
NMR SPECTROSCOPY

VOLUME **69**

Annual Reports on
NMR SPECTROSCOPY

VOLUME **69**

Edited by

GRAHAM A. WEBB

Royal Society of Chemistry
Burlington House
Piccadilly, London, UK

Amsterdam • Boston • Heidelberg • London • New York • Oxford
Paris • San Diego • San Francisco • Singapore • Sydney • Tokyo
Academic Press is an imprint of Elsevier

Academic Press is an imprint of Elsevier
Linacre House, Jordan Hill, Oxford OX2 8DP, UK
84 Theobald's Road, London WC1X 8RR, UK
Radarweg 29, PO Box 211, 1000 AE Amsterdam, The Netherlands
30 Corporate Drive, Suite 400, Burlington, MA 01803, USA
525 B Street, Suite 1900, San Diego, CA 92101-4495, USA

First edition 2010

British Library Cataloguing in Publication Data
A catalogue record for this book is available from the British Library

Library of Congress Cataloging-in-Publication Data
A catalog record for this book is available from the Library of Congress

ISBN: 978-0-12-381355-8
ISSN: 0066-4103

For information on all Academic Press publications
visit our web site at books.elsevier.com

Printed and bound in the United Kingdom

Transferred to Digital Print 2010

Working together to grow
libraries in developing countries

www.elsevier.com | www.bookaid.org | www.sabre.org

ELSEVIER   BOOK AID
              International   Sabre Foundation

CONTENTS

CONTRIBUTORS

Bernard Ancian
UPMC Univ. Paris 6, Laboratoire PECSA, Physicochimie des Electrolytes, Colloïdes et Sciences Analytiques (UPMC/CNRS/ESPCI), Case 51, 4 place Jussieu, Paris, France

Qun Chen
Department of Physics, Shanghai Key Laboratory of Magnetic Resonance, East China, Normal University, Shanghai, China

K. V. Romanenko
MRI Centre, Physics Department, University of New Brunswick, Fredericton, Canada

H. Van As
Wageningen University, Wageningen, The Netherlands

J. van Duynhoven
Unilever Research and Development, Vlaardingen, The Netherlands

A. Voda
Unilever Research and Development, Vlaardingen, The Netherlands

M. Witek
Wageningen University, Wageningen, The Netherlands

Yefeng Yao
Department of Physics, Shanghai Key Laboratory of Magnetic Resonance, East China, Normal University, Shanghai, China

PREFACE

Volume 69 of *Annual Reports on NMR Spectroscopy* consists of contributions from the usual medley of areas of science where NMR studies are of great significance. The volume commences with a contribution from K. V. Romanenko on '^{129}Xe NMR Studies of Xenon Adsorption'; this is followed by a chapter from B. Ancian on 'NMR Studies for Mapping Structure and Dynamics of Nucleosides in Water'; J. van Duynhoven, A. Voda, M. Witek and H. Van As report on 'Time-Domain NMR Applied to Food Products'; finally, Yefeng Yao and Qun Chen describe 'From Helical Jump to Chain Diffusion: Solid-State NMR Study of Chain Dynamics in Semi-Crystalline Polymers'.

It is a great pleasure for me to thank all these reporters for their very interesting and timely contributions.

G. A. Webb
Royal Society of Chemistry
Burlington House
London, UK

CHAPTER 1

^{129}Xe NMR Studies of Xenon Adsorption

K. V. Romanenko

Contents

Abstract

Original ideas based on NMR spectroscopy of the probe molecules have become indispensable for characterisation of porous materials and catalysts. The most important technique has been the ^{129}Xe NMR of adsorbed xenon. This method including MAS NMR and modern MRI facilities provide unique information on a wide range of objects from atomic to macroscopic scale. Many challenging and fascinating ^{129}Xe NMR studies, published in a few decades, referred to nanoscale properties of porous media. Basic approaches

MRI Centre, Physics Department, University of New Brunswick, Fredericton, Canada

Annual Reports on NMR Spectroscopy, Volume 69
ISSN 0066-4103, DOI: 10.1016/S0066-4103(10)69001-1

of ^{129}Xe NMR of adsorbed Xe developed in the model studies of idealistic, structurally and chemically pure materials should be subject to criticism when a more complex class of objects is considered. This chapter summarises these approaches and compares them with the recent results obtained for micro- and mesoporous carbon materials.

Key Words: Xe, NMR, Adsorption, Porosity, Micropores, Mesopores, Silica, Zeolites, Porous carbon, Activated carbon, Onion-like carbon, Diamond nanoparticle, Nanotube, Exchange spectroscopy, Pd, PdCl$_2$, Knight shift.

1. INTRODUCTION

Nuclear magnetic resonance (NMR) has evolved in many directions related to specific problems of modern science. At present, NMR is among the most informative means for characterisation of diverse industrially important materials.

The chemical industry is essentially dependent on the knowledge of the structural properties of complex porous materials that are either produced or used during production. The suitability of porous materials for various applications is related to a few important properties: porosity, surface area, pore structure and chemical composition of the surface. These properties are of vital importance for catalysis wherein porous materials are regularly used as supports of the active component. Porosity control, characterisation of surface morphology, determination of the nature of the surface sites and localisation and dispersion of supported catalyst species are the most challenging issues in catalysis research.

NMR of adsorbed xenon is a relatively new technique that provides independent information on the structure of adsorbents and supported catalysts. A number of papers concerning ^{129}Xe NMR applications have appeared over the past two decades. The uses and limitations of this technique have been discussed in several critical reviews.[1–4] The ^{129}Xe NMR technique as a tool for the characterisation of porous solids was for the first time proposed by Ito and Fraissard.[5] The central idea was to find a chemically inert molecule that is highly sensitive to its environment and physical interactions with other chemical species. Xenon seems to be an ideal choice since it is an inert gas with a large electronic environment and NMR-sensitive isotopes (^{129}Xe and ^{131}Xe). The isotope ^{129}Xe has a natural abundance 26.4, a spin quantum number $1/2$ and an absolute sensitivity 5.60×10^{-3}. The ^{129}Xe NMR chemical shift is the most informative characteristic ranging over 7500 ppm (over 1500 ppm for physically adsorbed xenon). Significant sensitivity improvement arises from the possibility of Xe nuclear polarisation by the optical method.[6] Dramatic signal-to-noise ratio, however, is not always as important as quantitative information. The relaxation behaviour of the polarised spin system is different from that of a system in the thermal equilibrium. The relative intensities of the observed ^{129}Xe NMR lines may not reflect the true occupancies of the corresponding adsorption sites. For quantitative Xe NMR studies of porous media, thermal equilibrium may seem preferable.

Despite its extensive use for characterisation of various microporous materials, particularly zeolites, ^{129}Xe NMR has been rarely used to probe the properties of carbon materials and carbon-based catalysts. Due in part to the variety of structural/surface properties, paramagnetism and different impurities, porous carbons are among the most complex materials. There have been many publications about carbon and its use as an adsorbent, catalyst support or catalyst in its own right.

The general purpose of this review chapter is to survey the potential of ^{129}Xe NMR spectroscopy of adsorbed xenon for the characterisation of micro- and mesoporous materials that are attractive for catalysis applications. Particular attention has been devoted to ^{129}Xe NMR studies of porous carbon materials— microporous activated carbons and mesoporous aggregates of various carbon nanoparticles.

Quantitative analysis of ^{129}Xe NMR data is of great practical significance. Relationships between parameters of pore structure and ^{129}Xe NMR parameters were the particular objectives of many studies. Section 2 describes the general approaches of the method developed for zeolites as well as mesoporous and amorphous silica. Section 3 describes the use of ^{129}Xe NMR for the characterisation of microporous amorphous carbon materials. The associated difficulties are pointed out. Particular attention is paid to the analysis of chemical shift versus local xenon density, $\delta(\rho)$. Section 4 concentrates on the examination of the surface properties and pore structure of aggregates formed by ordered carbon nanostructures (diamond nanoparticles, onion-like structures and various carbon filaments).

2. BASIC APPROACHES OF ^{129}XE NMR SPECTROSCOPY

2.1. Chemical shift of adsorbed xenon

The perturbation of the electronic shell changes the local magnetic field at the Xe nucleus and affects the NMR frequency. As discussed by Ramsey,[7] the chemical shift is composed of a diamagnetic contribution σ_d, which depends only on the fundamental state of the electronic system and a paramagnetic contribution σ_p, which depends on the excited states and the symmetry of valence orbitals:

$$\sigma = \frac{e^2}{3mc^2} \int \frac{\rho_e}{r} dr - \frac{4}{3\Delta E} \left\langle 0 \left| \sum_{j,k} L_j \frac{L_k}{r_k^3} \right| 0 \right\rangle \tag{1}$$

where ρ_e is the electronic density, ΔE is the average energy of the excited state, L_j and L_k are the angular moment operators and r_k is the distance between the nuclei and the electron. The screening constant of the isolated xenon atom is due to σ_d only.

For most xenon interactions with other species or in xenon compounds, the electron distribution is not spherically symmetric and σ_p varies over the wide range. The nature of the atom to which the xenon is bonded influences the Xe chemical shift. Except in a few cases, the ^{129}Xe chemical shift is positive, that is, the spectra manifest downfield with respect to the resonance of the isolated Xe atom.

Generally, the relaxation times of adsorbed xenon should provide useful information about local xenon environment and dynamics. However, this is expected for extremely pure solids only. Real catalysts, particularly, carbon-based catalysts, always contain a relatively high concentration of paramagnetic impurities, dramatically reducing the relaxation times.

In the case of zeolites, Fraissard and co-workers have shown that the chemical shift of adsorbed xenon measured at room temperature is essentially the sum of several terms[8,9]:

$$\delta = \delta_{ref} + \delta_S + \delta_{Xe} + \delta_{SAS} + \delta_E + \delta_M \tag{2}$$

δ_{ref} is the reference (gaseous xenon at zero pressure). The term δ_S arises from interactions between xenon and the surface of the zeolite pores, given that the surface of the adsorbent is free of electrical charges. In that case, it depends on the xenon–surface interaction, on the dimensions of the cages or channels and on the ease of xenon diffusion at long range. The term δ_{Xe} describes the perturbation posed by xenon collisions in the confined space, δ_{SAS} accounts for strong adsorption sites, δ_E is due to electric field created by cations and δ_M describes the magnetic properties of the solid.

Similar to xenon in bulk gas phase,[10] the isotropic chemical shift of Xe adsorbed in zeolites versus local Xe density ρ is given by[11]

$$\delta(T,\rho) = \delta_S(T) + \delta_{Xe-Xe}(T) \cdot \rho + \delta_{Xe-Xe-Xe}(T) \cdot \rho^2 + \cdots \tag{3}$$

The virial coefficients δ_S, δ_{Xe-Xe} and $\delta_{Xe-Xe-Xe}$ are temperature-dependent.[12] For the pure materials, δ_S is regarded as a surface characteristic dependent on Xe diffusion rate. The term $\delta_{Xe-Xe} \cdot \rho$ arises from binary collisions of xenon atoms in a confined space and depends on the pore size and shape. The term $\delta_{Xe-Xe-Xe} \cdot \rho^2$ is significant only at very high xenon density ($\gg 10\,mmol\,cm^{-3}$).

For a system of isolated pores, δ_S is the average value of the chemical shift of Xe in rapid exchange between pore volume and surface:

$$\delta_S = \frac{N_a \delta_a + N_V \delta_V}{N_a + N_V} \tag{4}$$

where δ_a and δ_V are the chemical shifts of adsorbed and gaseous xenon, respectively; N_a and N_V are the corresponding occupancies of adsorbed and gaseous (volume) states.

A quantitative 'rapid exchange model' proposed by Cheung et al.[13] describes the chemical shift of Xe adsorbed on zeolites as a function of density and temperature.

$$\delta(T,\rho) = \frac{N_{ads}\theta}{N}\delta_a + \frac{N - N_{ads}\theta}{N}\delta_g \tag{5a}$$

where θ is the fraction of the N_{ads} adsorption sites occupied on average and N is the total number of xenon atoms present in the available volume V. The term

δ_g is defined as $\delta_{Xe-Xe} \cdot \rho$.[8] An expression for θ was derived from a condition of adsorption equilibrium:

$$k_a(1 - \theta)\rho_g = k_d\theta$$

where $\rho_g = (N - N_{ads}\theta)/V$; k_a and k_d are the adsorption and desorption constants, respectively and ρ_g is the gas-phase xenon density. If we denote $n_a = N_{ads}/V$ and $b = k_a/k_d$, Equation (5a) can be rewritten as

$$\delta(T, \rho) = \frac{bn_a\delta_a + (1 + b\rho_g)\delta_{Xe-Xe}\rho_g}{1 + bn_a + b\rho_g} \tag{5b}$$

In the case of weak adsorption, $b\rho_g \ll 1$ and Equation (5b) reduces to

$$\delta(T, \rho) = \delta_S + \left(\frac{\delta_{Xe-Xe} - b\delta_S}{(1 + bn_a)^2} \right)\rho \tag{5c}$$

where $\delta_S = bn_a/(1+bn_a)$. In the case of a strong adsorption, $b\rho_g \gg 1$, one obtains

$$\delta(T, \rho) = \frac{n_a\delta_a + \delta_{Xe-Xe} \cdot n_a^2}{\rho} - 2n_a\delta_{Xe-Xe} + \delta_{Xe-Xe} \cdot \rho \tag{5d}$$

The initial decrease of the chemical shift with increasing density in the case of strong adsorption sites is described by the $1/\rho$ dependence (confirmed experimentally).

The rapid exchange model is not valid in the presence of strong chemical shift anisotropy (CSA). The CSA effects on the spectra usually appear for Xe in small micropores and decrease with the atomic motion rate. When the pore size is large enough, only isotropic shift is observed because of the effective averaging of the chemical shift tensor. The anisotropy effects and corresponding line-shape change upon Xe occupancy of micropores (6.7×4.4 Å) in crystalline aluminium phosphate ALPO-11 are reported in Refs. 14,15. The calculations of average ^{129}Xe chemical shielding tensor in nanochannels were proposed in Refs. 16,17.

2.2. Variation of the ^{129}Xe chemical shift due to porosity and temperature

The most important feature of the ^{129}Xe chemical shift is its dependence on pore dimensions. Successful prediction of the pore size from NMR data would be of great value in the characterisation of novel porous materials, as adsorption measurements are not always effective. In the initial studies,[18] δ_S was correlated with the mean free path of xenon $\bar{l}$ imposed by the micropore structure of zeolites and molecular sieves. In the case of zeolites, the empirical equation is given by

$$\delta_S = 243 \times \frac{0.2054}{0.2054 + \bar{l}} \tag{6}$$

Later, a similar relationship was proposed for porous silica-based materials—silica gels, porous glasses and porous organosilicates.[19] It is valid over the range 0.5–40 nm with a relatively large error.

In a model proposed by Liu et al.,[20] a virial expansion of the chemical shift is exploited,[9] with adsorbed xenon treated as two-dimensional gas. Derouane et al. considered the effects of surface curvature of the zeolite pores on the value of δ_S.[21,22] Other models are based on Lennard-Jones potential curve calculations.[23,24]

The use of existing chemical shift correlations for accurate measurements of pore diameters appeared questionable.[25] Ripmeester et al. have shown that prediction of the pore dimensions based on δ_S values is not sufficiently reliable. Numerical calculations showed that δ_S reflects the true void space of relatively narrow pores alone, that is, when their size is close to the van der Waals diameter of Xe atom. For large cages, the chemical shift was found to be a complicated function of void space, sorption energy and temperature. Thus, no unique correlation of δ_S with the pore size (D) should be expected.

A model describing the ^{129}Xe chemical shift as a function of temperature was proposed for zeolites[23,24] and mesoporous silica materials.[26,27] In the latter case, the chemical shift was proposed in a simple form derived from Equation (4), if $\delta_V \ll \delta_a$:

$$\delta_S = \frac{N_a \delta_a + N_V \delta_V}{N_a + N_V} = \frac{\delta_a}{1 + (D/\eta KkT)} \tag{7}$$

where $N_V = PV/kT$ is the amount of Xe atoms in the pore volume; P is the equilibrium xenon pressure; $N_a = KPS$ is the amount of surface Xe atoms; D is the effective pore size defined as $D = \eta V/S$, where η is a proportionality coefficient, depending on the geometry; and K is the Henry constant.

Cheung[28] showed that at 144 K pure amorphous materials (silica and alumina) may show an initial decrease in ^{129}Xe chemical shift with xenon loading due to pore sizes distribution.

The other model proposed by Cheung[24] describes the chemical shift of a single xenon atom confined in slit, cylindrical and spherical micropores:

$$\delta(T) = \frac{C\varepsilon}{1 + F \exp(-\varepsilon/T)} \tag{8}$$

where $F = (L' - D_{Xe})/(2lm' - 1)$; D_{Xe} is the xenon diameter (4.4 Å); $m' = 1, 2$ and 3 for slit, cylindrical and spherical pore geometries, respectively; L' (Å) is the free pore size for a considered pore geometry; and ε (K) and l (Å) are the potential well depth and width, respectively. The constant C (ppm K^{-1}) depends on ionisation potentials of xenon and the surface atom.[29] However, for Cheung's model where the square-well potential is used, C should be considered as a phenomenological parameter.

According to Cheung's calculations, the chemical shift can be an increasing or decreasing function of temperature depending on the pore size. If the pore size is close to the xenon atom radius, the chemical shift increases with increasing temperature. However, it decreases for sufficiently large pores, that is, $L' + 2R_a > 2.6(R_{Xe} + R_a)$, where R_{Xe} and R_a are the van der Waals radii of Xe and surface atoms, respectively.

The second virial coefficient δ_{Xe-Xe} in Equation (2) can also be considered as an important characteristic of the material studied. It describes the interaction of xenon atoms in pores and, therefore, depends on the pore size and shape which influence the xenon collision frequency.

The higher order chemical shift terms in the virial expansion obtained for the gas-phase xenon under standard conditions (298 K and 1 atm pressure)[10] are much more pronounced than those of ^{19}F or ^{1}H, due to xenon's much higher polarisability. The value of 0.55 ppm amagat^{-1} was obtained for δ_{Xe-Xe}. Theoretical calculations of the chemical shift carried out by Adrian[30] evidenced that during the lifetime of a binary xenon collision, short-range electron exchange forces were responsible for the large density-dependent paramagnetic resonance shift.

First attempts to analyse the experimental δ_{Xe-Xe} values with respect to the porosity looked promising. In the work of Julbe et al.,[31] the slope δ_{Xe-Xe} was correlated with the pore size of a sol–gel-derived amorphous microporous silica. The authors concluded that δ_{Xe-Xe} decreases with increasing pore size. These conclusions are questionable, since the discussed δ_{Xe-Xe} values were calculated from the chemical shift as a function of xenon pressure. By definition, δ_{Xe-Xe} is a slope of δ versus local Xe density ρ. The local Xe density ρ does not vary linearly with Xe pressure in the full pressure range. This is particularly true for microporous samples. Nevertheless, on the basis of numerical calculations, the authors emphasised that the term δ_{Xe-Xe} should be a better probe of the micropore size than δ_S.

An original approach to calculate the size of narrow one-dimensional channels was proposed by Chen et al.[32] It is based on the confined one-dimensional collision model. In this model, the channel diameter D_C (nm) is related to δ_{Xe-Xe} (ppm amagat^{-1}) as

$$D_C = 1.681 \times \delta_{Xe-Xe}^{1/2} \tag{9a}$$

Or using δ_{Xe-Xe} in ppm cm^3 mmol^{-1} (1 amagat $= 0.045$ mmol cm^{-3}):

$$D_C = 0.356 \times \delta_{Xe-Xe}^{1/2} \tag{9b}$$

It is remarkable that δ_{Xe-Xe} is the increasing function of the pore size. This dependence is related to the single-file diffusion (SFD) phenomenon.[33–35] For molecules in a channel with a diameter less than twice the molecular size, the diffusion character is different from that of diffusion in much wider pores. As δ_{Xe-Xe} is proportional to the xenon collision frequency (ν), in a one-dimensional channel it increases with increasing collision cross-section area (σ): $\delta_{Xe-Xe} \sim \nu \sim \sigma \sim D_C^2$.

Equation (9a) was tested on the experimental data published earlier for various microporous zeolites. The pore diameters were calculated using experimental δ_{Xe-Xe} values and compared with the channel characteristics measured by X-ray diffraction (XRD). In most cases, this equation provides a better estimate of the pore size than what is provided by Equation (6), given that the pore size calculated from XRD is very precise. For Y zeolite, Equation (9a) gives the pore size with a relatively large error (~ 0.3 nm), which is not surprising since the distribution of Xe–Xe collisions in the zeolite supercages is assumed to be isotropic.

2.3. Xenon exchange spectroscopy

The xenon diffusion in the porous media can be studied with ^{129}Xe NMR exchange spectroscopy.[36,37] This technique was extensively applied to study the structural heterogeneities and the dynamical features that are inherent in porous materials such as various microporous solids, liquid crystal systems and amorphous polymers.

The relevant timescale τ for ^{129}Xe NMR measurements is defined by the reciprocal frequency separation of resonance lines w_1 and w_2 arising from different adsorption sites:

$$\tau = (w_1 - w_2)^{-1}$$

Resolved peaks from two different Xe environments in the same sample are observed when the exchange time is lower than τ; faster exchange gives rise to a single NMR line.

The 2D exchange experiment is based on a stimulated echo pulse scheme (Figure 1). Initially, transverse magnetisation is prepared by a $\pi/2$ radio frequency (rf) pulse. During the evolution period t_1, magnetisation vectors acquire a frequency-dependent phase. After the second $\pi/2$-pulse (z-storage), the magnetisation is longitudinal. The exchange process takes place during the mixing period t_m. This delay is fixed within a 2D experiment. A third $\pi/2$-pulse rotates the longitudinal magnetisation into the $x-y$ plane for detection during the period t_2. The experiment has to be repeated for a number of equally spaced values of the evolution time t_1 leading to a data matrix $s(t_1, t_2)$. The desired 2D-exchange spectrum $S(w_1, w_2)$ is obtained by 2D Fourier transformation of $s(t_1, t_2)$. Off-diagonal peaks in the 2D spectrum appear as a result of exchange within the mixing period t_m.

Evolution of the longitudinal magnetisation $\mathbf{M}_z$ is described by the equation:

$$\frac{\mathrm{d}}{\mathrm{d}t}\Delta M_z(t) = \hat{L}\Delta M_z(t) \tag{10}$$

where $\Delta\mathbf{M}_z(t) = \mathbf{M}_z(t) - \mathbf{M}_0$, $\mathbf{M}_0$ is the vector of equilibrium magnetisation along magnetic field z-axis and $\hat{L}$ is the dynamic matrix presented by the sum of chemical exchange and relaxation matrices $\hat{L} = \hat{K} - \hat{R}$. Chemical exchange and relaxation processes are described by the kinetic matrix $\hat{K}$ and relaxation matrix $\hat{R}$, respectively. The longitudinal relaxation involves two different processes: the nucleus may relax independently of other nuclei with relaxation rate $1/T_1$ due to external relaxation; a direct dipolar interaction of nuclei causes dipolar relaxation.

For a one-spin system with two non-equivalent sites A and B, the first-order chemical exchange is assumed:

$$A \underset{k_{BA}}{\overset{k_{AB}}{\rightleftarrows}} B$$

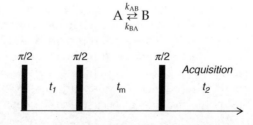

Figure 1   2D exchange spectroscopy pulse scheme.

$$\mathbf{L} = \begin{pmatrix} -k_{AB} - \dfrac{1}{T_{1A}} & -k_{BA} \\[2ex] -k_{AB} & -k_{BA} - \dfrac{1}{T_{1B}} \end{pmatrix}$$

where k_{AB} and k_{BA} are the rate constants, and T_{1A} and T_{1B} are the longitudinal relaxation times. Thus, the dynamic matrix $\mathbf{L}$ has eigenvalues $\lambda_{\pm}$:

$$\lambda_{\pm} = -\sigma \pm D$$

where

$$\sigma = \frac{1}{2}\left(k_{AB} + k_{BA} + \frac{1}{T_{1A}} + \frac{1}{T_{1B}}\right); \quad \delta = \frac{1}{2}\left(k_{BA} - k_{AB} + \frac{1}{T_{1A}} - \frac{1}{T_{1B}}\right);$$

$$D = (\delta^2 + k_{AB}k_{BA})^{1/2}$$

A ratio of integral intensities of the cross-peaks I_{AB} and diagonal peaks I_{AA} as a function of mixing period can be derived:

$$\frac{I_{AB}}{I_{AA}} = \frac{k_{AB}[1 - \exp(-2Dt_m)]}{(D - d) + (D + d)\exp(-2Dt_m)} \tag{11}$$

Using this equation the exchange dynamics parameters can be evaluated from the experimental values of I_{AB}/I_{AA} versus mixing period t_m. If $k_{AB} = k_{BA} = k$ and $T_{1A} = T_{1B} = T_1$, Equation (11) turns into a simple form, which can be used for the estimation of exchange rate constants:

$$\frac{I_{AA}}{I_{AB}} \approx \frac{1 - kt_m}{kt_m} \tag{12}$$

EXSY (Exchange Spectroscopy) was effectively used to study inter-crystalline xenon diffusion in a mixture of NaX and NaY zeolites.[38] From the experimentally measured exchange rate constants, the authors were able to estimate effective diffusion coefficients of 1.3×10^{-8} and 1×10^{-8} m^2 s^{-1} for NaX and NaY, respectively. The variable temperature measurements of exchange rate constants allowed the effective activation energies of exchange processes to be estimated at 3–4 kJ mol^{-1}. These results are consistent with PFG (pulse field gradient) NMR results.[39] For example, for NaX the diffusion coefficient value was 7×10^{-9} m^2 s^{-1} and the activation energy was 6 kJ mol^{-1}. A considerably larger value of activation energy was obtained for NaA zeolite, 65 kJ mol^{-1}.[40] This was explained by a smaller size of the window connecting two adjacent cavities.[41]

2.4. NMR of adsorbed species as a probe for metal clusters in porous materials

Supported metals constitute an important class of heterogeneous catalysts. Adsorption isotherm analysis,[42] IR of chemisorbed molecules,[43] electron microscopy,[44,45] Mossbauer spectroscopy[46] and EXAFS[47] are traditionally used to examine the state of supported metals.

Because of the technical difficulties associated with NMR detection of metals, the metallic properties were tested indirectly by NMR of the detectable chemisorbed phases (^{1}H, ^{13}C) or physically adsorbed probes (^{129}Xe). ^{1}H is the most NMR-sensitive nucleus and hydrogen is one of the most used gases in catalytic systems. Fraissard and co-workers[48,49] proposed the use of hydrogen as a universal probe of supported metal. Subsequent technical progress has enabled Slichter[50] and Van der Klink et al.[51] to run original experiments with chemisorbed molecules on supported platinum. Many authors have applied these methods to platinum and other metals.[52–56]

The distribution of fine particles (a few atoms) is hardly detectable by electron microscopy, but it can be studied by ^{129}Xe NMR of xenon co-adsorbed with hydrogen or other species.[57] The ^{129}Xe NMR technique is particularly interesting for determining the number of supported metal particles and the quantitative distribution of phases chemisorbed on them. The ability to explore NMR of adsorbed molecules is based on the Knight shift phenomenon in metals.[58–62]

Knight shift measures the magnetic hyperfine field at the nucleus produced by the polarised conducting electrons[58,59]:

$$K = \frac{8\pi}{3} \langle |\psi(0)|^2 \rangle_F \chi_P$$

where $\langle |\psi(0)|^2 \rangle_F$ is the average density of the conduction electron wave functions with the Fermi energy at the nucleus and ξ_P is the Pauli susceptibility of the conduction electrons. The Knight shift K of metal nuclei in the metal host is measured with respect to the NMR frequency of the same metal nuclei in a salt. The shift can be positive or negative depending on the polarisation of the conduction band by d-electrons. An NMR-active nucleus of a molecule adsorbed on a metal would show a Knight shift with respect to this nucleus in a non-magnetic environment. A survey of ^{1}H, ^{13}C, ^{2}D and ^{129}Xe NMR applications based on Knight shift phenomenon is given by Khanra.[60]

3. MICROPOROUS CARBON MATERIALS

Three forms of carbon are traditionally used as supports for precious-metal catalyst to provide optimum properties and performance in the chemical industry. These are activated carbon, carbon black and graphite or graphitised materials. Characterising amorphous microporous carbons is a classical problem in adsorption research. The particular difficulty with determining the micropore size is related to the very complex theoretical approaches to describe adsorption. Another challenge is the stringent requirements that adsorption equipment should meet.

^{129}Xe NMR is sensitive to the nature of organic molecules and their density inside zeolite channels.[63] Incomplete carbon oxidation resulting in carbonaceous deposits (coking) in zeolite pores has been discussed in Refs. 64,65. ^{129}Xe NMR studies of activated carbons and coals,[66–70] carbon black materials,[71,72] hard carbon,[73] carbon replicas of Y zeolite[74] and various carbon filaments including carbon nanotubes[75-79] have been reported. ^{129}Xe NMR of polyacenic semiconductor materials,[80] and optically polarised ^{129}Xe NMR tests of graphitised carbon surface[81] are worth mentioning as well. Most of these studies implied the strategies developed for ordered porous solids, particularly zeolites. A few studies were aimed at finding the relationships between ^{129}Xe NMR data and surface properties of carbon materials.

Microporous amorphous carbons, like many zeolites, are characterised by a linear relationship between ^{129}Xe chemical shift and local xenon density. Ryoo and co-workers[69] showed that the chemical shift extrapolated to zero density is dependent on the nature of the carbonaceous surface for several types of amorphous carbon. No correlations were found between pore size distribution in amorphous carbons and the ^{129}Xe chemical shift.

In contrast to pure silicates, the interpretation of ^{129}Xe NMR data obtained on porous carbon is complicated for several reasons: structural disorder—distribution of the crystallite size or presence of amorphous domain; heterogeneity of surface properties—presence of various surface groups and surface structures (basal and edge); conductivity; and, in some cases, strong paramagnetic sites. The paramagnetic properties of carbon materials originate from natural structural defects and paramagnetic impurities of inorganic nature.

3.1. Commercially available amorphous carbon materials

The following example demonstrates some difficulties in the analysis of ^{129}Xe NMR data obtained for amorphous carbons (AR). The commercially available amorphous carbon materials discussed in Ref. 82 are: AR produced from coconut shells (Sutcliffe Speakman Carbon, Ashton-in-Makerfield, UK); Nor produced from peat (Norit, Amersfoort, The Netherlands) and Sib produced by hydrocarbon pyrolysis followed by gas activation[83,84] (trademark: Sibunit, Russia). The textural parameters obtained by the analysis of N_2 adsorption data are given in Table 1. The presence of micropores in these samples is confirmed by the strong dependence of the chemical shift on Xe density ρ. The linear fit parameters δ_S and δ_{Xe-Xe} are given in the Table 1. In addition, the values of δ_S and δ_{Xe-Xe} adopted from Ref. 69 are also included. These values were obtained for samples denoted as SA (Strem Chem.), DA (Aldrich–Dargo G-60) and CA1 (Shin Ki Chem.). In contrast to most of these samples, two lines are resolved in the ^{129}Xe NMR spectra of Xe adsorbed on Nor (Figure 2).

It is hard to claim the presence of reliable correlations between ^{129}Xe NMR data and textural parameters. The chemical shift approximated to zero density, δ_S, is expected to correlate with the pore dimensions of amorphous carbons as it is for zeolites. The mean pore sizes of all these samples, except Sib, are very close. This could explain the similarity of the δ_S values. However, the δ_S value obtained for Sib seems to be too high for the mean pore size of 5.7 nm.

The ^{129}Xe NMR line-shapes observed for Nor (Figure 2A) and WA1 were very similar. Two ^{129}Xe NMR lines of WA1 were attributed to macroscopic inhomogeneity of the environment experienced by xenon on NMR timescale. Coalescence of the two separate lines occurred after careful grinding and mixing of the sample.

TABLE 1   Textural and [129]Xe NMR parameters of commercial amorphous carbons

Sample	D (nm)	W (cm³ g⁻¹)	S_{BET} (m² g⁻¹)	δ_S (ppm)	δ_{Xe-Xe} (ppm cm³ mmol⁻¹)
Sib[a]	5.7	0.53	415	47	15.4
Nor[a]	1.6	0.13	1119	46 (line 1)	2.1
				57 (line 2)	3.1
AR[a]	1.5	0.12	1456	43	1.5
SA[b]	1.74	0.427	1140	46	6.3
DA[b]	1.62	0.242	742	50	5.5
CA1[b]	1.48	0.256	527	49	8.1

D, the mean pore size; Wt, the total specific pore volume; S_{BET}, BET specific surface area; δ_S, the zero density chemical shift approximation; δ_{Xe-Xe}, the slope of the chemical shift versus density.
[a] Data from Ref. 82.
[b] Data from Ref. 69.

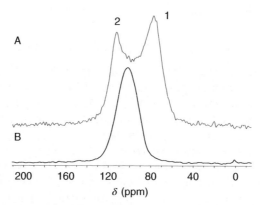

Figure 2   [129]Xe NMR spectra of xenon adsorbed on commercial activated carbons Nor (A) and AR (B) at a pressure of 100 kPa. Figure adapted from Ref. 82.

This explanation was reasonable; however, the dynamics of xenon exchange should be analysed for more reliable conclusions on the nature of different adsorption domains.

According to estimations done by N_2 adsorption analysis, the average pore size of Nor is 1.6 nm (Table 1). This could cause sufficiently high Xe mobility at room temperature to average Xe interactions over length scales on the order of micrometers. When Xe exchange between two sites is 'rapid' (as defined by their chemical shift difference), the observed chemical shift is a time-weighted average of the shifts in the absence of exchange. Thus, in order to ascertain whether peak shifting has occurred, it is necessary to know how rapid the exchange process is. This is conveniently done using two-dimensional exchange spectroscopy.

The off-diagonal peaks in the two-dimensional spectrum cannot be observed unless exchange takes place on the timescale of the mixing period used. The 2D ^{129}Xe EXSY spectra of Nor acquired at room temperature are shown in Figure 3. Weak intensities of the cross-peaks observed for $t_m = 10$ ms indicated a near-absence of xenon exchange on this timescale (Figure 3A). The rate of exchange between the two sites was lower than approximately $(10\ \text{ms})^{-1} = 100$ Hz. The intensive cross-peaks were observed for a mixing period of $t_m = 400$ ms (Figure 3B). Since the peak separation is $4830\ \text{Hz} \gg (400\ \text{ms})^{-1} = 2.5$ Hz, the peaks have not undergone appreciable shift due to xenon exchange.

The chemical shift δ_S depends essentially on the chemical state of the carbon surface and possible surface defects. Indeed, Ryoo and co-workers[69] demonstrated that δ_S varied depending on the surface state of a carbon sample. Acidic functional groups were introduced to the carbon surface by the treatment of samples with nitric acid. The oxidative treatment of the activated carbons resulted in the increase of chemical shift by 20 ppm without substantial change in the slope δ_{Xe-Xe}. The term δ_{Xe-Xe} should depend on pore connectivity and size. The use of δ_{Xe-Xe} for the characterisation of zeolites was reported by Chen et al.[32] It is remarkable that the slope δ_{Xe-Xe} varies at least by an order of magnitude from one sample to another (Table 1). From these data, it seems that δ_{Xe-Xe} is much more sensitive to the pore structure of amorphous carbons than δ_S.

The lines '1' and '2' in the ^{129}Xe NMR spectrum of Xe adsorbed on Nor are characterised by different δ_S and δ_{Xe-Xe} values (Table 1). Therefore, the presence of two domains different both in pore structure and chemical state of the surface could be predicted. The exchange between these domains is very slow: ~ 2.5 Hz.

The correspondence between ^{129}Xe chemical shifts and mean pore sizes obtained by traditional methods such as N_2 adsorption analysis has not been reported in the literature. This can be rationalised by the discrepancy of the pore

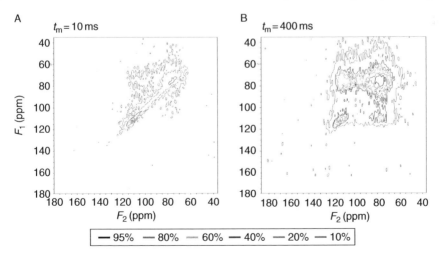

Figure 3   2D ^{129}Xe exchange spectra of Xe adsorbed on *Nor* acquired using mixing periods of 10 and 400 ms; xenon pressure ~ 100 kPa. Figures adapted from Ref. 82. (See Plate 1 in Color Plate Section)

structures and surface compositions as a result of different origins and activation procedures. In addition, N_2 adsorption analysis relies on various simplifications, which often give unreliable estimation of mean pore diameters.

3.2. Pitch-based activated carbon modified by air oxidation/pyrolysis cycles

Structural disorder of microporous carbon materials, their bad reproducibility and various impurities make the interpretation of ^{129}Xe NMR data ambiguous. A series of activated carbons (AC) obtained by successive air oxidation/pyrolysis cyclic treatments of a unique precursor were tested with ^{129}Xe NMR.[68] The virial coefficient δ_{Xe-Xe} arising from binary xenon collisions appeared to be a better probe of the microporosity than the chemical shift extrapolated to zero pressure. The first clear correlation between δ_{Xe-Xe} and the mean micropore size was proposed for a particular set of activated carbons. The preliminary studies of commercially available amorphous carbon materials and those obtained by air oxidation/pyrolysis cycles showed that the use of the ^{129}Xe chemical shift approximated to zero density is very limited. It is more sensitive to the chemical state of the surface and to the ease of Xe diffusion than to the pore size.

The N_2 adsorption analysis used for characterisation of the AC series is described in detail in Ref. 85. The AC series presents narrow pore size distributions (PSD) and is of particular interest for carbon molecular sieve (CMS) applications. The samples conserve the same basic carbonaceous structure and surface chemistry. The mean pore size increase per cycle was 0.16 nm on average (Figure 4A). The regular increase of specific total volume, microporous volumes, BET (Brunauer, Emmett, Teller theory) specific surface and external surface was observed as well, except at the last cycle when the thin micropore walls collapsed. The specific total and microporous volumes are displayed in Figure 4B.

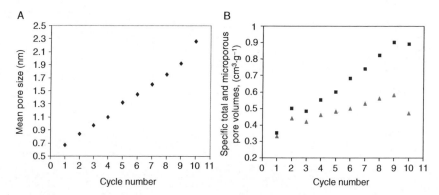

Figure 4   Evolution of (A) the average pore size (◆) and (B) specific total (■) and microporous (▲) pore volumes obtained by N_2 adsorption analysis versus the activation cycle number. Figures adapted from Ref. 68.

The ^{129}Xe NMR spectra of xenon adsorbed on AC series exhibited a single line over a wide range of xenon densities (ρ). The line-width narrowed with increased cycle number. The line-width and line-shape asymmetry increased with Xe pressure for the first four treatment cycles. The line-shape asymmetry features related to the CSA and the pore structure heterogeneity are not easily separated by line-shape analysis. The former contribution is expected for the narrowest micropores only.[86–89] When the pore size is large enough only, the isotropic shift is observed as a result of the effective averaging of the chemical shift tensor. The ^{129}Xe NMR spectrum of sample AC(10) was remarkably widened compared to those of the other samples. This clearly indicates some dramatic pore structure changes in the last treatment cycle.

The density dependences of the chemical shift were linear over a wide density range corresponding to Xe pressures from 0 to 100 kPa. The density ρ was calculated as the amount of adsorbed xenon divided by the total pore volume Wt of the sample considered. The calculated parameters δ_S and δ_{Xe-Xe} were compared with the corresponding mean pore sizes determined by N_2 adsorption, Figures 5 and 6, respectively. Surprisingly, no regular dependence of δ_S on the mean pore size was observed: δ_S varied within 10 ppm over the pore size range 0.6–2.3 nm. Although in the case of zeolites and various molecular sieves the variation of δ_S is not perfect either,[90] it is generally monotonic with the pore size. On the contrary, the slope δ_{Xe-Xe} varied very regularly with the pore size (Figure 6). A linear relationship between δ_{Xe-Xe} and D was used to describe these experimental data:

$$\delta_{Xe-Xe} = \eta + \kappa D \tag{13}$$

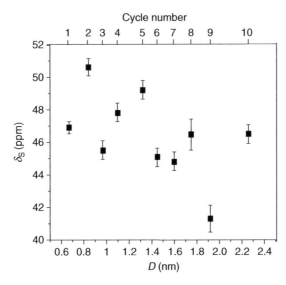

Figure 5 ^{129}Xe NMR chemical shifts approximated to zero xenon density versus mean pore size. The treatment cycle numbers are given on the upper axis. Figure adapted from Ref. 68.

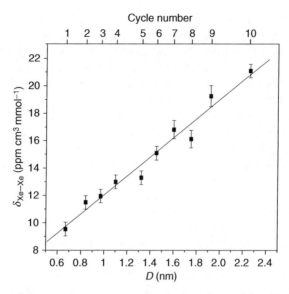

Figure 6 Experimental δ_{Xe-Xe} values versus mean pore size obtained for the series of activated carbons (■), the treatment cycle numbers are given on the upper axis. The linear fit of the experimental data $\delta_{Xe-Xe} = 5.1 + 7D$ is shown as a solid line (—). Figure adapted from Ref. 68.

where δ_{Xe-Xe} and D are used in units of ppm cm^3 mmol^{-1} and nm, respectively; κ is 7 ± 0.2 ppm cm^3 mmol^{-1} nm^{-1} and η is 5.1 ± 0.3 ppm cm^3 mmol^{-1}. The term δ_{Xe-Xe} is proportional to the xenon collision frequency and, therefore, to the molecular collision cross-sectional area σ. The latter is a geometrical factor that can be calculated according to the known pore structure. Therefore, the character of $\delta_{Xe-Xe}(D)$ should reflect the pore geometry. The positive slope led the authors to a logical conclusion that the Xe collision frequency increases with the mean pore size.

The pore structure of activated carbons is very complex and irregular.[91] It is, therefore, hard to compare the obtained δ_{Xe-Xe} with that measured for highly ordered materials like zeolites. For cylindrical pores, the term δ_{Xe-Xe} can be derived from Equation (9a):

$$\delta_{Xe-Xe} = 7.89D^2 \, (\text{ppm cm}^3 \, \text{mmol}^{-1}) \tag{14}$$

This equation is valid only in the range $D_{Xe} \leq D \leq 2D_{Xe} \sim 9$ Å, where D_{Xe} is the xenon diameter. Therefore, the maximum $\delta_{Xe-Xe}(D)$ for cylindrical channels is $\delta_{Xe-Xe}(2D_{Xe}) \sim 6.1$ ppm cm^3 mmol^{-1} (Equation (14)). The δ_{Xe-Xe} value established for bulk xenon[92] is 0.548 ppm amagat^{-1} = 12.2 ppm cm^3 mmol^{-1}.

It is doubtful that the $\delta_{Xe-Xe} - D$ correspondence discussed in Ref. 68 could be rationalised in terms of the simple Xe collision model. Xenon diffusion in activated carbons should be considered as a motion in a complex three-dimensional pore framework taking into account interaction with walls. The structural heterogeneity of the carbon pore walls should be taken into account as well.

The temperature dependence of the chemical shift, $\delta(T)$, is sensitive to the shape of the potential experienced by the xenon atom. Examination of the activated carbons pore geometry could be useful by means of ¹²⁹Xe NMR if one could find a suitable approach. For instance, the slit-like pore geometry, often considered for activated carbons, could be examined with the model proposed by Cheung.[24] However, the adequacy of this model would be questionable since the carbon pore structure is heterogeneous.

Variable temperature (VT) chemical shift measurements were performed for the AC series at a xenon density of 0.5 mmol cm⁻³ (3 atoms per 10 nm³). Low Xe density is commonly used in VT experiments in order to minimise the contribution of binary Xe collisions to the chemical shift. The local Xe density is kept constant by minimisation of the sample free volume.

The chemical shift changed by 2 and 9 ppm over the temperature range 150–304 K for samples AC(1) and AC(9), respectively (Figure 7). The observed chemical shift variations with temperature are qualitatively consistent with the Cheung's model.[24] The range of the chemical shift variation with temperature increases with the heterogeneity of the xenon potential inside the pores. Indeed, the chemical shift is almost independent of temperature for AC(1) representing the smallest micropores of 0.67 nm. This can be ascribed to high potential homogeneity inside the micropores, when no preferential localisation of xenon occurs at decreasing temperature. A similar observation was reported for the zeolite ZSM5. The chemical shift varied within 6 ppm over the temperature range 173–373 K.[93] The similarity of the results could be rationalised by the close pore sizes of AC(1) and ZSM5.

The temperature dependence of the chemical shift established for AC series is different from zeolites if wider pores are considered. Though the temperature influence exhibited for AC(9) is more pronounced than that for AC(1), it is

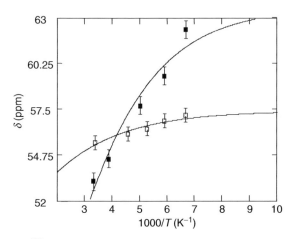

Figure 7 Experimental ¹²⁹Xe NMR chemical shifts versus reciprocal temperature: AC(1) (□) and AC(9) (■). The curves represent the least-squares fits of Equation (8) to the experimental data. Figure adapted from Ref. 68.

substantially weaker than those observed for microporous zeolites of a proximate pore size and mesoporous or amorphous silica.[26,27,93,94] For instance, a chemical shift variation of 50 ppm over the temperature range 173–373 K is reported for Y zeolite (diameter of supercages: $D = 1.3$ nm). This is remarkably greater than the range of 9 ppm observed for AC(9) ($D = 1.92$ nm).

Based on the Cheung's model, the authors excluded from consideration the slit-like pore geometry for the samples from AC series. The pore structures of the samples AC(1) and AC(9) were better described by cylindrical and spherical geometries, respectively. Indeed, the modelling performed by Py et al.[91] showed that the pore morphology of carbons obtained by air oxidation/pyrolysis cycles is not slit-like. The edge/edge and edge/face nanomorphologies of porosity available between disc-like basic structural units were proposed.

4. MESOPOROUS CARBON MATERIALS

4.1. Testing the nanoparticles surface: Diamond and onion-like carbon aggregates

Mesoporous carbons is an important class of materials with the pore size scaling from 2 to 50 nm. The ^{129}Xe NMR chemical shift of adsorbed xenon is known to be very sensitive to the presence of functional surface groups and strong adsorbtion sites.

Well-characterised, structurally similar nanodiamond (UDD) and onion-like carbon (OLC) samples present a good model system for ^{129}Xe NMR studies.[95] High concentration of different oxygen- and hydrogen-containing functional groups on the surface of diamond nanoparticles makes the essential difference between UDD and OLC. The OLC nanoparticles, as well as UDD, form aggregates. The pore structures responsible for the adsorption properties are very similar since the OLC sample is obtained by high temperature annealing of UDD in vacuum. The transmission electron microscopy (TEM) images of UDD and OLC are shown in Figure 8A and B, respectively. The surface of OLC is graphitised and similar to the surface of fullerenes, but not as perfect. In contrast to UDD, the OLC particles are free from functional groups and represent chemically pure carbon. To expose the graphite edge faces, the initial sample OLC was oxidised in a flow of O_2 and N_2. The oxidised sample is denoted OLC-ox. The parameters of the texture obtained with the N_2 adsorption analysis are given in Table 2. The PSD are shown in Figure 9. This data evidence that UDD and OLC aggregates has very similar PSD ranging from 3 to 20 nm. It is worth mentioning that the bulk structure of primary nanoparticles is not responsible for the adsorption properties of their aggregates. Therefore, the ^{129}Xe NMR technique is not sensitive to the bulk structure of nanoparticles of UDD and OLC samples since xenon adsorption occurs only in the aggregate pores, that is, between the particles.

The different surface compositions of OLC and UDD resulted in the sharp contrast of ^{129}Xe NMR data. The ^{129}Xe NMR spectrum of optically polarised xenon adsorbed on UDD powder is shown in Figure 10A. The ^{129}Xe polarisation method

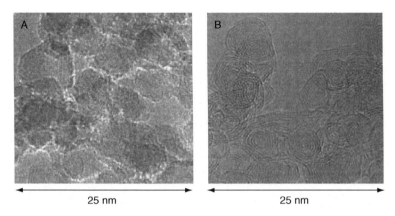

Figure 8   TEM images of aggregated nanoparticles UDD (A) and OLC (B). Figures adapted from Ref. 91.

TABLE 2   Textural properties of diamond and onion-like nanoparticles obtained by N_2 adsorption analysis

Sample	S_{BET} $(m^2\,g^{-1})$	D (nm)	Wt $(cm^3\,g^{-1})$
UDD	288	15.4	1.12
OLC	326	10.4	0.83
OLC-ox	440	9.7	1.04

S_{BET}, the BET specific surface area; D, the average pore diameter; Wt, the total specific pore volume; UDD, ultradispersed diamond; OLC, onion-like carbon.

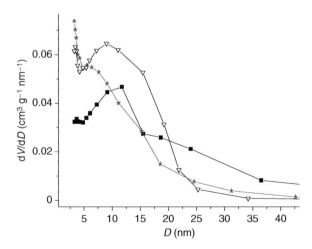

Figure 9   Pore size distributions obtained by N_2 adsorption analysis for UDD (■), OLC (⋆) and OLC-ox (△). Figure adapted from Ref. 91.

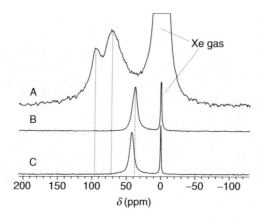

Figure 10 ^{129}Xe NMR spectra of Xe adsorbed on UDD (A) optical polarisation, OLC (B) and OLC-ox (C) at a pressure of 30 kPa. Figure adapted from Ref. 91.

was used to compensate for the low amount of the available UDD sample. An intense line near 0 ppm corresponds to polarised gaseous Xe. Two broad lines ($\sim$2000–3000 Hz) at 95 and 75 ppm correspond to adsorbed Xe. The chemical shift decrease of 10 ppm (higher field) was indicated with Xe pressure varied over the range from 1 to 70 kPa. The ^{129}Xe NMR spectra of Xe adsorbed on OLC and OLC-ox contain single, relatively narrow lines (Figure 10B and C), that shift downfield with Xe density. The narrow resonance lines are indicative of the fast exchange of xenon between the adsorption domains of OLC/OLC-ox. The chemical shifts approximated to zero density, δ_S, were 35.7$\pm$0.5 ppm (OLC) and 39.4$\pm$0.5 ppm (OLC-ox). The oxidation treatment of OLC led to an increased δ_S for OLC-ox. This is consistent with the increase of adsorption enthalpy. Enthalpies of xenon adsorption estimated from the adsorption isotherms were 16.3 and 24 kJ mol^{-1} for OLC and OLC-ox, respectively.

Despite the texture similarity of UDD and OLC, their ^{129}Xe NMR linewidths, chemical shifts and $\delta(\rho)$ dependences are remarkably different. These results led the authors to several conclusions. The chemical shift of adsorbed xenon (δ_S) cannot be unambiguously used to probe the pore size of mesoporous carbon materials with wide pore size distribution and heterogeneous surface composition. On the other hand, δ_S is very sensitive to the chemical state of the surface of mesoporous carbon materials that commonly present wide pore size distribution.

4.2. Filamentous carbon: Surface, porosity and paramagnetic impurities

The texture of most porous carbon adsorbents is heterogeneous, particularly because of the presence of several types of carbon surface, that is, different graphitic facets (basal and edge). The adsorption properties of these facets are considerably different. The chemical shift δ_S is expected to be an average characteristic of the surface structure.

The chemical shift δ_S was shown to depend on the relative contribution of edge and basal facets to the total surface area.[76] A series of catalytic filamentous carbons (CFC) obtained by methane decomposition on iron subgroup metal catalysts[96] was chosen for model ^{129}Xe NMR studies. The angle (α) between the constituent crystallite plane and the filament axis strongly depends on the catalyst composition. The filaments are built up by the crystallite planes, arranged as cones or tubes put one into one another (Figure 11). The ^{129}Xe NMR study revealed the dependence of ^{129}Xe chemical shift on the structure of carbon surface, particularly on the orientation of graphite crystallites. The CFC samples with angles (α) between graphitic planes and a filament axis of $0°$, $20°$, $45°$, $90°$ were denoted as CFC-0, CFC-20, CFC-45 and CFC-90, respectively. The chemical shift follows a general trend to increase with the adsorption potential of the surface. However, attention should be paid to paramagnetic impurities. As a result of the fibre formation, deactivated catalyst particles are embedded in carbon granules. Considerable paramagnetic contribution to the ^{129}Xe chemical shift was reported for carbon filaments with metal content as high as 2 wt%. Quantities of paramagnetic sites, *Nspin*, obtained by ESR are listed in Table 3.

Graphite edges do not form any slit pores with diameter $\sim$0.343 nm, as would be expected for graphite single crystal facets. The linkage of nearest or next-nearest surface edges is thermodynamically favourable. The Gibbs free energy of the edge face is roughly 40 times larger than that of the basal one, 6.3 and 0.16 J m^{-2}, respectively.[97] The so-called closed-layer structures are commonly observed with high-resolution TEM (HRTEM).

CFC are produced as a skein of chaotically interlaced filaments. The porous texture of CFC is presented mainly by mesopores. Bimodal PSD are composed of dense ($\sim$3–5 nm) and loose (10–30 nm) aggregates (Figure 12). The average pore size, total specific volumes (*Wt*) and micropore (*Wo*) volumes are listed in Table 3. The large contribution of loose aggregate pores to the total porosity is generally a characteristic for samples produced at lower carbon yield. The large free space is progressively filled by the filaments during the reaction, leading to the decrease of the average pore size and the dense pore texture.

The analysis of Xe adsorption isotherms (Figure 13), confirmed that the interaction of xenon with carbon surface depends on the surface structure. The parameter of the Langmuir's model of monolayer adsorption, $\gamma_{max}b$, describing the adsorption strength of the adsorbent, tends to increase with the angle between the crystallite planes and the filament axis, α.

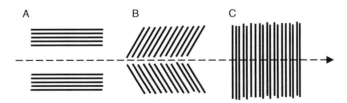

Figure 11 Various structures of catalytic filamentous carbons, CFC. The angles between constituent crystallite plane and filament axis α are $\alpha = 0°$ (A), $\alpha = 45°$ (B) and $\alpha = 90°$ (C).

TABLE 3 NMR, adsorption and EPR characteristics of filamentous carbons, CFC

Sample (angle α)	D (nm)	Wt (cm³ g⁻¹)	W_0 (cm³ g⁻¹)	S_{BET} (m² g⁻¹)	K, $\gamma_{max}b$ (m⁻² kPa⁻¹)	δ_S (ppm)	$\nu_{1/2}$ (Hz)	$Nspin$ (g⁻¹)	G (g/g$_c$)
CFC-0	17.5	0.663	0.011	151	1.14×10^{16}	24.6	3500	0.48×10^{20}	145
						35.9	1000		
CFC-20	20.1	0.585	0.001	116.4	0.78×10^{16}	77.6	5000	2.4×10^{20}	45
CFC-45	9.9	0.255	0.007	102.4	2.4×10^{16}	53	2400	1.1×10^{20}	100
CFC-90	6.1	0.466	0.007	307.5	3.1×10^{16}	67.9	860	0.34×10^{20}	220

α, the angle between the crystallite plains and the filament axis; D, the average pore diameter; Wt, the total specific pore volume; W_0, the specific micropore volume; S_{BET}, the BET specific surface area; K, Henry constant; γ_{max}, the monolayer capacity; b, the adsorption equilibrium constant; δ_S, the chemical shift approximated to zero loading; $\nu_{1/2}$, full width at half maximum; $Nspin$, the number of paramagnetic sites; G, catalyst efficiency or carbon capacity, weight of produced carbon per gram of catalyst.

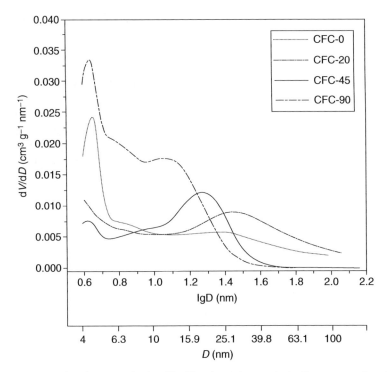

Figure 12 Pore size distributions obtained by N$_2$ adsorption analysis. Figure reproduced from Ref. 76.

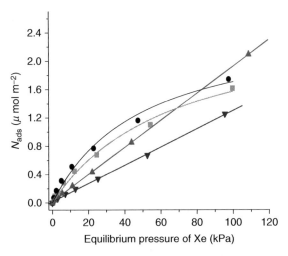

Figure 13 Isotherms of Xe adsorption (amount per square metre): (▲) CFC-0, (▼) CFC-20, (■) CFC-45, (●) CFC-90, lines correspond to the least-square data fit according to the Henry's and Langmuir's laws.

The ^{129}Xe NMR spectra of Xe adsorbed on CFC-0, CFC-20, CFC-45 and CFC-90 at a pressure of 100 kPa are shown in Figure 14. Table 3 contains the line-widths and δ_S values extracted from experimental $\delta(\rho)$ dependences.

The CFC samples have small microporous volume and the average pore sizes are in the range 6–20 nm. In the mesopore range (pore diameter > 2 nm), both the surface type and the porosity influence the chemical shift. However, any correlation of the chemical shift with the mean pore size would be doubtful, since the samples present wide bimodal PSD. The surface structure as well as the chemical composition of the surface should affect the ^{129}Xe chemical shift of adsorbed xenon.

The correspondence between the surface structure, adsorption potential of the surface and the chemical shift was indicated for CFC-0, CFC-45 and CFC-90 (Table 3, Figure 15). The authors rationalised this result in terms of the adsorption

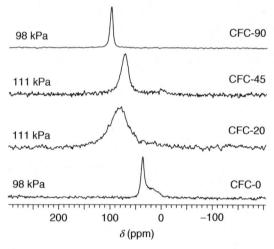

Figure 14 ^{129}Xe NMR spectra of xenon adsorbed on CFC-0, CFC-20, CFC-45, CFC-90. Figure adapted from Ref. 76.

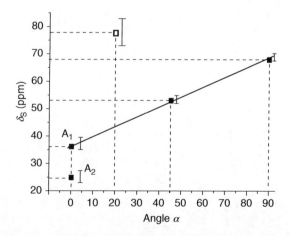

Figure 15   The chemical shift approximations to zero loading, δ_S, versus angle α.

properties of basal and edge faces. The adsorption potential of the edge faces is larger than that of the basal face.[98] If α is between $0°$ and $90°$, the filament surface structure can be considered as a combination of basal, edge and closed-layer edge faces.[99,100] Therefore, under the fast exchange condition, the chemical shift δ_S is a measure of the average surface structure of carbon filaments. This assumption is confirmed by the chemical shift obtained for CFC-45, $\delta_S = 53$ ppm, which is close to the average of the values obtained for $\alpha = 0°$ and $90°$, respecively (Table 3). The sample CFC-20 fell out of the observed trend. Most likely it was because of the high concentration of paramagnetic impurities. Strongest paramagnetic contribution is confirmed by EPR, $Nspins$, full-width at half-maximum of the NMR line, $v_{1/2}$ and carbon capacity of a catalyst, G (Table 3).

Similar results were reported by Simonov et al.[101] The chemical shifts obtained for graphitised carbon black, step-like roughened surface and filamentous carbons with exposed edge faces were 14, 50 and 100 ppm, respectively. Although these values were measured at a fixed Xe density ~ 100 μmol g^{-1}, that is, without extrapolation of chemical shift to zero density, they support the relationship between the chemical shift and the surface structure.

The $\delta_{Xe-Xe} \cdot \rho$ term depends not only on the density of adsorbed xenon but also on the probability of Xe–Xe collisions and therefore on the distribution of Xe atoms in the pores and their diffusion. The coefficient δ_{Xe-Xe} was shown to increase with the mean pore size in the case of micropores.[68] In the case of mesopores, the opposite dynamics was reported[76,95] (Figure 16). A simple model describing these data has been proposed.[102]

The number of Xe atoms in the pore volume, N_v, is negligible compared to the number of atoms on the surface, N_S. Thus, the collision frequency at the surface, v_S, is much greater than that in the pore volume. Consideration of the chemical shift as a function of the surface Xe density, γ, would seem logical:

$$\delta(\gamma) = \delta_S + \delta_{Xe-Xe}^{Surf} \cdot \gamma \tag{15}$$

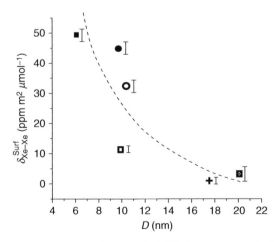

Figure 16 δ_{Xe-Xe}^{Surf} versus average pore diameter: (+) CFC-0, (⊠) CFC-20, (□) CFC-45, (■) CFC-90, (○) OLC, (●) OLC-ox. Dashed curve is for convenience only. Figure adapted from Ref. 102.

This expression is very similar to Equation (3). The term $\delta_{Xe-Xe}^{Surf} \cdot \gamma$ describes the binary xenon collision at the carbon surface. If Xe atoms spend times t_S and t_V on the surface and in the pore volume, respectively, the time-averaged collision frequency is given by Equation (16):

$$\bar{v}_S = \frac{v_S t_S}{t_S + t_V} = \frac{v_S}{1 + t_V/t_S} \quad (16)$$

where t_V is the average time between two adjacent collisions with the surface, that is, t_V is proportional to the mean free path imposed by the structure, $\bar{l}$. $\bar{l}$ is proportional to the pore diameter D. Finally, one obtains an expression for δ_{Xe-Xe}^{Surf} with two phenomenological parameters φ and ϑ:

$$\delta_{Xe-Xe}^{Surf} = \frac{\vartheta}{1 + \xi D} \quad (17)$$

4.3. Multi-wall carbon nanotubes

4.3.1. Identification of adsorption domains

The adsorption on carbon nanotubes is commonly addressed in the literature as a problem of identification of the adsorption domains: inter-crystalline or aggregate pores corresponding to the external surface and one-dimensional channels. The important role of aggregate pores in adsorption, capillarity, or other physico-chemical properties was accentuated in Ref. 103. This type of pore appears to have a large adsorption capacity responsible for 78.5% of the total adsorbed amount. Understanding the aggregation mechanisms and inter-crystalline pore structure is necessary for development of efficient gas-storage materials. Different sorbates were used for volumetric measurements in order to identify the adsorption sites of single-wall carbon nanotube bundles—interstitial channels, grooves and remaining outer parts.[104] Carbon nanotubes may seem to be an ideal geometrical model of cylindrical pores.[105] However, analysis of the inner cavity filling from the usual adsorption measurements is quite complicated.[106] Porosity of as-made nanotubes can hardly be ascertained using N_2 adsorption.[107] Adsorption data seem to be more informative with regard to the surface characteristics than to the porosity. Several studies report the native nanotubes as being closed.[107,108] The carbon nanotube channels become accessible to guest molecules only after chemical or mechanical treatments: uniform burning-out of the tips by mild oxidation,[109] ball milling,[110] or thermal activation.[111] The purification procedures also may result in the removal of the nanotubes tips.[112]

A few attempts have been made to characterise multi-wall carbon nanotubes (MWCNT) with ^{129}Xe NMR.[77,78] The reported ^{129}Xe NMR results were somewhat similar to those obtained for mesoporous silica.[19,26,27] This similarity is based on the validity of the fast exchange approximation for Xe interacting with the MWCNT surface. Xenon was shown to be a reliable probe of accessible nanotubes channels and aggregate texture of MWCNT. Adsorption of Xe in aggregate (or inter-crystalline) pores and inside MWCNT was distinguished by ^{129}Xe NMR chemical shift and exchange dynamics analysis.

Two MWCNT samples (referred to as CNT and CNT(O)) were obtained in the catalytic decomposition of C_2H_4 over an iron–cobalt catalyst at 700 °C.[113] A thorough analysis of the TEM images of the sample CNT(O) (not shown) revealed nanotubes with open tips. The sample CNT presented only closed nanotubes. As result of the ball-milling procedure, the average nanotubes length reduced from 5 to 0.5 μm.

The following results were adapted from Ref. 77. Figures 17A and 18A display the [129]Xe NMR spectra of Xe adsorbed on as-made carbon nanotubes CNT and CNT(O). The [129]Xe NMR spectra of ball-milled nanotubes, CNT-BM and CNT(O)-BM are displayed in Figures 17B and 18B. After ball milling, the spectra changed dramatically for both samples.

Two principal adsorption domains were considered: (1) the aggregate porosity corresponding to the external nanotube surface and (2) the interior of the nanotubes (nanotube channels) accessible for adsorption after chemical or mechanical treatments. The NMR lines ($x_{0,1,2}$, $y_{1,2}$ and $z_{1,2}$) were assigned to these adsorption domains based on the general trend of the chemical shift to decrease with the pore size. This correspondence was supported by exchange dynamics studies, variable temperature chemical shift measurements, N_2 adsorption analysis and electron microscopy.[77]

The relatively low chemical shift values ($\sim$10 ppm) correspond to Xe in fast exchange between the surface of loosely packed nanotube (large aggregate mesopores) and the gas phase. Dense nanotubes packing could be distinguished from loose packing by the additional low-field shift of 10–20 ppm. Since some fraction of the as-made nanotubes CNT(O) had open tips, additional resonance lines manifested at 40–50 ppm. These results are summarised in Table 4.

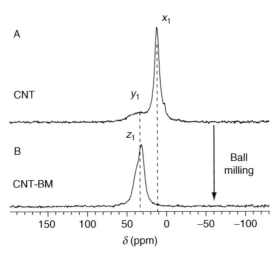

Figure 17 [129]Xe NMR spectra of Xe adsorbed on CNT (A) and CNT-BM (B) at 100 kPa. Figure adapted from Ref. 77.

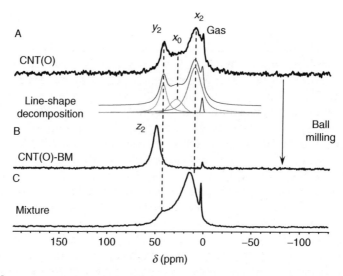

Figure 18 ^{129}Xe NMR spectra of Xe adsorbed on CNT(O) (A), CNT(O)-BM (B), blend of CNT(O) and CNT(O)-BM (C, D) at 100 kPa; the signal-to-noise ratios are different due to the differences in accumulation number and adsorption. Figure adapted from Ref. 77.

TABLE 4   The ^{129}Xe NMR chemical shifts assignment to the carbon nanotube adsorption sites

	CNT	CNT (O)	Assignment
As-made	11 ppm (x_1)	10 ppm (x_2)	Inter (loose aggregates)
	35 ppm (y_1)	30 ppm (x_0)	Inter (dense aggregates)
		43 ppm (y_2)	Intra (long tubes)
Ball-milled	35 ppm (z_1)	50 ppm (z_2)	Exchange inter–intra-edge

The chemical shift changed by 10 ppm over the wide range of Xe density (0–0.2 mmol g^{-1}). These observations were indicative of fast exchange between the pore volume and the carbon surface.

The exchange between nanotube channels and aggregate pores of CNT(O) was slow and two lines, x_2 and y_2, were distinguished (Figure 18A). The exchange rate between these domain sites was estimated with EXSY. The 2D ^{129}Xe EXSY spectrum obtained at a mixing time of 10 ms is shown in Figure 19.

The effects of ball milling on nanotube pore structure and Xe exchange dynamics were examined. The nanotube channels became accessible to Xe as indicated by xenon volumetric adsorption studies and chemical shift increase. A single line (z_1 or z_2) observed in the spectra (Figures 17B and 18B) was indicative of fast exchange between nanotube channels and aggregate pores. The average nanotube length after ball milling was 0.5 μm. The Xe residence times inside ball-milled nanotubes estimated with Einstein's diffusion equation was 10^{-7} s, a negligible quantity on the timescale of NMR experiment. The low-field contribution

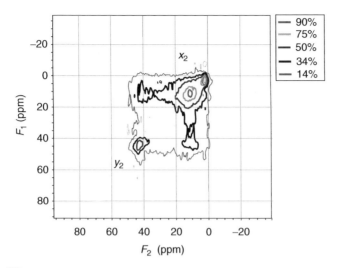

Figure 19   2D ^{129}Xe exchange spectrum of CNT(O) using a 10-ms mixing period. Figure adapted from Ref. 77. (See Plate 2 in Color Plate Section)

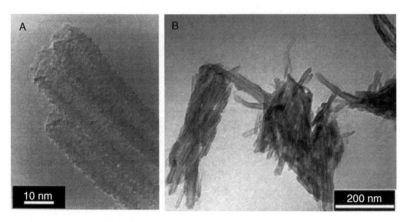

Figure 20   HRTEM images of CNT(O)-BM: (A) nanotube sections, (B) agregates of ball-milled nanotubes. Figures adapted from Ref. 77.

to the chemical shift of the exchange line comes from the nanotube sections formed by the milling procedure. Short nanotubes with side sections caused by ball milling are visible on the HRTEM images of ball-milled samples (Figure 20). Densifying of the overall MWCNT texture after ball milling was expected to cause additional low-field shift. The aggregates of ball-milled nanotubes are shown in Figure 20B.

The rates of Xe exchange between adsorption domains of as-made and ball-milled nanotubes were examined. For that purpose, the samples CNT(O) and CNT(O)-BM were mixed in a weight proportion of 1:0.2. This ratio was adjusted experimentally in order to make the intensities of the lines a_1 and c_1 similar. The ^{129}Xe NMR spectrum of Xe adsorbed on the mixture of CNT(O) and CNT(O)-BM is displayed in Figure 18C. Its line-shape could not be reproduced by linear superposition of the individual spectra displayed in Figure 18A and B. This confirmed the fast Xe exchange between the channels of ball-milled nanotubes and the aggregate pore volumes.

4.3.2. Variable temperature measurements

The variable temperature ^{129}Xe NMR measurements confirmed the validity of the fast exchange approximation. The temperature decrease forces the occupancies of different adsorption domains to change. If the amount of Xe atoms is constant, the relative occupancy of a stronger adsorption site increases at the expense of a weaker one. The difference of adsorption potentials could be estimated from ^{129}Xe NMR spectra acquired at different temperatures.

Variable temperature ^{129}Xe NMR spectra of Xe adsorbed on as-made CNT(O) are shown in Figure 21. Cooling causes the relative intensities of the NMR lines to change, reflecting the population of adsorption domains. All the lines moved downfield. At temperatures below 150 K, these lines merge into a single, relatively narrow line at 128 ppm. This chemical shift could be due to both Xe interaction with the carbon surface and Xe–Xe collisions. The former contribution dominated since the Xe density was low, $\rho \sim 15\ \mu\text{mol g}^{-1}$.

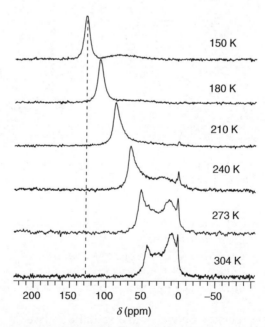

Figure 21   Variable temperature ^{129}Xe NMR spectra of Xe adsorbed on CNT(O). The Xe loading is 0.15 mmol g^{-1}. Figure adapted from Ref. 77.

The chemical shift variation with temperature seems to be helpful for the pore geometry characterisation, Equation (7). The authors considered the geometry of non-intersecting cylindrical pores to describe the variable temperature Xe NMR data. The local Xe density was sufficiently low to neglect the Xe–Xe interaction. In the case of cylindrical pores, $\eta=4$, the chemical shift is given by

$$\delta(\beta) = \frac{\delta_a}{1 + [(D\beta/4K_0k)\exp(|\varepsilon|\beta)]} \tag{18}$$

Here the Henry constant K is expressed as $K_0 \exp(-\varepsilon\beta)$, where K_0 is the temperature-independent constant; $\beta = T^{-1}$ and $\varepsilon = H_{ads}/k$, where H_{ads} is the enthalpy of Xe adsorption. Experimental data obtained for the ball-milled MWCNT sample (CNT(O)-BM) and the fitting curve are shown in Figure 22. The pore diameter, D, was fixed at 4 nm. The best fit parameters were $\delta_a = 107.4$ ppm, $H_{ads} = 11.9$ kJ mol^{-1}, $K_0 = 0.23 \times 10^{10}$ Pa^{-1} m^{-2}. The Henry constant K corresponding to these H_{ads} and K_0 was 0.29×10^{12} Pa^{-1} m^{-2} at $T = 296$ K. This could be compared to a K value obtained from Xe adsorption isotherms, 5×10^{12} Pa^{-1} m^{-2}. The difference in the K-values is acceptable in view of all the simplifications made. The model (Equation (18)) corresponds to the channel-like pores of nanotubes only. The difference in binding energies of inner and outer nanotube surfaces[114,115] has a strong effect on the relative amount of adsorbed xenon at low temperatures.

4.3.3. Evidence for localisation of Pd-containing particles supported on MWCNT

Carbon-supported metal catalysts are advantageous from economical and ecological standpoints. Metal recovery, refining and recycling procedures are very convenient and economical, since the carbon support can be burnt off, leading to highly concentrated ashes.

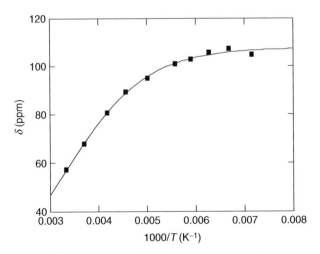

Figure 22 Experimental ¹²⁹Xe NMR chemical shifts versus reciprocal temperature measured for the sample CNT(O)-BM (■). The solid curve represents the least-squares fit of Equation (18) to the experimental data. Figure adapted from Ref. 77.

   This technology does not produce large amounts of solid waste that need to be land-filled. Porous carbon materials are stable in both acidic and basic media, which is not true for alumina or silica supports. The catalysts most frequently used in fine chemicals production are metals. About 30% of the catalysts in this sector are supported palladium catalysts, and most of them are used for hydrogenation reactions, for example, the synthesis of amines from nitro compounds or the saturation of carbon–carbon and carbon–heteroatom multiple bonds.

   As-made and ball-milled carbon nanotubes were used as model supports of the $PdCl_2$ particles.[116] The rate of Xe exchange between channels and aggregate pores of ball-milled nanotubes was strongly influenced by Pd-containing particles localised inside nanotubes.

   The $PdCl_2$ particles were supported on MWCNT by the incipient wetness preparation method, Table 5.

   According to studies of Simonov et al.,[117] both metallic (Pd^0) and oxidised ($PdCl_2$–Pd^{2+}) Pd states are formed. Adsorption of H_2PdCl_4 from the aqueous solution takes place on the carbon surface by two mechanisms. Reduction results in formation of metallic Pd^0 particles with a wide size distribution of 6–100 nm. The $PdCl_2$ clusters are deposited in the form of surface π-complexes with $>C=C<$ fragments of the carbon matrix. The quantitative ratio of the adsorbed Pd^{2+} and Pd^0 species, dispersion and morphology of the metallic particles depend on various factors like size of carbon particles, state of the carbon surface, conditions of drying, etc. Simonov et al. showed that the first process is localised near the external surface of the porous carbon particles (Sibunit). The second process takes place on the whole surface of these particles.

   Formation of Pd-containing particles on the surface of CNT and CNT-BM was confirmed by TEM images. In agreement with the results of Simonov et al.,[117,118] the $PdCl_2$ particles were present mainly in a highly dispersed state (<3 nm). Rare presence of coarse Pd metal particles 10–80 nm in size was indicated. Initiation of $PdCl_2$ clusters occurred at the numerous micro-defects present on the outer surface and on the side sections of the nanotubes. The presence of fine $PdCl_2$ clusters inside or at the sections of ball-milled nanotubes could be suggested from the TEM images (Figure 23). Because of the appropriate size, these particles could partially or completely block the nanotube channels. HRTEM images, however, were ambiguous since they do not allow distinguishing between the particles inside and outside the nanotubes.

   Figure 24A and B display the ^{129}Xe NMR spectra of xenon adsorbed on CNT and on the sample A_1 (CNT-supported $PdCl_2$). Despite the presence of Pd-containing particles on the surface of A_1, the ^{129}Xe NMR spectra are very

TABLE 5 The samples of MWCNT-supported $PdCl_2$

Sample	A_1	A_2	A_3	A_4
Support	CNT	CNT-BM	CNT-BM	CNT-BM
[$PdCl_2$] (mmol g^{-1})	1.025	0.1	0.425	1

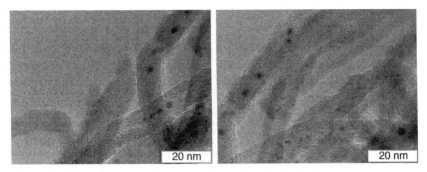

Figure 23   TEM images of the sample A_3 (0.425 mmol g^{-1} PdCl$_2$/CNT-BM). Figure adapted from Ref. 77.

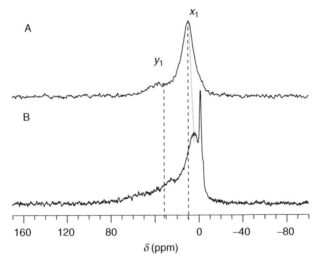

Figure 24   The [129]Xe NMR spectra of Xe adsorbed on CNT (A) and A_1 (B) at a pressure of Xe 100 kPa. Figure adapted from Ref. 77.

similar to those obtained for the bare support CNT. There was no Xe access inside channels of CNT and A_1. The NMR lines were attributed to Xe in fast exchange between the external surface of nanotubes and the volume of aggregate pores. These observations indicated relatively weak interaction of Xe with Pd-containing particles. The blocking of wide aggregate mesopores of A_1 by the Pd-containing particles (mainly fine PdCl$_2$ clusters) did not occur.

Figure 25 displays the [129]Xe NMR spectra of Xe adsorbed on CNT-BM and MWCNT-supported PdCl$_2$ samples A_2, A_3 and A_4. The single line (z_1) (CNT-BM) split into two well-distinguishable lines w_1 and w_2. The sensitivity of [129]Xe NMR to the presence of nanotube sections and accessible nanotube channels was demonstrated. Nanotube sections contained a considerable amount of >C=C< fragments necessary for the growth of PdCl$_2$ clusters. Hydrogen adsorption on

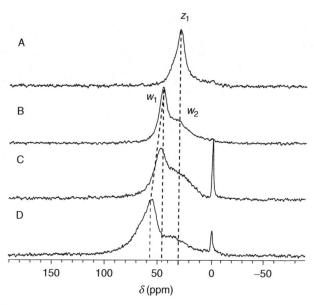

Figure 25   The ^{129}Xe NMR spectra of Xe adsorbed on CNT-BM (A), A_2 (B), A_3 (C) and A_4 (D) at a Xe pressure of 90 kPa. Figure adapted from Ref. 77.

metal clusters is known to reduce the spin density of the conduction electrons.[52] Using this idea, the authors confirmed that the chemical shifts of the lines w_1 and w_2 were free of the Knight contribution. The lines w_1 and w_2 therefore were atributed to nanotubes channels partially blocked by $PdCl_2$ and aggregate pores, respectively. The Xe exchange rate decreased sufficiently to make these adsorption domains distinguishable by NMR. The effect of supported $PdCl_2$ on Xe exchange is opposite to the effect of ball-milling discussed above. The variation of the ^{129}Xe NMR line-shape with concentration of adsorbed $PdCl_2$ was strongly pronounced. The line w_1 shifted to low field as a result of free space reduction inside the nanotubes channels.

5. CONCLUSIONS

Demands for robust analytic techniques that enable characterisation of complex porous media expand continuously. Catalysis and petroleum engineering are the areas where expectations from magnetic resonance-based methods are particularly high. Porous carbon materials are very attractive to adsorption, catalysis and other relevant areas due to their great variety of textures and surface properties. Characterisation of amorphous microporous carbons has been much discussed in the literature. Traditional adsorption techniques used for determination of parameters of micropore structure are based on very complex theoretical approaches and stringent technical requirements. A relatively new class of carbon nanomaterials has given rise to an enormous amount of scientific literature within a few years. Despite its extensive use for the characterisation of various porous materials, particularly zeolites, Xe NMR has been rarely used to probe the carbonaceous solids. The uses and limitations of Xe NMR for characterisation of porous carbon materials and carbon-based catalysts have been elucidated. Although Xe NMR exhibits definite advantages over conventional adsorption and difraction techniques, it can hardly be referred to as an all-sufficient method. Its use as a supplementary tool, however, will be highly appreciated.

REFERENCES

1. J. L. Bonardet, J. Fraissard, A. Gedeon and M. A. Springuel-Huet, *Catal. Rev. Sci. Eng.*, 1999, **1999**(41), 115.
2. P. J. Barrie and J. Klinowski, *Prog. NMR Spectrosc.*, 1992, **24**, 91.
3. D. Raftery and B. F. Chmelka, Xenon NMR spectroscopy, *in: NMR Basic Principles and Progress*, 1994, Vol. 30, p. 112. Springer, Berlin, Heidelberg.
4. D. Raftery, *Annu. Rep. NMR Spectrosc.*, 2006, **57**, 205.
5. T. Ito and J. Fraissard, *in: Proceedings of the 5th International Zeolite Conference, Naples*, (L. V. C. Rees ed.), 1980, p. 510. Heyden, London.
6. B. M. Goodson, *J. Magn. Reson.*, 2002, **155**, 157.
7. N. F. Ramsey, *Phys. Rev.*, 1950, **78**, 699.
8. J. Fraissard and T. Ito, *Zeolites*, 1988, **8**, 350.
9. Q. Chen, M. A. Springuel-Huet and J. Fraissard, 129Xe NMR of adsorbed Xenon for the determination of void spaces, *in: Catalysis and Adsorption by Zeolites*, J. Holmann, H. Pfeifer, and R. Fricke (eds.), 1991, Vol. 65, p. 219. Elsevier, Amsterdam.
10. A. K. Jameson, C. J. Jameson and H. S. Gutowsky, *J. Chem. Phys.*, 1970, **53**, 2310.
11. T. Ito and J. Fraissard, *J. Chem. Phys.*, 1982, **76**, 5225.
12. C. Dybowski, N. Bansal and T. M. Duncan, *Annu. Rev. Phys. Chem.*, 1991, **42**, 433.
13. T. T. P. Cheung, C. M. Fu and S. Wharry, *J. Phys. Chem.*, 1988, **92**, 5170.
14. M. A. Springuel-Huet and J. Fraissard, *Chem. Phys. Lett.*, 1989, **154**, 299.
15. J. A. Ripmeester and C. I. Ratcliffe, *J. Phys. Chem.*, 1995, **99**, 619.
16. C. J. Jameson, A. C. Dios, *J. Chem. Phys.*, 2002, **116**(9), 3805.
17. C. J. Jameson, *J. Chem. Phys.*, 2002, **116**(20), 8912.
18. J. Demarquay and J. Fraissard, *Chem. Phys. Lett.*, 1987, **136**, 314.
19. V. Terskikh, I. L. Moudrakovski, S. R. Breeze, S. Lang, C. I. Ratcliffe, J. A. Ripmeester and A. Sayari, *Langmuir*, 2002, **18**, 5653.
20. S. B. Liu, C. S. Lee, P. F. Shiu and B. M. Fung, *Stud. Surf. Sci. Catal.*, 1994, **83**, 233.
21. E. G. Derouane, J. M. André and A. A. Lucas, *Chem. Phys. Lett.*, 1987, **137**, 336.
22. E. G. Derouane and J. B'Nagy, *Chem. Phys. Lett.*, 1987, **137**, 341.
23. J. A. Ripmeester and C. I. Ratcliffe, *J. Phys. Chem.*, 1990, **94**, 7652.
24. T. T. P. Cheung, *J. Phys. Chem.*, 1995, **99**, 7089.
25. M. J. Annen, M. E. Davis and B. E. Hanson, *Catal. Lett.*, 1990, **6**, 331.
26. W. C. Conner, E. L. Weist, T. Ito and J. Fraissard, *J. Phys. Chem.*, 1989, **93**, 4138.
27. V. Terskikh, I. Mudrakovski and V. J. Mastikhin, ^{129}Xe nuclear magnetic resonance studies of the porous structure of silica gels. *Chem. Soc., Faraday Trans.*, 1993, **89**, 4239.
28. T. T. P. Cheung, *J. Phys. Chem.*, 1989, **93**, 7549.
29. R. A. Kromhout and B. J. Linder, *J. Magn. Reson.*, 1969, **1**, 450.
30. F. J. Adrian, *Phys. Rev.*, 1964, **136**, A980.
31. A. Julbe, L. C. Menorval, C. Balzer, P. David, P. Palmeri and C. J. Guizard, *Porous Mater.*, 1999, **6**, 41.

32. F. Chen, C. Chen, S. Ding, Y. Yue, C. Ye and F. Deng, *Chem. Phys. Lett.*, 2004, **383,** 309.
33. K. Hahn, J. Kärger and V. Kukla, *Phys. Rev. Lett.*, 1996, **76,** 2762.
34. G. Roedenbeck, K. Hahn and J. Kärger, *Phys. Rev. E*, 1997, **55,** 5697.
35. P. Demontis, J. G. Gonzalez, G. B. Suffritti and A. Tilocca, *J. Am. Chem. Soc.*, 2001, **123,** 5069.
36. B. H. Jeneer, P. Bachmann and R. R. Ernst, *J. Chem. Phys.*, 1979, **71**(11), 4546.
37. M. Mansfeld and W. S. Veeman, *Chem. Phys. Lett.*, 1993, **213**(1–2), 153.
38. I. L. Moudrakovski, I. C. Ratcliffe and J. A. Ripmeester, *Appl. Magn. Reson.*, 1995, **8,** 385.
39. W. Heink, J. Karger, H. Pfeifer and F. Stallmach, *J. Am. Chem. Soc.*, 1990, **112,** 2175.
40. M. Janicke, B. F. Chmelka, R. G. Larsen, J. Shore, K. Schmidt-Rohr, L. Emsley, H. Long and A. Pines, *Proceedings of the 10th International Zeolite Conference, Germany*, 1994.
41. R. G. Larsen and B. F. Chmelka, *Chem. Phys. Lett.*, 1993, **214,** 220.
42. J. H. Sinfelt, *Annu. Rev. Mater. Sci.*, 1972, **2,** 641.
43. L. H. Little, *Infrared Spectra of Adsorbed Species*. Academic Press, London, New York, 1966.
44. C. R. Adams, H. A. Benessi, R. M. Curtis and R. G. Meisenheimer, *J. Catal.*, 1962, **1,** 336.
45. G. R. Wilson and W. K. Hall, *J. Catal.*, 1970, **17,** 190.
46. J. A. Dumesic and H. Topsoe, *Adv. Catal.*, 1977, **26,** 121.
47. F. W. Lytle, G. H. Via and J. H. Sinfelt, *in: Synchrotron Radiation Research*, H. Winick and S. Doniach (eds.), 1980, p. 401. Plenum, New York.
48. J. L. Bonardet, J. P. Fraissard and L. C. De Menorval, *in: Proceeding of the VIth International Congress on Catalysis*, 1977, Vol. 2, p. 633. The Chemical Society, London.
49. L. C. De Menorval and J. P. Fraissard, *Chem. Phys. Lett.*, 1981, **77**(2), 309.
50. C. P. Slichter, NMR and surface structure, *in: The Structure of Surfaces*, M. A. Van Hove and S. Y. Tong (eds.), p. 84. Springer, Heidelberg.
51. J. J. Van der Klink, J. Buttet and M. Graetzel, *Phys. Rev. B*, 1984, **29,** 6352.
52. D. Rouabah and J. Fraissard, *Solid State Nucl. Magn. Reson.*, 1994, **3,** 153 and references therein.
53. T. Sheng and J. D. Gay, *J. Catal.*, 1981, **71,** 119.
54. J. Sanz and J. H. Rojo, *J. Phys. Chem.*, 1985, **89,** 4974 and references therein.
55. X. Wu, B. C. Gerstein and T. S. King, *J. Catal.*, 1989, **118,** 238.
56. M. Polisset and J. Fraissard, *Colloids Surf A: Physicochem. Eng. Asp.*, 1993, **72,** 197.
57. L. C. De Menorval, J. Fraissard and T. Ito, *J. Chem. Soc., Faraday Trans. I*, 1982, **78,** 403.
58. W. D. Knight, Nuclear magnetic resonance shift in metals. *Phys. Rev.*, 1949, **76,** 1259.
59. C. P. Slichter, *Principles of Magnetic Resonance*. Springer, New York, 1978.
60. B. C. Khanra, *Int. J. Modern Phys. B*, 1997, **11**(14), 1635.
61. T. H. Chang, C. P. Cheng and C. T. Yeh, *J. Phys. Chem.*, 1991, **95,** 5239.
62. T. H. Chang, C. P. Cheng and C. T. Yeh, *J. Phys. Chem.*, 1992, **96,** 4151.
63. L. C. deMenorval, D. Raftery, S.-B. Liu, K. Takegoshi, R. Ryoo and A. Pines, *J. Phys. Chem.*, 1990, **94,** 27.
64. J. T. Miller, B. L. Meyers and G. J. Ray, *J. Catal.*, 1991, **128,** 436.
65. C. Tsiao, C. Dybowski, A. M. Gaffney and J. A. Sofranko, *J. Catal.*, 1990, **128,** 520.
66. N. Bansal, H. C. Foley, D. S. Lafyatis and C. Dybowski, *Catal. Today*, 1992, **14,** 305.
67. C. C. Tsiao and R. E. Botto, *Energy & Fuels*, 1991, **5,** 87.
68. K. V. Romanenko, X. Py, J.-B. d'Espinose de la Caillerie, O. B. Lapina and J. Fraissard, *J. Phys. Chem. B*, 2006, **110**(7), 3055.
69. D. J. Suh, T. J. Park, S. K. Ihm, R. J. Ryoo, *J. Phys. Chem.*, 1991, **95**(9), 3767.
70. P. C. Wernett, J. W. Larsen, O. Yamada and H. J. Yue, *Energy Fuels*, 1990, **4**(4), 412.
71. K. J. McGrath, *Carbon*, 1999, **37,** 1443.
72. K. Saito, A. Kimura and H. Fujiwara, *Magn. Reson. Imaging*, 2003, **21,** 401.
73. K. Gotoh, T. Ueda, H. Omi, T. Eguchi, M. Maeda, M. Miyahara, A. Nagai and H. Ishida, *J. Phys. Chem. Solids*, 2008, **69**(1), 147.
74. A. Garsuch, W. Böhlmann, R. R. Sattler, J. Fraissard and O. Klepel, *Carbon*, 2006, **44**(7), 1173.
75. P. A. Simonov, S. V. Filimonova, G. N. Kryukova, H. P. Boehm, E. M. Moroz, V. A. Likholobov and T. Kuretzky, *Carbon*, 1999, **37,** 591.
76. K. V. Romanenko, J.-B. d'Espinose, J. Fraissard, T. V. Reshetenko and O. B. Lapina, *Micropor. Mesopor. Mater.*, 2005, **81,** 41.

77. K. V. Romanenko, A. Fonseca, S. Dumonteil, J. B'Nagy, J.-B. d'Espinose, O. B. Lapina and J. Fraissard, *Solid State NMR*, 2005, **28**, 135.
78. C. F. M. Clewett and T. Pietra, *J. Phys. Chem. B*, 2005, **109**, 17907.
79. J. M. Kneller, R. J. Soto, S. E. Surber, J.-F. Colomer, A. Fonseca, J. B. Nagy, G. Van Tendeloo and T. Pietra, *J. Am. Chem. Soc.*, 2000, **122**, 10591.
80. H. Ago, K. Tanaka, T. Yamabe, T. Miyoshi, K. Takegoshi, T. Terao, S. Yata, Y. Hato, S. Nagura and N. Ando, *Carbon*, 1997, **35**, 1781.
81. D. Raftery, H. Long, T. Meersmann, P. J. Grandinetti, L. Reven and A. Pines, *Phys. Rev. Lett.*, 1991, **66**, 584.
82. K. V. Romanenko, O. B. Lapina, X. Py and J. Fraissard, *Russ. J. Gen. Chem.*, 2008, **LII**(1), 3.
83. V. A. Likholobov, V. B. Fenelonov, L. G. Okkel, O. V. Goncharova, L. B. Avdeeva, V. I. Zaikovskii, G. G. Kuvshinov, V. A. Semikolenov, V. K. Duplyakin, O. N. Baklanova and G. V. Plaksin, *React. Kinet. Catal. Lett.*, 1995, **54**(2), 381.
84. US Patent No. 4978649, 1989.
85. B. Cagnon, X. Py, A. Guillot, J. P. Joly and R. Berjoan, *Micropor. Mesopor. Mater.*, 2005, **80**, 183.
86. M. A. Springuel-Huet and J. Fraissard, *Chem. Phys. Lett.*, 1989, **154**, 299.
87. J. A. Ripmeester and C. I. Ratcliffe, *J. Phys. Chem.*, 1995, **99**, 619.
88. C. J. Jameson and A. C. Dios, *J. Chem. Phys*, 2002, **116**(9), 3805.
89. C. J. Jameson, *J. Chem. Phys.*, 2002, **116**(20), 8912.
90. J. L. Bonardet, J. Fraissard, A. Gedeon and M. A. Springuel-Huet, *Catal. Rev.*, 1999, **41**, 115.
91. X. Py, A. Guillot and B. Cagnon, *Carbon*, 2004, **42**, 1743.
92. A. K. Jameson, C. J. Jameson and H. S. Gutowsky, *J. Chem. Phys.*, 1970, **53**, 2310.
93. Q. J. Chen and J. Fraissard, *J. Phys. Chem.*, 1992, **96**, 1814.
94. I. L. Moudrakovski, V. V. Terskikh, C. I. Ratcliffe, J. A. Ripmeester, L.-Q. Wang, Y. Shin and G. J. Exarhos, *J. Phys. Chem. B*, 2002, **106**, 5938.
95. K. V. Romanenko, O. B. Lapina, V. L. Kuznetsov and J. Fraissard, *Kinet. Catal.*, 2009, **50**(1), 26.
96. T. V. Reshetenko, L. B. Avdeeva, Z. R. Ismagilov, V. V. Pushkarev, S. V. Cherepanova, A. L. Chuvilin and V. A. Likholobov, *Carbon*, 2003, **41**, 1605.
97. J. Abrahamson, *Carbon*, 1973, **11**, 337.
98. N. N. Avgul, A. V. Kiselev and D. P. Poshkous, Адсорбция газов и паров на однородных поверхностях *Adsorption of Gases and Steams on Uniform Surfaces, M.: Chemistry*. Moscow, Russia, 1975 (in Russian).
99. V. B. Fenelonov, A. Y. Derevyankin, L. G. Okkel, L. B. Avdeeva, V. I. Zaikovskii, E. M. Moroz, A. N. Salanov, N. A. Rudina, V. A. Likholobov and S. K. Shaikhutdinov, *Carbon*, 1997, **35**(8), 1129. http://www.sciencedirect.com/science?_ob=ArticleListURL&_method=list&_ArticleListID=1150935364&_sort=r&view=c&_acct=C000050221&_version=1&_urlVersi on=0&_userid=10&md5=c989dd664138f95058677ed6801e3b4b.
100. P. A. Simonov, S. V. Filimonova, G. N. Kryukova, H. P. Boehm, E. M. Moroz, V. A. Likholobov and T. Kuretzky, *Carbon*, 1999, **37**, 591.
101. P. A. Simonov, S. V. Filimonova, G. N. Kryukova, H. P. Boehm, E. M. Moroz, V. A. Likholobov and T. Kuretzky, *Carbon*, 1999, **37**, 591.
102. K. V. Romanenko, J.-B. d'Espinose de Lacaillerie, O. Lapina and J. Fraissard, *Micropor. Mesopor. Mater.*, 2007, **105**, 118.
103. Q.-H. Yang, P.-X. Hou, S. Bai, M.-Z. Wang and H.-M. Cheng, *Chem. Phys. Lett.*, 2001, **345**, 18.
104. M. Muris, N. Dupont-Pavlovsky, M. Bienfait and P. Zeppenfeld, *Surf. Sci.*, 2001, **492**, 67.
105. G. Stan and M. W. Cole, *Surf. Sci.*, 1998, **395**, 280.
106. S. Inoue, N. Ichikuni, T. Suzuki, T. Uematsu and K. Kaneko, *J. Phys. Chem. B*, 1998, **102**, 4689.
107. C. Gommes, S. Blacher, N. Dupont-Pavlovsky, C. Bossuot, M. Lamy, A. Brasseur, D. Marguillier, A. Fonseca, E. McRae, J. B. Nagy and J.-P. Pirard, *Colloids Surf. A Physicochem. Eng. Asp.*, 2004, **241**, 155.
108. S. Talapatra, A. Z. Zambano, S. E. Weber and A. D. Migone, *Phys. Rev. Lett.*, 2000, **85**, 138.
109. H. Z. Geng, X. B. Zhang, S. H. Mao, A. Kleinhammes, H. Shimoda, Y. Wu and O. Zhou, *Chem. Phys. Lett.*, 2004, **399**, 109.

110. N. Pierard, A. Fonseca, Z. Konya, I. Willems, G. Van Tendeloo and J. B. Nagy, *Chem. Phys. Lett.*, 2001, **335**, 1.
111. A. Kuznetsova, D. B. Mawhinney, V. Naumenko, J. T. Yates, Jr., J. Liu and R. E. Smalley, *Chem. Phys. Lett.*, 2000, **321**, 292.
112. P. X. Hou, S. Bai, Q. H. Yang, C. Liu and H. M. Cheng, *Carbon*, 2002, **40**(1), 81.
113. N. Nagaraju, A. Fonseca, Z. Konya and J. B. Nagy, *J. Mol. Catal. A: Chem.*, 2002, **181**, 57.
114. M. W. Maddox and K. E. Gubbins, *J. Chem. Phys.*, 1997, **107**, 9659.
115. Q. Wang and J. K. Johnson, *J. Chem. Phys.*, 1999, **110**, 577.
116. K. V. Romanenko, P. A. Simonov, O. G. Abrosimov, O. B. Lapina and J. Fraissard, *React. Kinet. Catal. Lett.*, 2007, **90**(2), 355.
117. P. A. Simonov, A. V. Romanenko, I. P. Prosvirin, E. M. Moroz, A. I. Boronin, A. L. Chuvilin and V. A. Likholobov, *Carbon*, 1997, **35**(1), 73.
118. P. A. Simonov, E. M. Moroz, A. L. Chuvilin and V. A. Likholobov, *in: Preprints of the 6th International Symposium Sci. Bases for the preparation of heteroeneous catalysts.* Louvain-la Neuve, 5–8 September 1994, Vol. 3, p. 201.

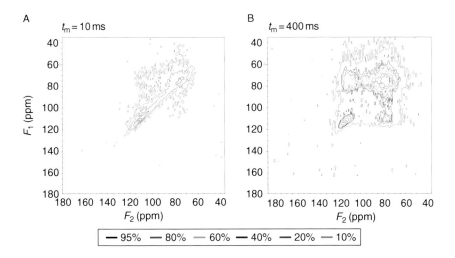

Plate 1 2D ^{129}Xe exchange spectra of Xe adsorbed on *Nor* acquired using mixing periods of 10 and 400 ms; xenon pressure $\sim$100 kPa. Figures adapted from Ref. 82. (For B/W version, refer page 13, Figure 3)

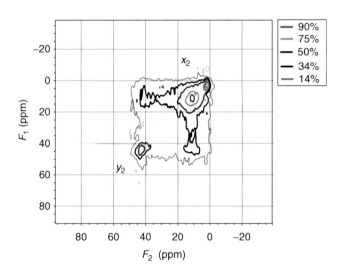

Plate 2 2D ^{129}Xe exchange spectrum of CNT(O) using a 10-ms mixing period. Figure adapted from Ref. 77. (For B/W version, refer page 29, Figure 19)

CHAPTER 2

NMR Studies for Mapping Structure and Dynamics of Nucleosides in Water

Bernard Ancian

Contents

Abstract

An overview of the literature on structure of the nucleosides for the last 40 years is presented in order to gain a better understanding of their specific and non-specific interactions with water. The emphasis has been on NMR intermolecular interactions to probe the hydration of the compounds,

UPMC Univ. Paris 6, Laboratoire PECSA, Physicochimie des Electrolytes, Colloïdes et Sciences Analytiques (UPMC/CNRS/ESPCI), Case 51, 4 place Jussieu, Paris, France

Annual Reports on NMR Spectroscopy, Volume 69
ISSN 0066-4103, DOI: 10.1016/S0066-4103(10)69002-3

especially homonuclear and heteronuclear Overhauser spectroscopy. New aspects as given by well-designed diffusion experiments are investigated. Dynamics and microdynamics, e.g. lifetimes of the hydrates, structure of the first hydration shell and chemical exchange of the amide protons, are fully discussed on the example of uridine in water.

Key Words: Nucleosides, Overhauser effect, Water interaction, HOESY, Intermolecular dipole–dipole interactions, Diffusion, DOSY, Dynamics, Microdynamics, Uridine.

1. INTRODUCTION

Water is generally considered to be an integral part of nucleic acids and proteins and plays an important role in biochemical and biophysical processes such as transcription and translation, recognition and replication, processing, transport, stability and editing.[1-4] The complex network of hydrogen-bond interactions between biomolecular structures and water determines to a large extent their sizes and structures as well as their chemical and physical properties.[3-6] The question of protein and/or DNA and RNA hydration, viz. the interaction of protein and/or DNA molecules with water and the involvement of water molecules in their reactions, has been a focus of NMR attention in the last few years.[7-9] By the even essence of the physical phenomenon of hydration, the concept of "rigid shell" of water around a protein or a nucleic acid remains an elusive one because there is always a fluctuating cloud of water molecules which are thermodynamically affected more or less strongly by the solute.[6] Nevertheless, transport measurements, spectroscopic techniques, molecular dynamics (MD) and Brownian dynamics (BD) have allowed gaining a clearer understanding of the aqueous environment, first near the macromolecular surface, and second, far from the surface. They all lead to a better clarification of the hydration shell, giving at least two time scales for the water: one, relatively long on the nanosecond scale for the water which is in close contact with the mosaic-like surface, and the other short, on the picosecond range for the bulk water.[7-14]

Most studies of protein–water interactions are concerned with the collection of water molecules at the surface of the protein, which influences their rotational and translational dynamics.[15-19] On the other hand, water can also be an integral part of the protein three-dimensional (3D) structure,[7-14,19-22] and these internal water molecules can exchange with bulk water, typically on a the microsecond time scale but sometimes for much longer.[7-14,17-20] The surface of a protein is highly heterogeneous in which invaginations formed by the hydrophilic and hydrophobic side chains of the amino acids give a unique "roughness" that may favour water molecular interactions.[6] Moreover, an analysis of the fluctuations in the positions of the protein atoms shows that the addition of water molecules makes the protein more flexible and this increased flexibility appears to be due to an increased length and weakened strength of protein–protein hydrogen bonds, thereby helping cavity formation.[21,22] Water–protein surface interactions that persist for times of the order of nanoseconds are revealed by ^1H, ^2H and ^{17}O magnetic relaxation dispersion (MRD) techniques.[20] These experiments show that rotational and translational diffusion of waters in direct contact with the protein surface are typically slowed by a factor of 2–5 relative to their behaviour in bulk water. These conclusions are generally supported by neutron scattering experiments and MD calculations.[23–25]

In DNA and RNA, X-ray crystallographic and NMR spectroscopic measurements have shown the presence of an ordered spine of hydration, at the nanosecond scale, in the narrow minor groove.[7,9,26–31] Furthermore, structural fluctuations in DNA at this time scale result in less structured local hydration patterns around the base, giving a heterogeneous environment, including also polar, non-polar and stacking interactions between adjacent bases and sugars.[29–31] The sites of hydration have been extensively studied in the 1970s by the Pullmans' group by using molecular orbital self-consistent field (MO-SCF) quantum calculations.[32] These calculations are today currently run by using MD and Monte Carlo (MC) approaches[33–37] and naturally at the level of the density functional theory (DFT).[38] All these interactions are naturally implied in the much slower time scale of base opening—at about the millisecond range—where the biochemical processes begin.[39] Hydration processes are also directly involved in antisense therapy in order to understand how antisense modifications affect the structure and stability of nucleic acid fragments and single strand mRNA requiring the estimation of modification-induced hydration changes, both at the global and the local levels.[40]

Concerning the hydration structure of a nucleoside, the situation is completely different. Such a simple molecule has neither an internal cavity nor a groove which can host the water, in contrast to the precedent macromolecules. Here, water is necessarily maintained at the external molecular surface and the hydration should be investigated at the atomic level. Besides the two almost perpendicular base and furanose type-sugar cycles, the surface roughness is mainly given by the amide group, the alcohol function, the hydroxymethyl group and perhaps the lone electronic pairs on the heteroatoms. Knowing that water plays simultaneously the role of a hydrogen-bond donor or an acceptor,[5,41] hydrophilic interaction can be expected between all these functional groups and water. These hydrogen bonds, which are mainly electrostatic with only a small covalent part,[42–44] are weak (3–6 kcal mol^{-1}) and only weakly directional as compared to purely covalent ones. Their lengths are very sensitive both to the electronegativity of the heteroatoms X and Y they share and to their geometrical characteristics, ranging from 1.7 to 2.6 Å for an angle X–H$\cdots$Y ranging from 180° to 120°, respectively.[5] In addition, hydrophobic interactions[45–48] between apolar groups and water, generally of less than 2 kcal mol^{-1} (van der Waals repulsion, dispersion forces, surface tension), should also be accounted for. Nevertheless, hydrophobicity is clearly different for small apolar groups in water from those between large assemblies like macromolecules, as shown by Lum, Chandler and Weeks[49] in a general theory of solvation at small and large length scales. This is another distinction between the hydration of nucleosides which are smaller than the nanometer scale, and proteins, DNA and RNA which are beyond this crossover limit.[49,50]

Because of the nucleoside hydration, the magnetic properties of the nuclear environment (mainly ^{1}H, ^{13}C, ^{14}N, ^{15}N and ^{17}O nuclei) change, leading to an observable change in the structural NMR parameters such as chemical shifts and scalar coupling constants.[9,51–56] At the same time, hydration also acts as a local "structure making" and "structure breaking" mechanism with characteristic life times that modulate the transport characteristics (rotation and translation) of the solute in the solvent.[57] The resulting dynamical properties are thus also altered and can be adequately probed by nuclear relaxation, Overhauser effect and diffusion experiments.[58–63].

As an example, we present in Figure 1 the two-dimensional (2D) ^{13}C, ^{1}H correlation spectrum of uridine (hereafter abbreviated as U) in water obtained in the carbon-13 direct detection mode with proton broadband decoupling. As can be seen, the partial overlap between H5 of the uracil moiety and H1' of the ribose can create a problem for a-selective analysis of water interaction (see below). Attempts to solve this ambiguity in nucleosides constitute a large part of the literature and are discussed in the following sections.

2. OVERVIEW, TRENDS AND OPINIONS

NMR in water is now a continuously developing field in all aspects of physics, chemistry and, last but not least, biology. With the constant puzzling water NMR response, the adapted methodology for water suppression[65,66] and/or elimination

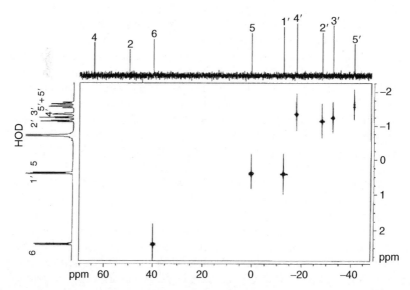

Figure 1   2D ^{13}C, ^{1}H correlation spectrum of uridine U in water ($\approx$1 mol l^{-1}) obtained in the carbon-13 direct detection mode with WALTZ-16 proton broadband decoupling at room temperature. Note the partial overlap between the H5 and H1' resonances. Only the direct correlations between the carbons and their attached protons are shown (from Reference 64).

of radiation damping[67–71] is currently well documented and is clearly not of concern in this review. On the other hand, study of water interaction with nucleotides, oligonucleotides and polynucleotides as well as natural and modified DNA and RNA, which is also a burgeoning area growing at the same speed as the high-field technology, is also not covered by this chapter. This aspect will be only discussed as an aid for understanding the special water interaction with a given nucleoside, if any. An effort has been made to include citations of all papers from the last 30 years (prior to January 2009) and to only recall earlier key works, in the author's opinion, in order to obtain the best representation and not totality. Indeed, this broad and important subject has produced such a large volume of literature that even modern search tools will miss many important and relevant papers. For example, in the book edited by Townsend[72] in 1994, more than 2000 nucleosides are reported! How many more since this date? The author apologises in advance to any researchers whose work has been overlooked.

The goal has been to provide a digestible and didactic summary of modern NMR techniques that provide information about the specificity, strength and duration of interactions of nucleosides—*and only nucleosides*—with water, as this solvent is of prime importance in living organisms and finally in life. The review will contain some comments on trends and some personal opinions of the author which will be labelled as such, but the reader should be aware. There are several high-level specialised reviews that cover part of this field, but it seems completely unrealistic and illusory to extensively quote all of them as they concern biology, chemistry, physics, computing science and analytical instrumentation! The reader is directed to consult them, according to his proper preoccupations, choices and interest. As an example, we will quote an excellent Web site, maintained by Martin Chaplin (http://www.lsbu.ac.uk/water/sitemap.html) which carefully describes hydration of proteins, nucleic acids, sugars and polysaccharides and many other macromolecules and which provides up-to-date literature references to all aspects of the chemistry of water. It also gives an outstanding clarification of "chaotropes" and "kosmotropes" which increase and decrease, respectively, the solubility of hydrophobic aggregates in water.

Finally, the author emphasises that all the advances in the understanding of these intimate interactions between water and life constituents which are only at their very start are clearly dependent of all the progress in cutting-edge technology which pushes ahead the limits of molecular sensitivity, atomic resolution and observation time scale. This is particularly acute for NMR, which is certainly one of the most powerful techniques, but with the concomitant problems of limited sensitivity and a dramatically reduced time scale of observation. Much progress has to be made in the future in these directions.

3. NOMENCLATURE AND SYMBOLS

For atom numbering and torsion angle definition in nucleosides, we follow the IUPAC/IUB guidelines.[41,73] Accordingly, the chemical structure and atom numbering of the most common nucleosides, viz., the four purine nucleosides, abbreviated hereafter Pus, adenosine A, guanosine G, Inosine I, xanthosine X and the four pyrimidine nucleosides, abbreviated hereafter Pys, cytidine C, thymidine T, orotidine O and uridine U, are given, respectively, in Figures 2 and 3. Notation

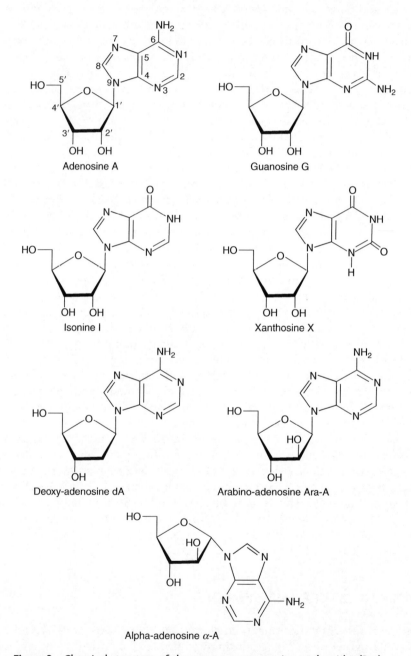

Figure 2   Chemical structure of the most common purine nucleosides (Pus).

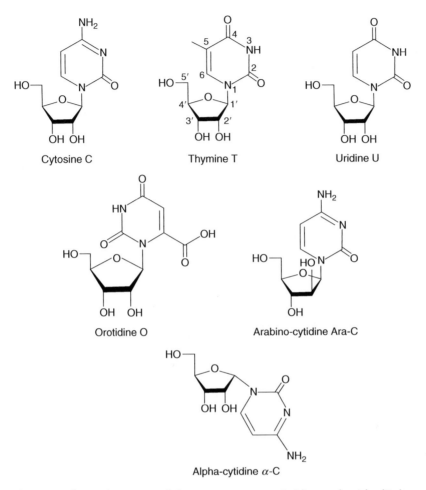

Figure 3   Chemical structure of the most common pyrimidine nucleosides (Pys).

of the furanoside-type sugar (β-D ribose r in RNA and β-D-2′ deoxyribose d in DNA) is also pictured on A in Figure 2 and on C in Figure 4.

The nucleoside flexibility is fully characterised by three internal modes of motion:

— The glycosidic linkage torsion angle χ, O4′–C1′–N1–C2 (χ, Pys) and O4′–C1′–N9–C4 (χ, Pus), is pictured in Figure 4A, Section 5.1.1.1, for U. According to this definition, the *syn* conformation is in the range $\chi = 0 \pm 90°$, whereas the *anti* conformation is in the range $\chi = 180 \pm 90°$ (Figure 4B).

— The pseudo-rotation of the furanose ring or sugar puckering mode is illustrated in Figure 5A, with the two most common states of the ribose cycle, the C2′ *endo* (referred to as ^2E, ^2_3T or S-type) and the C3′ *endo* (referred to as ^3E, ^3_2T or N-type) represented in Figure 5B. Endocyclic torsion angles of the sugar are denoted v_0 to v_4, P is the pseudo-rotation phase angle and τ_m is the maximum torsion angle which describes the maximum out-of-plane pucker, according to the usual convention.[41,73] They are exemplified on U in Figure 4A.

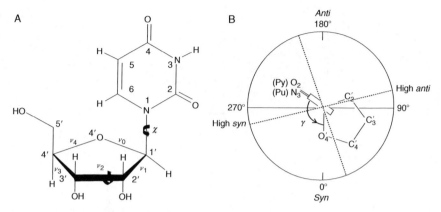

Figure 4 (A) The torsion angle χ in uridine U and the definition of the torsion angles v_0, v_1, v_2, v_3 and v_4 in the ribose ring. (B) Conformational model for the glycosidic torsion angle χ in nucleosides for Pys (O_2) and for Pus (N_3) showing the *anti* and *syn* ranges and denoting the four symmetric quadrants (dashed lines).

— The three main rotamers in staggered conformations gg, gt and tg as obtained by rotation about the exocyclic C4′⁻C5′ bond in the ribose moiety are drawn in Figure 6, as Newman projections about this bond.

Other more detailed nomenclature and symbols about inter-atomic bond distances, hydrogen bonding and base stacking, bond angles and torsion angles can be found in original publications,[74–78] in classical texbooks[41,79] and in previous NMR reviews in this field.[80–84] So far, in this review, and in order to avoid possible misunderstandings, all the (ribo)nucleosides are quoted by a one letter symbol (A, C, G, I, O, T, U, X,...), whereas for the deoxy(ribo)nucleosides the first lower-case letter d indicates the β-D-2′ deoxyribose-type sugar and where the second capital letter (A, C, G, I, O, T, U, X,...) refers to the nucleobase. For example, dT means the deoxythymidine, which is unfortunately and often yet named as thymidine, whereas T is the "true" thymidine which is also named ribothymidine to avoid confusion with deoxythymidine dT because this latter was discovered before the "true" thymidine T. and was erroneously named "thymidine". As stated above, there is a large number of common nucleosides and we describe below only the most frequently reported, namely:

(a) C-Nucleosides such as pseudo-uridine (Ψ uridine, symbolised as ΨU hereafter) which occurs ubiquitously as a minor component in various tRNAs.[85–87] and a large number of C-analogues which have been the cornerstone of antiviral and anticancer chemotherapy over the past three decades.[72,86–88]
(b) Azanucleosides which are also powerful chemotherapeutic agents with, for example, anti-human immunodeficiency virus (HIV) activity.[89–91]
(c) Thionucleosides which are often found in the wobble position of transfer RNA anticodon and occur exclusively in the N *anti* form.[92,93]
(d) Halogeno-nucleosides which have anti-herpes virus activity and are also used in anticancer chemotherapy.[94,95]

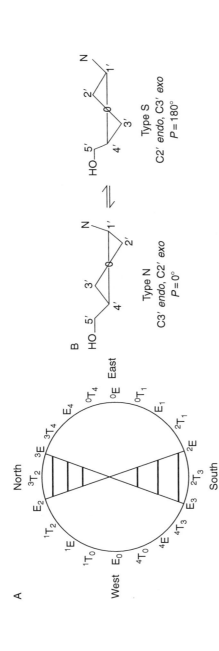

Figure 5 (A) The pseudo-rotational wheel of the ribose sugar in nucleosides. The dashed angle represents the phase angle of 36°. Envelope E and twist T alternate every 18°. After rotation by 180°, the mirror image of the starting position is found as schematised for the North position and the South position in the hatched part. (B) Schematic representation of the ribose equilibrium between the two states N and S: C3′, $endo$ (³E, N) ⇌ C2′, $endo$ (2E, S). $P = 0°$ and $P = 180°$ are the phases along the pseudo-rotation cycle.

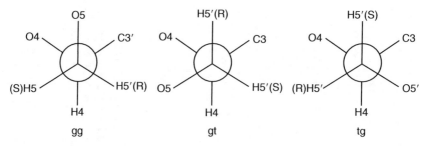

Figure 6   Newman projections showing the three main staggered conformations about the C4'–C5' exocyclic bond.

(e) 2',3'-Dideoxynucleosides such as 3'-azido-2',3'-dideoxyribosylthymine (AZT) with also anti-HIV activity and cytopathic effect on human T-lymphotropic virus type.[96–98]

(f) Bicyclic heterocyclic nucleosides which possess a base ring fused to various membered heterocyclic systems and are known for their reported *in vitro* and *in vivo* inhibition of various tumour cell lines.[99–101]

(g) 5'-O-Amino-2'-deoxy-nucleosides which are building blocks for antisense oligonucleotides and have recently gained much attention for their usefulness in antisense therapy.[102]

(h) α-Nucleosides which are conformational enantiomers of the common nucleosides with the anomeric carbon C1' in an inverted configuration and are found in vitamin B-12 and in arabino-nucleosides which differ from their ribo analogues in the altered configuration at C2' and exhibit broad antiviral activity against DNA-containing viruses as well as against RNA tumour viruses.[103]

From these preliminary considerations, it emerges that most nucleosides are generally in the *anti* conformation around the glycosidic linkage with the ribose in the N state, C3' *endo* form and the gg arrangement around the C4'–C5' exocyclic bond.[7–10] Nevertheless, it should be stressed that the presence of electronegative substituents either on the ribose moiety[76,94,104–110] or/and on the aromatic ring[76,89,109–118] as well as the presence of bulky groups on either one or the other cycle[76,112,113] can dramatically change the *anti/syn* geometry and/or the N/S puckering with also important modifications in the proportions of the gg, gt and tg rotamers. In some well-defined cases, hydrogen bonding between the C5' hydroxyl group and a particular acceptor group on the ribose or on the base can increase the proportion of the *syn* conformation.[76,109,113,116,119–126] In other well-defined cases, changes in the solvent or pH can also change the conformation and the furanose pucker.[114,121,127] Finally, it has also been shown that the S (C2' *endo*) form appears to be greater in deoxynucleoside as compared to ribonucleoside. This variability in flexibility, structure shaping and biological functions has been tentatively estimated, mainly, by Pullman and co-workers[128,129] in the 1970s by using (old) semi-empirical MO-SCF calculations and also by others[130–136] with often some recourse to molecular mechanics (MM). Today, this theoretical field has opened the way to the state of the art for the sugar puckering and the conformation simulated by MD and/or Car–Parrinello molecular dynamics (CPMD) and/or BD,[37,137,138] empirical potential functions (PF),[78,122,139–148] post-Hartree–Fock (HF) methods at the Möller–Plesset second-order perturbation level (MP2)[149–153] and naturally the DFT.[151–158] These aspects, which are also very dependent on the solvent, will be discussed in more detail below with respect to their structural and dynamics implications.

As a final point, it is well known that the nucleic bases are prone to tautomerism equilibrium between an amide form and an imide form. Nevertheless, with a usual pK_a of ~9–10, the equilibrium constant for aqueous conditions (pH ~6–8) favours the amide form very strongly so that the imide form, much less than 1%, can be safely neglected in this neutral water.[79]

4. ABOUT STACKING OF NUCLEOSIDES IN WATER

Before going into the details of water interaction at the atomic level of nucleosides, the problem of vertical stacking—a term introduced by Ts'o[159] and now currently in use—requires some comments. Such an important point has been largely investigated in the literature during the last 40 years, so that it is not possible to discuss here about the many published interpretations, sometimes completely contradictory.[160–166] The interested reader is referred to the many good relevant reviews on this aspect.[163–167] Nevertheless, since this stacking, if any, may compete with hydration for nucleosides in water, a brief survey and the personal point of view and experience of the author are given below.

According to Ts'o and many others,[159–167] the bases, the nucleosides, the nucleotides and the polynucleotides all associate extensively in an aqueous solution by an enthalpy-driven but entropy-unfavourable process where the bases pile up in parallel planes.[163–167] NMR evidence of these non-specific interactions is given by an upfield chemical shift of the base protons with an increase in concentration.[159] There is no doubt that the nucleic bases are so strongly hydrophobic that they repel the water molecules and tend to stack with one aromatic cycle under another at the van der Waals distance of about 3.5 Å. The origins and details of this arrangement, which is diffusion controlled, still remain a subject of debate, but it appears that intra-strand base–base stacking and inter-strand base–base stacking involve distinct, elaborate patterns.[163–172] In a single-stranded nucleotide as well as in nucleic bases and nucleosides, a head-to-tail structure seems to better conform to electrochemical diffusion measurements and to electrical energy minimisation.[163,172] This stacking mainly depends on London dispersion forces, so that the purines with large electronic polarisabilities and often large permanent dipole moments are more likely to stack than pyrimidines.[173] This effect, which largely depends upon the presence of carbonyl and/or amino groups, is pronounced in halogenated and methylated bases. Moreover, it is now clear that stacking is always much weaker than hydrogen bonding, and it has also recently been shown by quantum mechanical (QM) and DFT calculations that π–π sandwich and π induction do not play any role in the base stacking,[165,172] thus invalidating some previous claims.[162]

In an extensive analysis of the whole link between stacking, hydration, volumetric properties and adiabatic compressibilities in water, Chalikian and Breslauer[4] have shed very interesting light on this effect. Base stacking results in a volume decrease and a positive increase in adiabatic compressibility, reflecting thus "smaller hydration". In contrast, a volume increase that is accompanied by an increase in the water-accessible surface area of polar groups enhance hydration and results in a more negative compressibility and a "greater hydration". Within this framework, pyrimidine and purine bases are only poorly hydrated, cytosine exhibits the strongest hydration, but pyrimidine with a small dipole moment ($\mu = 2.3$ D) shows a less hydrated structure than purine ($\mu = 4.3$ D). On the other hand, nucleosides have large partial molar volumes without any sensitive difference between ribonucleosides and their 2'-deoxyribo counterparts, with large negative adiabatic compressibilities, mainly for the ribose. Indeed, it has been shown by many researchers, including Ts'o himself,[163] that there is no stacking for nucleosides in an organic solvent[104,138,163,174–188] and that this effect is also completely negligible in water, not to say non-existent,[174–188] at concentration less than 1 mol l^{-1}. Indeed, the pioneering work of Ts'o was undertaken with an ancient spectrometer with a poor magnetic field stability and a poor spectral resolution at a very low frequency, compared to modern instruments. Using changes in linewidths and chemical shifts less than 5 Hz with temperature and/or concentration at 40 MHz for proton resonances seem nowadays very adventurous to set up a stacking theory. It is our own experience that for ribo and C, T and U, the absence of any cross-correlation peaks between relevant nuclei in two distinct planes (protons for an inter-ring NOESY experiment, proton and carbon in an inter-ring HOESY experiment) is another proof of the lack of any sufficient long-lived stacked structure in water.[189–191] Others have also arrived at the same conclusions by relaxation data and intermolecular 2D NOE investigations.[192–194] Due to many conflicting data,[177,184,195–199] the situation is more puzzling for G, which is the most insoluble nucleoside in water, because it self-associates with its edges having self-complementary hydrogen-bond donors and acceptors.[195–199] This structure is also predicted as ideal for stacking with its polarisable aromatic surface with a strong molecular dipole.[195–199] For clearing the way, self-diffusion measurments,[75,76,200–203] which are still missing, are naturally welcomed since this extremely attractive technique should be very informative about the lifetime and the chemical structure of the postulated molecular aggregates.[204–216]

5. STRUCTURE IN WATER

Ever since the beginning of NMR, it has been shown by Ramsey[217] that the screening constant of a given nucleus mainly consists of two parts: a diamagnetic Lamb contribution associated with the free circulation of s electrons around the nucleus, and a paramagnetic term resulting from the non-spherical electronic distribution due to low-lying energy np states. Roughly speaking, chemical shifts are directly related to the electron density around the relevant nucleus, which is essentially due to the diamagnetic contribution for the proton and to the paramagnetic contribution for the heteroatoms (carbon-13, nitrogen-14 and nitrogen-15 and oxygen-17). For nuclei belonging to aromatic compounds such as the nucleic base ring, the approximate Karplus–Pople[218] Equation (1) relating the chemical shift $\Delta\delta$ and the electronic density Δq, both reported from a reference molecule, is frequently used to get a qualitative idea of the electronic delocalisation from the NMR chemical shifts.

$$\Delta\delta = -k\Delta q \qquad (1)$$

In Equation (1), k is a constant depending on the observed nucleus (e.g. ca. 10, 160, 850 and 540 ppm/electron for, respectively, proton,[219] carbon-13,[220] nitrogen-15[221] and oxygen-17[222] but with more uncertainty for the two latter nuclei). Hydration of the nucleoside may affect this atomic electron density by several effects[219–222] such as hydrogen bonding, electrostatic field, magnetic anisotropy interaction, dispersion forces and, naturally for the labile protons, chemical exchange with the water.

In the same manner, indirect nuclear spin–spin coupling constants, which are mediated via the bonding electrons, largely depend on the electron density at the relevant nuclei by the Fermi contact term[223] and may also probe intermolecular interactions. Indeed, these solvent effects are generally very low for a molecule with a rigid and unique conformation and their quantitative estimation by the usual theories such as reaction field, cluster model or rotating point dipole is always highly speculative. When considering compounds that can accommodate many conformations, it should be borne in mind that the solvent-induced changes in relative conformer populations can give rise to large effects on the measured coupling constants. In this way, the Karplus relation,[224] obtained by an approximate valence bond theory for a six-electron fragment, gives a useful information about the dihedral $N-X-X''-N'$ angle χ from the coupling constant $^3J_{NN'}$ and constitutes the most important, successful application of the early theoretical methods to the conformational analysis, Equation (2).

$$^3J_{NN} = A\cos^2\chi + B\cos\chi + C \qquad (2)$$

Extensive use of this relation has been made, often by an adequate re-parameterisation of the constants A, B and C and, for example, Rüterjans and co-workers[225] have reviewed 25 Karplus-type equations corresponding to the six χ-angle-related couplings in proteins. In this way, the empirical generalisation of Altona and coworkers.[226] which superpose in Equation (2) a linear dependency on the electronegativity of the substituents attached to the $H-C-C-H$ fragment under study is supported by some electronic justifications and seems to give better results for the analysis of pseudo-rotation of the sugar ring in a large population of nucleosides.

The author is not an expert in this field of theoretical calculations of chemical shifts and coupling constants, so it is hoped that the reader can pick up some papers previously quoted in the text and refer to the early reviews by Kowalewski[227], Webb et al.[228], de Dios[229] and Fukui[230] as well as to the two more recent papers by Jaszusńki and co-workers[231] using the modern concepts of DFT calculations.[232]

It appears that the Jardetzkys[233] were the first in the 1960s to report the chemical shifts of nucleosides in D_2O at acidic and basic pHs. After some controversial assignment of the H2 and the H8 of purine,[234,235] it has been clearly established that, in A, H8 is shifted downfield by about 0.2–0.3 ppm as compared to H2. With the low magnetic field in use at that time (40 MHz) and poor homogeneity, neither the ribose protons were resolved nor the diastereotopic exocyclic methylene protons were separated. However, by comparing the H2', H3' and H4' proton shifts in U, C, A, I and X, it was shown that the pyrimidine resonances were less well resolved and always at a higher field than those of the purine ones.[232] Another important point is that there did not seem to have any noticeable stacking of the nucleosides in water. From these observations and a crude analysis of spin–spin coupling constants, Jardetzky[233] concludes in the next paper that the D-ribose takes a specific conformation in the Pys, another one in the Pus and yet another one in nucleotides. Ts'o et al.[236–238] observed a differential effect on the proton chemical shifts of both Pys and Pus in D_2O which they attributed to a greater solvent–solute interaction of the water molecules with the H6 or H8 protons than with the H5 or H2 proton in Pys and Pus, respectively. This assertion seems to be supported from a comparison with chemical shifts observed in anhydrous dimethylsulfoxide (DMSO) by the same group. Finally, the authors also confirm that in water all the nucleosides have the *anti* conformation with respect to the sugar–base torsion angle.

Determining the *syn* and *anti* conformations of the glycosidic bond of nucleosides as well as the N or S puckering of the ribose ring in water is fundamental in nucleoside chemistry because the structural geometry is often a critical factor in biological activity (e.g. antiviral) of these compounds. Multinuclear NMR is undoubtedly one the best ways for tackling the problem of nucleoside conformation in water. To this end, we report in Tables 1 and 2 the proton chemical shifts of the most common ribonucleosides and deoxyribonucleosides, respectively. Table 3 gives proton chemical shifts of some particular nucleosides (both ribo and deoxyribo) which might be of interest in biological and pharmaceutical fields. These include (i) arabinonucleosides (Figures 2 and 3)—quoted hereafter with the prefix ara—which are epimers of β-nucleosides on the carbon C2', (ii) α-nucleosides which are epimers of β-nucleosides on the carbon C1' (Figures 2 and 3), (iii) some rigid cyclic nucleosides with specific bonding between the base and a ribose hydroxyl group—quoted with the prefix c—and (iv) other aza, thio or C-nucleosides. All these compounds may be of interest for their antiviral activity and/or useful for determining the preferential geometry of the most common β-nucleosides. Tables 4 and 5 are concerned with the carbon-13 chemical shifts of the most common ribo and deoxyribonucleosides, respectively. The nitrogen-15 chemical shifts of these common nucleosides are reported in Table 6. Finally, all the most useful coupling constants (J_{HH} and J_{CH}) are given in Tables 7 and 8 for ribo and deoxyribo nucleosides, respectively. As far as possible, all data have been selected from the literature for nucleosides dissolved in water at room temperature. But due to the low solubility of some of them in this solvent, we have also reported

	H2	H5	H6	H8	H1'	H2'	H3'	H4'	H5'	H5''
A[a]	8.20			8.30	6.06	4.80	4.44	4.31	3.93	3.87
A[b](d)	8.23				5.92	4.62	4.17	4.33	3.91	3.91
8-BrA[c]	8.04				6.14	5.07	4.52	4.30	3.92	3.92
8-NHMeA[c]	8.10			2.97*	5.98	4.80	4.44	4.29	3.90	3.90
8-NMe$_2$A[c]	8.17			3.02*	5.95	5.17	4.47	4.24		
iPA[d](d)		1.34*	1.56*	8.34	6.14	5.35	4.98	4.24	~3.56	~3.56
6-NMe$_2$iPA[e]	7.76	Me endo: 1.36*	Me exo: 1.64* NMe$_2$: ~3.52*	8.28	5.84	5.24	5.11	4.50	3.95	3.77
C[f]		6.11	7.93		5.91	4.34	4.23	4.16	3.96	3.85
5-MeC[g]		2.06*	7.69		~6.02	~4.33	~4.33	~4.33	~4.07	~3.97
6-MeC[g]		~5.93	~2.47*		~5.77	~4.92	~4.38	~4.10	~3.98	~3.88
G[h](d)				~8.55	5.67	4.40	4.11	3.88	3.64	3.52
isoG[i](d)				8.38	5.81	5.46	4.52	4.12	3.94	3.94
8-BrG[i](d)						4.95				
8-NH$_2$G[i](d)						4.53				
8-OHG[i](d)					5.56	4.83	4.07	3.77	3.57	3.43
8-SHG[i](d)					6.25	4.95	4.22	3.79	3.66	3.50
8-OMeG[i](d)						4.70				
8-SMeG[i](d)						4.83				
8-OCH$_2$ΦG[i](d)						4.70				
8-SOMeG[i](d)						4.96				
8-ThioG[i](d)						4.92				
8-OxoG[i](d)						4.81				
8-tBuG[h](d)				1.39*	5.95	5.06	4.17	3.87	3.68	3.54
6-ThioG[k](d)						4.38				
G[c]				8.00	5.91	4.73	4.41	4.24	3.87	3.87
8-BrG[c]					6.00	5.09	4.53	4.25	3.89	3.89
8-NH$_2$G[c]					5.86	4.86	4.43	4.21	3.89	3.89
8-NHMeG[c]				2.92*	5.85	4.79	4.41	4.22	3.89	3.89

(continued)

TABLE 1 (continued)

	H2	H5	H6	H8	H1'	H2'	H3'	H4'	H5'	H5''
8-NMe₂G^c				2.89*	5.84	5.13	4.50	4.21	3.89	3.89
I^f(d)	8.07			8.33	5.90	4.51	4.16	3.88	3.68	3.58
iPI^f(d)	8.08			8.30	6.10	5.26	4.94	4.22	~3.56	~3.56
O^m		5.77			5.56	4.75	4.34	3.95	3.87	3.76
T^n(est)		1.90*	7.70		~5.91	~4.24	~4.24	~4.24	~3.94	~3.84
U^o		5.90	7.89		5.92	4.36	4.24	4.14	3.92	3.82
U^p(d)		5.67	7.88		5.92	4.37	4.24	4.15	3.93	3.82
5-BrU^q(est)			8.36		~5.90	~4.38	~4.25	~4.38	~4.00	~4.00
5-FU^q(est)			~8.25		~6.02	~4.38	~4.26	~4.38	~4.00	~4.00
5-ClU^q(est)			8.24		5.89	~4.17	~4.05	~4.17	~3.77	~3.77
5-IU^q(est)			8.44		~5.81	~4.09	~4.02	~4.09	~3.74	~3.74
5-OHU^r			7.45		5.94					
5-OmeU^r			7.87		5.91					
5-NH₂U^r			7.32		5.95					
5-NMe₂U^r			7.52		5.96					
5-CNU^r			8.71		5.87					
5-NO₂U^r			9.66		5.92					
iPU^f(d)		5.66 / Me endo:1.32* / Me exo: 1.50*	7.85		5.88	5.00	4.72	4.08	3.58	3.58
5-CNiPU^s(d)			8.67		5.76	4.90	4.71	4.17	3.59	3.59
3-MeU^t	3.33*	5.96	7.88		5.94	4.34	4.22	4.14	3.93	3.82
6-MeU^u(est)		5.76	~2.41*		5.67	~4.82	~4.40	~4.01	~3.90	~3.80
X^v(est)				~8.53	~6.36	~4.81	~4.81	~4.81	4.47	4.47
X^w				7.93	5.89				3.93	3.93
Tph^x	Me-N1: 3.53* / Me-N3: 3.34*			8.33	6.21	4.54	4.31	4.21	3.96	3.82
W^y(d)	8.3	Me-N4: 4.0* / Me-C6: 2.2*		H7:7.5	6.1	4.4	4.1	4.0	3.7	3.6

Notes: (d) means that shifts are reported in DMSO-d₆ as solvent and from internal TMS

TABLE 2 Proton chemical shifts of some common deoxyribonucleosides in water and in ppm from internal TSPA

	H2	H5	H6	H8	H1'	H2'/H2''	H3'	H4'	H5'	H5''
dA[a](d)	8.38			8.51	6.57	2.76 / 2.36	4.44	4.08	3.80	3.90
dA[b]	8.04			8.21	6.36	2.75 / 2.54	4.63	4.18	3.84	3.78
(S,S)-isoddA[c]	~8.02			~8.19	α 4.18 / β 4.23		α 2.75 / β 2.02	4.26	3.87	3.67
dC[d](d)		6.64	8.69		6.47	2.55 / 2.31	4.68	4.26	4.11	3.94
dC[e]		6.06	7.84		6.28	2.45	4.45	4.07	3.85	3.77
5-MedC[t](d)		7.50	7.60							
6-AzadC[t](d)										
dG[f]				7.98	6.30	2.79 / 2.52	4.62	4.13	3.81	3.76
dG[g](d)				8.08	6.16	2.66 / 2.30	4.26	3.79	3.54	3.48
8-OHdG[h](d)					5.56	2.97 / 1.91	4.32	3.74	3.43	3.56
α-dG[i](d)				8.00	6.09	2.67 / 2.21	4.33	4.06	3.43	3.40
dI[i](d)	8.10			8.32	6.30	2.60 / 2.32	4.26	3.87	3.56	3.56
dT[k]		1.90*	7.67		6.28	2.38 / 2.40	4.48	4.03	3.85	3.78

(continued)

TABLE 2 (continued)

	H2	H5	H6	H8	H1'	H2' H2''	H3'	H4'	H5'	H5''
3-MedT[l]	3.31*	1.93*	7.67		6.33	2.37 2.41	4.47	4.04	3.86	3.78
α-dT[m]		1.91*	7.66		6.24	2.75 2.20	4.47	4.22	3.79	3.71
3'-Azido dT[n] (AZT)		1.74*	7.49		6.06	2.36 2.36	4.21	3.87	3.71	3.64
4'-Azido dT[o] (AZT)		1.97*	7.59		6.48	2.64 2.61	4.72		4.00	3.94
dU[p]		5.89	7.86		6.29	2.40 2.30	4.46	4.05	3.84	3.76
3-MedU[j]	3.29*	5.96	7.85		6.30	2.33 2.44	4.46	4.06	3.85	3.76
5-FdU[q]					6.27	2.33 2.41	4.46	4.01	3.83	3.76
6-MedU[r]	2.39*	5.71			6.19	2.27 2.96	4.54	3.93	3.86	3.76
6-Meα-dU[s]	2.37*	5.72			6.10	2.19 2.74	4.45	4.42	3.73	3.65
dX										

Notes: (d) means that shifts are reported in DMSO-d_6 from internal TMS. H2' is the proton over the ribose ring whereas H2'' is the proton under the ribose ring on the same side of the base in β nucleosides.

References: [a]244,256; [b]257,258; [c]259; [d]244; [e]243,257; [f]260; [g]244,256,261; [h]256; [i]261; [j]258,261; [k]103,117,257; [l]257; [m]103,117; [n]98; [o]90; [p]105,257; [q]103; [r]103,117,257; [s]117; [t]262.

TABLE 3 Proton chemical shifts of some other particular nucleosides (cyclic, aza, dihydro, thio, ara and α) in water and in ppm from internal TSPA

	H2	H5	H6	H8	H1′	H2′	H3′	H4′	H5′	H5″
8-AzaA[a]	8.23				6.43	5.10	4.72	4.38	3.94	3.83
β-araA[b]				8.37	6.45	4.58	4.38	4.10	3.99	3.91
5-AzaC[c](d)			8.58							
6-AzaC[c](d)		7.53								
β-araC[d]		6.05	7.82		6.21	4.42	4.14	4.03	3.93	3.85
β-aracC[e]		6.53	8.06		6.59	5.51	4.65	4.39	3.55	3.55
cA[f]	8.18				6.19	4.78	4.56	4.79	4.72	4.32
α-C[e]		5.98	7.68		6.08	4.41	4.26	4.18	3.86	3.66
α-cC[e]		6.53	8.06		6.47	5.53	4.36	3.83	3.88	3.69
FoA[g](d)	8.16			8.16	4.97	4.51	4.12	3.95	3.66	3.52
FoA[g]	8.20			8.20	5.24	4.62	4.38	4.25	3.93	3.82
FoB[h]					5.25	4.60	4.38	4.26	3.95	3.84
FoB[h](d)										
8-AzaG[a]					6.19	5.01	4.62	4.31	3.91	3.80
8-AzidoG[w](d)					5.72	~5.51	~4.78	3.82	~3.62	~3.38
cI[e](d)					5.94	4.46	4.27		3.91	3.80
8-AzaI[a]					6.37	5.07	4.68	4.34	3.91	3.80

(continued)

57

TABLE 3 (continued)

	H2	H5	H6	H8	H1'	H2'	H3'	H4'	H5'	H5''
dihydroU[i]	5a: 2.75 5b: 2.75			6a: 3.59 6b: 3.53	5.86	4.30	4.16	4.03	3.80	3.73
6-AzaU[j]		7.42			6.15	4.60	4.39	4.11	3.84	3.71
2-ThioU[k]		6.03	8.00		6.49	6.27	4.06	4.06	3.89	3.74
4-ThioU[l]		6.58	7.77		5.88	4.37	4.24	4.16	3.96	3.83
β-araU[m]		5.88	7.88		6.20	4.43	4.15	4.01	3.94	3.86
β-aracU[n]		6.22	7.86		6.48	5.51	4.61	4.34	3.51	3.51
β-ΨU[o]			7.68		4.69	4.29	4.16	4.02	3.86	3.74
3-Meβ-ΨU[p]		Me-N3: 3.23*	7.61		4.67	4.23	4.10	3.97	3.82	3.68
α-ΨU[q]			7.58		5.01	4.37	4.34	4.01	3.90	3.72
α-U[r]		5.81	7.73		6.09	4.42	4.26	4.23	3.84	3.65
5-CN				8.35	6.87	4.81	4.81	4.45	3.60	3.60
iPα-U[s](d)		Me endo: 1.27* Me exo: 1.34*								
cU[t]		6.21	7.92		6.54	5.48	4.67	4.40	3.58	3.57
α-cU[n]		6.18	7.85		6.36	5.43	4.30	3.81	3.88	3.69
β-aracO[n]		7.80			7.06	5.39	4.70	4.31	3.52	3.52
β-CAR[u]					6.12	4.73	4.38	3.97	3.87	3.74
6-Thio		Me-N1: 3.68*		8.50	6.94	4.25	~4.15	~4.05	~3.95	3.85
Tph[v] (d)		Me-N2: 3.50*								

Note: (d) means that shifts are reported in DMSO-d$_6$ as solvent and from internal TMS.

References: [a]239; [b]127; [c]262; [d]127,117; [e]114; [f]110; [g]263; [h]263; [i]249; [j]257; [k]93; [l]257; [m]117,264; [n]114; [o]249,265; [p]266; [q]265; [r]114; [s]251; [t]267; [u]249,265,248; [v]254.

TABLE 4 Proton–proton and carbon-13–proton coupling constants of some selected Ribonucleosides in water and in Hz

	HI'/H2'	H2'/H3'	H3'/H4'	H4'/H5'	H4'/H5''	H5'/H5''	C2/HI'	C6/HI'	C4-HI'	C8/HI'	CI'/H6	CI'/H8
A[a]	6.0	5.0	3.4	3.0	3.4				3.6	3.9		
2-FA[b]									3.2	4.1		
8-BrA[c]	7.1	5.1	~2.3	~5.3	~4.7	−12.1			~5.0	3.7		
8-NHMeA[d]	7.6											
8-NMe$_2$A[d]	7.6											
iPA[e]	2.8	6.2	2.4	4.7	4.7							
6-NMe$_2$ iPA[f]	4.7	5.9	1.2	1.5	2.0	−12.6			4.9	3.0		
araA[g]	5.5	5.5	6.3	2.8	4.6							
cA[h]	~0	6.3	~0	2.0	1.5	−13.5						
8-AzaA[i]	4.2	5.1	4.9	2.7	4.7	−12.7						
C[j]	3.6	5.0	5.8	2.8	4.2	−12.7	1.9	3.3				
5-FC[b]	3.3						1.7	3.4				
5-MeC[k]	3.3											
6-MeC[k]	3.7						~6.0					
araC[l]	4.9	3.9	4.9	3.0	5.7							
aracC[m]	6.0	0.7	1.8	3.4	3.4	−12.0						
G[n]	5.5	5.1	3.9	3.2	3.3				2.5	4.5		
8-BrG[o]	6.1	5.1	1.5	3.3	3.4	−12.5			4.8	4.6		
8-NH$_2$G[p]	7.0								3.0	5.5		
8-NHMeG[d]	7.0											
8-NMe$_2$G[d]	6.7											
8-OMeG[q]									4.1	4.5		
8-SMeG[q]									4.5	4.4		
8-OxoG[q]									4.1	5.3		

(continued)

TABLE 4 (continued)

	HI'/H2'	H2'/H3'	H3'/H4'	H4'/H5'	H4'/H5''	H5'/H5''	C2/HI'	C6/HI'	C8/HI'	C4-HI'	CI'/H6	CI'/H8
8-SO$_2$MeG[q]									4.4	4.5		
8-OCH$_2$ΦG[q]									4.2	3.9		
8-tBuG[r]	6.1	5.8	3.2	4.2	5.0	−12.3						
iPG[s]	2.9	6.3	2.3	~5.1	4.7	−11.8						
8-AzaG[i]	4.7	5.1	4.6	3.4	5.0	−12.7						
6-ThioG[q]									4.2	2.5		
8-ThioG[q]									4.0	5.4		
I[t]	5.8	5.0	3.6	3.2	3.9	−12.1						
iPI[k]	2.9	6.2	2.6									
8-AzaI[i]	4.5	5.1	4.9	3.2	5.0	−12.7						
O[u]	3.6	6.3	7.0	3.0	6.1	−12.2						
β-aracO[v]	5.9	1.9	3.9	6.1	6.1	−12.0						
T[w]	5.0	5.3	5.3	3.0	4.3							
2-ThioT[x]	2.7		7.6	2.7	3.3							
U[y]	4.3	5.3	5.6	3.0	4.3	−13.0	2.3	3.6	0.4*			
5-NH$_2$U[z]	4.7											
5-NMe$_2$U[z]	3.7											
5-OHU[z]	4.5											
5-OMeU[z]	3.6											
5-BrU[α]	3.3	4.5		3.0	3.0	−12.6						
5-FU[β]	4.0						2.1	3.6				
5-ClU[β]	3.7											
5-IU[χ]	4.0											
5-CNU[z]	2.8											
5-NO$_2$U[z]	1.7											
3-MeU[δ]	4.1	5.4	5.9	2.9	4.5							
6-MeU[ε]	3.3	6.2	6.6	6.0								

dihydroU$^\varphi$	6.3	6.0	3.6	3.6	4.8	−12.6	
6-AzaU$^\kappa$	3.2	5.5	5.5	3.6	5.5	−12.6	1.0
2-ThioU$^\lambda$	2.5	4.0	6.0	1.6	3.0	−13.5	
4-ThioU$^\mu$	3.9	5.4	5.6	3.0	4.3	−12.7	0.4
β-ΨU$^\nu$	5.0	5.0	5.2	3.2	4.6	−12.7	
α-ΨU$^\nu$	3.3	4.2	7.9	2.4	5.7	−12.7	
iPU$^\pi$(d)	2.6						
5-CNiPU$^\theta$	3.2	6.1	3.0	~4.0	~4.0		
5-CNiPα-U$^\theta$	3.9	6.3	~0	~3.0	~3.0		
cU$^\rho$	5.8	0.9	1.9	4.1	4.2		
araU$^\sigma$	5.0	4.6	5.6	3.1	5.5	−12.6	
aracU^m	5.9	0.7	1.7	4.0	4.0	−12.0	
X$^\tau$	5.7	4.9	3.8	3.0	3.3	−12.4	
β-CAR$^\omega$	3.9	6.4	6.6	3.2	6.2	−12.8	
Tph$^\xi$	3.5	5.6	5.6	2.4	4.4	−12.7	

References: ^{a}104,110,115,138,239,240,268; ^{b}138; ^{c}110,240,252; ^{d}240; ^{e}252; ^{f}241; ^{g}127; ^{h}110; ^{i}239; ^{j}112,117,127,134,138; ^{k}112,247; ^{l}117,127,134; ^{m}114; ^{n}110,240,268; ^{o}115,184,240; ^{p}184,240; ^{q}184; ^{r}115; ^{s}252; ^{t}110,268,269; ^{u}248; ^{w}114; ^{w}117,138; ^{x}92; ^{y}105,111,138,242,248,249,257; ^{z}118; $^\alpha$111,242; $^\beta$111,138; $^\gamma$242; $^\delta$117; $^\epsilon$242; $^\zeta$93; $^\eta$247; $^\theta$251; $^\iota$249; $^\kappa$249; $^\lambda$257; $^\mu$257; $^\nu$249; $^\xi$242,257,270; $^\rho$267; $^\sigma$117,264; $^\tau$271; $^\omega$248,249; $^\xi$254.

TABLE 5 Proton–proton and carbon-13-proton coupling constants of some selected deoxyribonucleosides in water and in Hz

	H1'/H2'	H1''/H2''	H2'/H3'	H2''/H3'	H2'/H2''	H3'/H4'	H4'/H5'	H4'/H5''	H5'/H5''	C2/H1'	C6/H1'	C4/H1'	C9/H1'	C1'/H6	C1'/H8
dA[a]	7.1	6.0	5.8	3.0	−13.9	3.2	3.0	4.0	−12.6						
8-BrdA[b]															
8-NHMedA[b]															
8-NMe₂dA[b]															
2'-FdA[c,*]	2.5	F 17.1	4.5	F 18.3	F 52.0	7.3	2.2	3.5	−13.0						
(S,S)-isoddA[d]	4.7	6.0*	6.9	6.7*	−11.2*	c 4.0 t 6.7	2.4	4.0	−10.2						
dC[e]	6.9	6.3	6.6	3.9	−14.0	4.0	3.7	5.0	−12.5						
5-MedC[f]															
6-MedC[g]															
2'-FdC[n,*]	1.2	F 17.0	5.0	F 22.5	F 52.9	7.2	2.3	3.5	−12.1						
dG[h]	7.4	6.5	6.3	3.6	−13.9	3.4	3.5	4.8	−12.5						
8-BrdG[i]															
8-NH₂dG[i]															
8-NHMedG[i]															
8-NMe₂G[i]															
α-dG[j]	7.7	2.8	7.1	3.0	−14.2	2.9	4.4	4.8	−11.7						
dI[l]															
dO[k]															
dT[l]	6.9	6.9	6.0	5.0	−14	3.9	3.7	5.0	−12.3						
3-MedT[m]	6.5	6.7	6.9	3.8		4.3	3.5	5.1		2.0					
2-ThiodT[n]	6.5		6.5			4.5	3.3	4.7							
α-dT[o]	7.2	3.3	6.5	3.0	−14.8	2.8	3.9	5.5	−12.4						

3'-Azido dT[p] (AZT)	6.4	6.7	7.2	5.2	-13.9	5.5	3.5	4.6	-12.6	2.3	3.9
4'-Azido dT[q] (AZT)	3.8	8.2	8.1	8.3	-14.2				-12.6		
dU[r]	6.4	6.5	6.5	4.0	-14.1	4.0	3.4	5.1	-12.6		
5-FdU[o]	6.6	6.2	6.7	3.9	-14.2	3.8	3.5	4.9	-12.5		
5-Fα-dU[o]	7.1	2.4	5.9	2.3	-14.9	2.2	3.7	5.4	-12.5		
5-CN, iPdU[s](d)											
3-MedU[t]	6.6	6.6	6.8	4.1		3.9	3.6	5.2			
6-MedU[t]	5.2	8.6	8.2	5.7	-13.9	5.8	3.4	6.7	-12.0		
6-MeαdU[u]	7.5	7.5	7.4	7.4	~-14	7.0	2.3	4.6	-12.5		
dTph											
dX											

Notes: For (S,S)-isoddA, 6.0* in column 3 is JH2'H1'', 6.7* in column 5 is JH2'H3'', c 4.0 and t 6.7 in column 7 are, resp., JH3'H4' and JH3''H4'.

References: [a]104,107,110,115,239,257; [b]240; [c]115; [d]259; [e]117,127,257; [f]117,242; [g]242; [h]240,260,261; [i]240; [j]261; [k]248; [l]103,117,257,272; [m]257; [n]92; [o]103,117,257; [p]98; [q]90; [r]105,249,257; [s]251; [t]117,257; [u]117.

TABLE 6 Carbon-13 chemical shifts of some selected ribonucleosides in water in ppm and from internal TMS

	C2	C4	C5	C6	C8	C1′	C2′	C3′	C4′	C5′
A[a]	153.3	149.2	119.9	156.4	141.4	89.2	74.7	71.5	86.6	62.9
2-FA[a]	159.6	150.8	118.2	158.1	141.5	89.0	74.5	71.3	86.4	62.3
A[b](d)	153.3	150.1	120.4	157.1	141.0	89.0	74.5	71.7	86.9	62.7
8-BrA[c](d)	153.4	150.9	120.6	156.1	128.1	91.4	72.1	71.9	87.7	63.1
8-ClA[d](d)	153.6	150.7	119.0	156.2	137.9	90.4	74.7	71.8	87.7	63.1
8-MeA[d](d)	152.3	150.6	119.1	156.4	150.0	86.7	72.0	71.4	86.5	63.4
8-OHA[d](d)	151.6	147.6	104.6	148.2	152.5	86.8	72.0	71.4	86.5	63.4
8-OMeA[d](d)	151.5	149.8	113.8	155.0	155.4	87.8	72.1	71.9	87.0	63.3
8-SMeA[d](d)	152.2	151.7	120.6	155.4	150.7	89.9	72.3	72.0	87.6	63.2
2-ClA[e](d)	154.4	151.5	119.5	157.8	141.4	89.1	75.0	71.6	86.9	62.7
6-NMe$_2$ iPA[f]	151.2	148.9	121.7	155.3	138.1	94.3	82.8	81.8	86.1	63.6
C[g]	158.7	167.3	97.3	142.8		91.4	75.1	70.4	84.9	61.9
C[h](d)	156.7	166.5	95.4	142.6		89.9	74.9	70.3	85.1	61.6
5-AzaC[e](d)	154.4	166.6		157.3		90.4	74.8	69.9	85.1	61.1
5-FC[a]	156.6	159.3	138.4	126.6		91.1	75.1	70.0	84.7	61.4
6-MeC[h](d)	158.5	167.4	97.6	156.3	22.2*	93.8	73.4	72.5	87.2	64.6
6-OHC[e](d)	155.9	165.0	119.9	152.3		88.5	76.1	71.8	85.7	64.0
2-ThioC[e](d)	181.1	161.3	99.2	142.8		94.6	76.2	69.5	85.3	61.0
β-araC[i]	158.6	167.4	96.5	144.0		87.2	76.8	76.8	84.5	62.1
FoA[j](d) [9]=C9	151.5	138.7	122.6	151.2	143.8[9]	78.4	75.2	72.4	86.1	62.6
FoA[k] [9]=C9	148.7	137.7	132.1	154.4	140.6[9]	78.2	75.0	71.8	84.9	62.1
FoB[l](d) [9]=C9	143.3	136.7	128.0	153.6	147.7[9]	77.7	74.9	72.1	85.7	62.5
G[m](d)	154.7	152.3	117.8	157.8	136.6	87.5	74.8	71.4	86.3	62.5
8-BrG[d](d)	154.4	153.0	113.6	156.4	122.1	90.8	71.5	71.5	86.9	63.1
8-MeG[d](d)	153.9	152.7	116.3	157.2	145.9	88.7	72.3	71.4	86.5	63.8
8-OHG[d](d)	153.6	148.5	99.7	155.1	153.0	86.7	71.9	71.0	86.1	63.4
8-SMeG[d](d)	154.0	153.7	118.2	156.7	147.2	89.3	71.7	71.7	86.8	63.2

isoG[n]	152.1	142.6	110.7	148.7	139.0	89.5	73.7	70.5	85.9	60.6
6-ThioG[e](d)	154.5	149.1	129.6	176.3	139.7	88.1	71.1	71.6	86.6	62.6
I[o](d)	148.7	146.4	124.9	157.1	139.7	88.0	74.6	70.8	86.1	62.0
8-BrI[d](d)	147.2	150.2	126.4	156.2	127.0	91.5	72.3	71.5	87.3	62.9
8-MeI[d](d)	145.9	149.8	124.4	157.1	149.3	89.4	73.0	71.5	87.1	62.8
8-SmeI[d](d)	157.8	149.6	122.2	158.7	139.1	88.5	75.0	71.4	86.5	62.4
6-ThioI[e](d)	146.3	144.9	136.5	176.9	142.1	88.7	75.3	71.0	86.6	62.1
cI[p](d)	152.9	148.7	115.4	155.7	154.4	89.7	78.1	72.0	89.0	75.6
T[i]	153.1	167.7	112.7	138.7	12.8*	90.2	74.8	70.7	85.4	62.0
U[q]	153.8	168.1	104.0	143.5		91.4	75.8	71.5	86.6	62.8
U[e](d)	152.2	164.5	102.8	142.0		88.9	74.6	70.9	85.7	62.0
5-FU[a]	151.0	160.2	141.6	126.4		90.2	74.7	70.1	85.2	61.4
5-OHU[e](d)	150.8	161.9	121.2	133.6		88.8	74.3	71.5	86.0	62.5
5-BrU[e](d)	151.1	160.2	96.9	141.5		89.8	75.2	70.5	85.9	61.5
5-NH$_2$U[e](d)	150.8	162.1	117.0	124.2		88.8	74.2	71.7	86.0	62.9
3-MeU[i]	153.3	166.7	102.7	140.9	29.0*	91.7	75.1	70.6	85.3	62.0
Me endo* iPU[r] C quat* Me exo*	152.0	165.3	103.2	143.5	26.5* 114.8* 28.4*	92.8	85.2	82.0	87.8	61.9
6-AzaU[e](d)	149.2	157.4	137.3			90.4	73.3	71.2	85.5	62.8
2-ThioU[s]	176.8	163.9	107.3	142.5		94.2	75.4	69.3	84.6	60.7
4-ThioU[e](d)	149.3	191.3	113.9	137.1		89.8	75.1	70.6	86.2	61.9
2,4-di ThioU[e](d)	174.0	187.2	138.5	135.7		94.5	75.1	69.8	85.8	60.8
5-CN iPU[t](d)	150.5	160.0	92.5	148.9	60.9[CN]	88.1	87.7	84.5	80.2	60.9
5-CN iPα-U[t](d)	150.1	160.0	87.4	148.0	62.1[CN]	87.1	83.6	81.4	79.0	62.1

(continued)

TABLE 6 (continued)

	C2	C4	C5	C6	C8	C1'	C2'	C3'	C4'	C5'
3-isoU[e](d)	152.3	164.5	156.1	142.6		88.6	72.2	71.4	85.7	63.6
β-ΨU[u]		165.7	110.9	141.9						
3-Meβ-ΨU[v]	153.2	164.8	110.1	139.3	27.2*	83.2	73.5	70.8	79.8	61.5
β-araU[i]	152.7	167.5	102.4	144.3		86.5	76.8	76.2	84.3	61.8
β-CAR[e](d)	150.5	149.5			150.5	89.6	72.6	71.3	85.8	63.5
Theoph[w]	153.6	150.3	107.8	156.7	142.1	91.3	75.9	70.0	85.4	61.7
Me-N1: 31.0*										
Me-N3: 29.2*										
6-Thio	149.3	145.3	116.9	175.6	143.0	89.6	75.3	67.9	83.9	59.6
Theoph[w] (d)										
Me-N1 33.9*										
Me-N3 30.1*										
X[e](d)	159.1	154.5	117.4	164.7	137.7	90.0	75.0	71.7	87.4	62.3
6-ThioX[e](d)	158.6	149.6	127.4	183.2	137.7	90.1	75.2	71.8	87.3	62.3
W[x](d)	135.4	140.0	C7:	137.3	C9:	88.7	74.8	69.7	85.6	60.6
4-Me-33.9	C3a:		105.7		C9: 151.4					
6-Me:14.0	140.0				C9a: 115.7					

Notes: (d) means that shifts are reported in DMSO-d_6 from internal TMS.

References: [a]138; [b]107,110,273; [c]107,110; [d]107; [e]273; [f]241; [g]117,138; [h]112,242; [i]117; [j]250,274,275; [k]250; [l]274,275; [m]107,110; [n]245; [o]107,269,273; [p]110; [q]105,138,276; [r]personal data; [s]93; [t]251; [u]277; [v]266; [w]254; [x]278.

5-CN, iP-U (Reference 251); Me endo: 25.1; Me exo: 26.9; C quat: 112.7.
5-CN, iP2x-U (Reference 251); Me endo: 23.7; Me exo: 25.1; C quat: 112.3.
6-NMe2, iPA (Reference 241); Me endo: 25.3; Me exo: 27.7; C quat: 113.9; NMe2: ~38.7.

TABLE 7 Carbon-13 chemical shifts of some selected deoxyribonucleosides in water in ppm from internal TSPA

	C2	C4	C5	C6	C8	C1'	C2'	C3'	C4'	C5'
dA[a](d)	149.5	152.8	119.9	156.7	140.4	84.8	40.0	71.4	88.4	62.2
(S,S)-isoddA[b]	154.5	150.8	121.5	157.4	142.2	74.5	57.3	36.2	82.3	65.0
dC[a](d)	157.2	167.1	96.0	142.6		86.8	40.8	71.9	88.8	62.8
dG[c](d)	151.3	154.2	117.9	157.7	136.0	82.8	39.8	71.0	86.1	62.4
8-BrdG[d](d)						85.0	36.4	70.9	87.8	
8-CldG[d](d)						83.9	36.6	70.9	87.8	
8-IdG[d](d)						87.3	36.7	71.2	88.0	
dT[a](d)	146.3	148.7	123.9	157.1	139.0	84.0	40.1	71.1	88.4	62.0
dT[e]	152.5	167.3	111.7	138.3	12.5*	86.0	39.4	71.3	87.4	62.1
3-MedT[f]	152.8	166.7	110.7	136.2	29.2*	87.0	39.6	71.2	87.5	62.1
4-ThiodT[g](d)	148.9	191.7	119.0	134.4	13.4*	86.0	40.1	71.3	88.0	62.3
4'-AzidodT[h]	152.3	167.3	112.4	138.5	12.3*	85.5	36.6	71.3	99.9	63.2
α-dT[i]	152.0	166.9	110.9	138.4	12.1*	87.2	39.9	71.1	88.9	61.9
dU[j]	152.4	167.0	103.1	142.8		86.4	39.6	71.4	87.1	62.3
5-BrdU[a](d)	150.8	160.2	96.7	141.4		86.1	40.7	71.0	88.6	62.0
3-MedU[f]	152.8	166.3	102.3	140.5	29.0*	87.3	39.9	71.3	87.2	62.2
6-MedU[i]	152.2	166.3	103.1	157.2	20.7*	86.5	37.8	71.6	87.0	62.6
6-Me αdU[i]	152.3	166.2	103.0	156.8	20.3	86.7	36.6	70.9	85.9	61.4
3-isodU[g](d)	152.2	142.2	101.4	164.4		81.7	37.1	72.1	87.0	62.6
dX										

Note: (d) means that shifts are reported in DMSO-d$_6$ from internal TMS.
References: [a]273; [b]259; [c]273,279; [d]107; [e]117,257; [f]257; [g]273; [h]90; [i]117; [j]105,117,257.

TABLE 8 Nitrogen-15 chemical shifts of selected nucleosides in water and in ppm from NH_3

	N1	N3	N7	N9	$NH_2(2)$	$NH_2(4)$	$NH_2(6)$
A^σ (d)	236.2	223.1	241.1	170.2			82.0
A^υ(d)	224.1	237.2	242.0	170.9			82.3
A^κ(d)	215.3	202.6	219.5	150.3			62.5
8-BrA^κ(d)	218.6	201.9					65.4
iPA^κ(d)	217.6	204.1	222.6	151.7			63.3
8-Br,iPA^κ(d)	218.4	203.3					64.6
dA^σ (d)	236.7	223.7	241.5	173.7			82.1
$C^{\iota,\rho}$(d)	152.9	209.6				94.0	
C^υ	153.3					93.6	
5-AzaC^σ(d)	167.9	191.4	216.6			97.4	
$G^{\rho,\sigma,\clubsuit}$(d)	148.1	166.6	247.8	170.8	74.1		
G^υ	148.8	167.6	248.5	171.6	74.9		
1-MeG^ψ(d)	143.4	164.8	247.3	168.3	80.2		
7-Me$G^{\psi,\clubsuit}$(d)	~192.1	167.0	159.0	170.7	80.6		
T							
dT^υ	145.3	156.9					
dT^σ	144.5	156.3					
U^σ (d)	142.5	157.6					
U^υ	146.6	158.9					
$I^{\tau,\omega,\sigma}$ (d)	175.2	212.3	249.4	175.3			
I^υ	176.0	215.3	249.9	176.0			
2-NH$_2$I$^\omega$(d)	156.7	165.2	246.7	169.3	72.8		
X^υ	155.1	215.1	250.3	167.4			

Notes: (d) means that shifts are reported in DMSO-d_6 as solvent.

some data in fully deuterated DMSO-d_6 as an essential condition for giving clear insights into the general trends. These data are quoted by (d) in the first column of all the relevant chemical shifts. As coupling constants seem to appear less sensitive to solvent effects than chemical shifts, the data shown in Tables 7 and 8 concern all the usual NMR solvents (D_2O, DMSO-d_6, $CDCl_3$, dioxane-d_6, etc.).

With the exception of the rigid cyclic nucleosides and to a lesser extent of the isopropylidene-substituted sugar ones, abbreviated iP, all nucleosides exhibit a great flexibility, pictured by the three interrelated and strongly dependent glycosidic torsion, sugar pucker and rotation about the exocyclic C4′–C5′ bond.[114] Such type of correlation between these structural parameters became clear since the 1970s when Prestegard and Chan[280] described correlations between the H6 chemical shift and the $^3J_{H'H2'}$ coupling constant of Pys in water. A mutual relation between ring puckering and the nucleoside conformation about the exocyclic C4′–C5′ bond has also been suggested by Wilson and Rahman.[140] Confronted with many contentious views on these aspects, researchers have then tried to put forth more or less empirical rules based mainly on chemical shifts and coupling constants of the sugar protons. The most remarkable contribution in this way is certainly the concept of pseudo-rotation of the sugar as proposed by Sundaralingam.[76,119] Its foundations and the general ideas have been briefly described above and despite its large application, it seems sometimes very cumbersome with some ambiguities. Nevertheless, it is now well recognised as very useful by the scientific community. More details about it can be found in the original papers[76,119] or in Saenger's book.[41]

5.1. Proton and carbon-13 data

Shieldings, homonuclear as well as heteronuclear coupling constants and relaxation times of these two nuclei are certainly the most useful—and the most used—parameters to obtain reliable structural information of nucleosides in water. All these aspects are discussed below, as well as the most important trends from a great number of publications since the 1960s. For clarity, we deliberately discuss separately the three dependent conformational aspects (glycosidic torsion, sugar puckering and rotation about the exocyclic bond) in an attempt to give the clearest insights of the most marked features for the reader.

5.1.1. Glycosidic torsion

5.1.1.1. Nuclear Overhauser effect   There are many spectroscopic and theoretical methods for approaching the torsion glycosidic torsion angle χ and its potential modifications with the solvent. From the NMR point of view, the most appealing method appears to be the nuclear Overhauser effect (NOE).[60] Because the fractional enhancement of the resonance of spin d when the resonances of spins s are saturated is related to the inverse sixth power of their distance r^{-6} if they mainly relax by an intra-molecular dipole–dipole mechanism, valuable information between their relative proximity may be gained. Qualitative application of this effect has been largely used in the past for the determination of molecular geometry of nucleosides,[109–111,122,125,158,176,188,241,255,256,259,261,266,268,269,271,280–293] nucleotides[169,294–298] and polynucleotides.[81–83,299–305]

By way of example, consider U. In the *anti*, C3' *endo*, N conformation, the protons H6, H1' are about 3.8 Å apart, whereas in the *syn*, C2' *endo* S, they are only distant from 2.4 Å, so that we should expect some NOE in the *syn* and not in the *anti*. Experiments by selective irradiation on the H1' and H2' protons of U in D_2O give, respectively, NOE enhancements $f_6(1')$ and $f_6(2')$ of $\approx 0\%$ and 23%, respectively, for the H6 proton,[282,283,285] thus ascertaining preferentially the *anti* form in which the H2', H6 protons are about 2.5 Å away.

β-pseudouridine β-ΨU

On the contrary, the β pseudo-uridine, β-ΨU, which seems to be essential in the anticodon of certain tRNAs for the genetic expression of the operon, should be in the atypical *syn* conformation according to Hurd and Reid[306] on the basis of the anomalous upfield shifts of the NH and the H6 resonances. By comparing the two independent NOE results of Guéron et al.[285] with the previous works of Smith and coll.,[284] the situation seems a bit confusing: the first authors advocate a *syn* form, whereas the latter ones do not exclude a rapid *syn–anti* conformational equilibrium in water! Moreover, the latter authors rule out the possibility of a hydrogen bond between the amide oxygen O4 and the hydroxyl OH5' which is predicted by CNDO calculations as a stabilisation of the *syn* form.[9,307,308] As a counter-example, we can quote the work by Chow et al.[266] on 3-methyluridine (3-MeU) and 3-methylpseu-douridine (3-MeΨU) in water: Here, 1D NOE difference spectroscopy tends towards an *anti* conformation for the first compound and a *syn* form for the second one, another very striking result! Indeed, such contradiction between many experiments and many authors is easily explained when it is considered that a large flexibility in the glycosidic bond rotation together with even small errors like 2–3% in the NOEs can result in an error of much as a factor of 2 for the estimation of the distance between the interacting spins.[309] In addition, for the special case of U, a partial overlap exists between the protons H5 and H1' and hydration of the amide group can seriously impede this intra-molecular rotation so that obtaining a reliable NOE between these two protons is highly challenging, even with the best attainable resolution in a NOESY experiment (see below).[190,191] This means that other ways should be explored to overcome these difficulties and that the NOE approach is not *per se* an alternative method, but should be considered only as a further proof, knowing that, in the point of view of this reviewer, 2D NOESY correlations should be preferred to the tedious and erratic 1D difference NOE experiments. Nevertheless, Hart and Davis[283] have measured a nearly constant NOE $f_6(2') \approx 0.2$–0.3 in a conformationally homogeneous series of 5-halogeno U in D_2O as in DMDSO-d_6, thus showing a solvent-independent *anti* conformation for these compounds.

Two independent studies of the conformational analysis in solution of the four common Pus A, G, I and X in DMSO-d_6 on one hand[90] and in D_2O and ND_3 on the other,[271] are extremely instructive of the limits of the 1D NOE in structural chemistry. Both groups arrive at the same conclusion that for all these Pus there is a rapid equilibrium *syn* $\rightleftarrows$ *anti* in close correlation with an even more rapid equilibrium S, C2' *endo* $\rightleftarrows$ N, C3' *endo* states for the ribose moiety. Moreover, results obtained in aqueous solutions are very similar to those found in deutero-ammonia, thereby proving that there are no conformational changes from one solvent to the other. Note that this appears like an *a posteriori* justification for using ND_3 as the solvent having similar hydrogen bonding capacity as water, according to the German group.[271] In contrast, comparative conformational analysis of guanosine 5'-monophosphate (5'-GMP) and G in DMSO-d_6 containing small traces of water is very illustrative of the potentialities of the method.[197] Whereas 5'-GMP is usually shown to adopt the *syn* form, some noticeable cross peaks between H8 and H2', OH5', H5' and H5'', respectively, give evidence of a preferential *anti* form.[197] Some interactions between NH_2 and H1' and H2', respectively, seem also to favour of dimer, while at the same time some intermolecular inter-actions can been observed between water and H8, (N)H1 and the two NH$_2$ protons.[197] Another enlightening example is given by the study in water of ribavirin Rb, an antiviral nucleoside with a triazole ring as the base instead of a pyrimidine ring.[176]

Ribavirin **Rb**

Irradiation of the H1′ ribose proton gives only one important nuclear Over-hauser enhancement equal to 22% on the H6 proton of the triazole as compared to the slight increase of 6% on the *trans* H2′ proton on the ribose. This results allows unambiguously assigning a *syn* glycosidic torsion for this compound in water, as opposed to the *anti* form observed for its inactive isomer.[176]

Finally, the most reliable NOE-based techniques for elucidation of nucleoside conformation are either based on a calibration graph from rigid molecules,[109] generally cyclic nucleosides, or on a population analysis accommodating multiple conformations and thus avoiding some biases inherent to the crude method.[310] In the example of 2′,3′-isopropylideneinosine (iPI), this last method puts forth three significantly populated conformers, mainly a *syn* form, then a high-energy stag-gered conformation and a minor *anti* form, the populations and distributions of which are more or less dependent on the rotational correlations used. These two points were not expected in the crude steady-state NOE analysis.[247]

2′, 3′-isopropylideneinosine **iPI**

As an indirect proof of the *anti* conformation of U in water, we will comment on our own work in water. Specific hydration occurs at the O2 and N3 of the amide groups of U. Because of steric hindrance by the ribose, this interaction of water can only be possible in an *anti* conformation where the carbonyl group C2=O2 is directed toward the exterior and then freely accessible. We have checked this conformation by running a phase-sensitive NOESY, which has shown two strong negative cross-peaks between the H6 proton of the uracil moiety and the H2' and H3' protons of the sugar (respectively, at distances 2.92 and 2.88 Å in the *anti* form, while these are 4.24 and 5.77 Å apart in the *syn* conformation). In this *anti* form, the ribose proton H1' and the pyrimidine carbon C2 are about 2.65 Å apart, so that a heteronuclear Overhauser effect (HOE) between them is expected, but seems to be completely negligible in the *syn* in which this separation becomes about 3.40 Å (smaller by a factor of about 6–7). The difficulty of the experiment lies in a partial overlap of the proton resonances H1' of the ribose and H5 of the base, which precludes a clear selective irradiation of the H1' proton (Figure 1).

To overcome this difficulty, we have developed a TOCSY–HOESY experiment which includes three distinct steps.[190] First, a selective E-BURP 2 excites the H2' proton of the ribose which is completely isolated and well separated from all the other protons (see Figure 1). Second, a TOCSY mixing transfers the H2' magnetisation to all the H1', H3' and H4' ribose protons engaged in the same J-coupling system. Third, the classical 1D HOESY follows with a 90° hard pulse on the protons followed by a mixing time $t_m = 2.5$ s between the two carbon-13 and proton longitudinal magnetisations. Neat HOE carbon-13 longitudinal magnetisations are finally read under broadband proton decoupling by a hard 90° read pulse on the carbon. The sequence and the neat 13C, {^{1}H} spectrum are given in Figure 7.

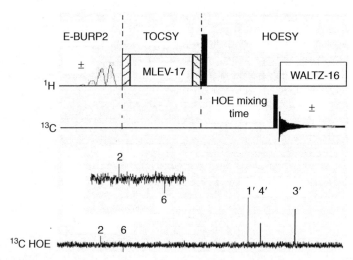

Figure 7 The selective TOCSY–HOESY sequence (top) and the HOE spectrum obtained for U after selective excitation of the proton H2' on the ribose by a E-BURP2 pulse (bottom). The insert (middle) represents an expansion of the response of the uracil carbons C2 and C6. The carbon C2 positive peak arises from a direct HOE transfer from the ribose proton H1' whereas the negative carbon C6 results from a NOE relay between H3' and H6 (adapted from Reference 190).

5.1.1.2. Chemical shifts on the ribose and deoxyribose   By looking at Tables 1–5, the reader should be surprised not to observe any significant changes on the chemical shifts of the base whatever the conformation *syn* or *anti* of the nucleoside. Indeed, H5 and H6 of Pys, like H8 of Pus, seem to be ambiguous monitors of the conformation about the glycosidic linkage[267] and the best sensors of this rotation are the nuclei on the furanose,[197,242,257] a situation which should be a little disappointing at first sight! This strange evidence has emerged after a large number of different authors studied a large family of nucleosides in a rigidly fixed *syn* or *anti* conformation.[104,114,141,306] It includes appropriate cyclo[109,110,119,267,287] and anhydronucleosides,[114,115] arabinonu-clesoides,[119,264] Pys or Pus substituted at carbon C6 or C8, respectively[115,240,260,279] and many others.[109] In some cases, an appropriate intra-molecular hydrogen between a ribose hydroxyl group and an aromatic nitrogen atom ensures an additional stability to the unique form.[109,110] It has then been shown that the protons H2', H3' and H4' of the *syn* compounds are all shifted downfield by 0.60–0.30 ppm and that the proton H1' is, on the contrary, shifted upfield by about 0.1–0.2 ppm as compared to their analogues in the *anti* forms.[242,257,311] The exocyclic protons H5' and H5'' are mostly not affected. But the story does not end here: going on, when it has also been observed that the corresponding carbons C2', C3' and C4' are all shifted upfield in an extent ranging from 1 to 3 ppm in these *syn* conformers,[107] another somewhat curious trend! Naturally, the C1' carbon generally appears shifted downfield and the C5' is essentially invariant. Such an apparent contradiction is well resolved if we accept the existence of an electrostatic field effect along the C–H bond which pushes away the bond electrons from the hydrogen atom, thereby deshielding the hydrogen nucleus and, correspondingly, shielding the attached carbon-13 nucleus.[312–314] This effect should be due to the lone electron pair of the nitrogen N3 of the bases, including probably the N1 of the Pus, which are just above the ribose in this *syn* conformation. In addition, in some pyrimidines like O or the β-cyanuric acid riboside, β-CAR, the magnetic anisotropy influence of the C2=O carbonyl group should also be accounted for.[248,257]

β-cyanuric acid riboside β-**CAR**

As this electric field is rapidly decreasing with the inverse square power of the interaction distance, its effect is more pronounced at the H2'' proton than at the H3' and H4' protons, respectively. The opposite trend observed for H1' and C1' as compared to the other furanose nuclei is no more puzzling since in the *syn* form the nitrogen N3 lone pair is closer to the carbon C1' than to the H1' proton and thus now pushes away the bond electrons from the carbon to the proton! A more detailed comparison of the data for O or β-CAR reveals other interesting trends consistent with the existence of the former as a *syn* rotamer in which the 2-keto is located above the sugar.[248,249] In the *syn* conformation, only H1′ and H2′ are in the vicinity of the negatively charged 6-substituent and are strongly affected by its presence. Relative to the corresponding shifts in β-CAR, a shielding of 0.547 ppm is observed for H1′, whereas a deshielding of 0.154 ppm is observed for H2′.[248] The relative magnitudes of the carboxylate contribution to the overall HI′ and H2′ magnetic shielding constants are reasonable in view of the larger distance of the latter hydrogen from the 6-position in the *syn* conformation. The fact that the contributions are of opposite signs may be surprising at first sight, but is readily rationalised by resorting to the electric field associated with the 6-substituent. The most favoured orientation of the carboxyl moiety which minimises steric interactions with the H1′ hydrogen appears to be that in which the plane of the carboxyl is essentially perpendicular to the uracil moiety. Assuming this relative orientation of the base and the carboxyl in an overall *syn* conformation for the molecule, electric field calculations according to Buchingham[312] predict a diamagnetic contribution for H1′ and a somewhat attenuated paramagnetic contribution to H2′. Moreover, the opposite result is predicted if an *anti* model is chosen.[248] Another direct experimental proof has recently be given by an analysis at low temperature (in $CDCl_3$) of an acetylated derivative of theophylline Tph where the *syn* H1′ proton appears about 1.3 ppm at higher field as compared to the *anti* H1′ on the same spectrum.[254] This through-space interaction is thus expected to be weaker in a polar solvent like water than in other less polar or apolar media.

Theophylline **Tph**

Incidentally, we must observe that this field effect may also play an important role in saline solution. For example, it is certainly responsible of the upfield shifts reported by Prestegard and Chan[280] for U protons in D_2O upon addition of a number of various salts, the shifts increasing with an increasing concentration in salt. The downfield effect produced upon addition of tetrabutylammonium chloride to the same solution results probably from other subtle causes and always remains without a realistic explanation to date.

By introducing many bulky substituents (van der Waals'radi of about 3.5–5. Å)—like a *t*-butyl group—at the C8 position of G, Shugar and co-workers[115] have also observed important downfield shifts in water for the H2' and H3' protons (respectively, 0.65 and 0.07 ppm) which are consistent with an exclusive *syn* conformation. In addition, the *t*-butyl substituent also induces a large downfield shift (≈ 0.30 ppm) on the H1' proton, which is probably due to a very specific van der Waals dispersion interaction. The authors also question the conformation of the 8-BrG, not exclusively in the *syn* form in water according to their results and in some disagreement with previous claims. In a next paper in which they examine the *syn* $\rightleftarrows$ *anti* dynamics of the Pus in water, Stolarski et al.[110] confirm upfield shifts by about 2–3 ppm for the C2' shielding and by about 0.5–1 ppm for the C3' and C4' shieldings, respectively, and for the same compounds as above. As expected, C1' is deshielded to an extent dependent primarily of the nature of the C8 substituent (1.2 ppm for the i-Pr group, 2.5 ppm for the Br substiuent and 3.0 ppm for the *t*-Bu group). The same trends are also observed for 8-substituted Gs and inosines[110] as well as for 8-halogeno-substituted dGs: the more bulky the substituent, the more shielded the C2' carbon and the more deshielded its H2' attached proton.[279]

Follmann and Gremmels[104] have analysed the proton NMR spectra of 30 adenine nucleosides substituted at the C5' position in water with respect to structure-dependent variations of the purine proton chemical shifts. It is observed that the resonance signal of H8 shows a marked dependence upon the nature of substituents linked to C5' of adenine ribofuranoside derivatives and upon the presence of a group iP group, while H2 is relatively unaffected. Therefore, the chemical shift difference, Δ(H8–H2), may serve as an indicator of intra-molecular ribose–base interactions. This varies with the chemical nature of the C5' substituent and it seems to increase from a positively charged substituent like an ammonium group to a hydrogen atom and is the greatest with a carboxylate group (0.4 ppm), but decreases with increasing chain length of these substituents. These results suggest the existence of electrostatic and dipolar repulsion or, in the majority of cases, attractive forces between the C5' and H8 regions of adenine nucleosides. This also implies a strong preference of most adenosine derivatives for the *anti* conformation range and an unusual gg conformation around the C4'–C5' bond.

The behaviour of the 8-amino Gs and 8-methylamino As in water and with temperature described by Jordan and Niv[240] appears more complex. Firstly, the very low upfield and downfield shifts experienced, respectively, for the H8 proton of unsubstituted G and A with increased temperature cannot be attributed to some kind of solvent effect due to its smallness, being in the range of the measurement accuracy! On the contrary, we are inclined to think that this observation is contradictory to the early reports by Ts'o and Jardetzky in the 1960s[233–234] of a possible hydrogen bond of this proton with water at neutral pH. Nevertheless, the comparison of the behaviour of identical substituted G and substituted A with temperature in water is very interesting and illustrative of the trends. Secondly, in all these Pus, upfield shifts of varying magnitudes are found with increasing temperature, while destacking is expected to lead to deshielded shifts of H1' resonances. From our personal view, this should be another proof of the minor importance of nucleoside stacking in water. Thirdly, according to the authors,[240] the difference in H1' and H2' chemical shifts, Δ(H1'–H2') seems to be a good index for the *anti/syn* ratio, a lower value (approx. 0.7 pip) suggesting a larger *sin* population and a larger one (>1.0 ppm) with predominantly *anti* conformers. On one hand, in all the temperature range studied in water (27–90 °C), G and A follow the same trend, both remaining in the *trans* conformation without any sensitive temperature effect on the N/S ratio. There is also no observable effect of a temperature increase on 8-N(CH$_3$)$_2$ G and 8-N(CH$_3$)$_2$ A and also on 8-BrG, all remaining in the *syn* form, whatever the temperature. On the other hand, 8-NHCH$_3$ A, 8-NHCH$_3$ G and 8-NH$_2$ G seem to have distinct relative conformers populations, but always insensitive to a temperature change. For the first conformer, a slight excess in the *anti* forms appears probable, while for the two others a similar proportion of *syn* and *anti* conformers seems more realistic. Finally, all these observations are at variance with any hydrogen bonding of water with the A and/or G substrates, but support the existence of a highly restricted rotation around the exocyclic C–N bond in the amino-substituted Pus and Pys.[241,315–321] Plochocka et al.[241] have also observed coalescence of the two magnetically non-equivalent N(CH$_3$)$_2$ in N(6)-dimethy-1',2',3' iPA at 250 K in CDCl$_3$, a nucleoside exclusively maintained in the *syn* conformation by an intra-molecular hydrogen bond between the N3 and the exocyclic hydroxyl OH5'. Indeed, this hydrogen bond is very weak since it is easily broken in polar solvents like water as indicated by a sudden change in the Cotton effect (CE) going from an undetectable absorption in cyclohexane to a strong positive circular dichroîsm (CD) in water. The same kind of "anomalous" *anti* conformations has also been reported[322] for 8-butylamino A as well as for its epimer 8-butylamino ara-A in dry DMSO-d$_6$ where the 8-alkyamino and the 5'-hydroxy groups are hydrogen bonded. According to these authors, this should explain the low-field resonance of the 5' hydroxyl proton (in the range 6.6–7.0 ppm) as compared to the "normal" range of 5.2–5.9 ppm for the *syn* form. Reese and Safhill[322] suggest that the chemical shift of 5'-hydroxyl proton might provide evidence regarding nucleoside conformations, a point much disputed by many other authors and a really useless probe in water! Finally, we can infer from these results that the height of this rotation barrier should depend on both the extent of the donating effect of the nitrogen lone electron pair to the aromatic ring and also to the amino hydrogen exchange to the water and that its magnitude is certainly close to the *syn* $\rightleftarrows$ *anti* barrier. These points would be discussed more extensively below.

Based on the same previous idea of Uesugi and Ikehara,[107] Nair and Young[245] have also empirically shown in 43 purine nucleosides in DMSO-d$_6$ or CDCl$_3$ as solvent that the differences in ^{13}C chemical shifts between the ribose carbons C2' and C3', $\delta_{C1'} - \delta_{C2'}$, is very sensitive to the glycosidic angle: compounds in the *syn* form exhibit a $\delta_{C1'} - \delta_{C2'}$ value of less than 0.5 ppm, whereas nucleosides in the *anti* conformation show a value of greater than 2.8 ppm. Wyosine W is one of the fluorescent hyper-modified T nucleosides found in tRNAs and has particular distinctive chemical properties and consequently is of biological interest. So far, by using the previous criterion, Agris et al.[255] have determined the glycosidic torsion angle of W and two of its isomers as indicating a preponderant *anti* conformation. Nevertheless, the method curiously fails for the 1-methyl and 5-methyl isomers where other NMR experiments have led to approximately equal amounts of *syn* and *anti* conformers.[255]

Wyosine **W**

Among the numerous hetero-nucleosides revealing a substantial chemotherapeutic activity, aza-nucleosides compose an especially interesting class. As examples, 6-azaC, 6-azaU, 8-azaA, 8-aza and formycin A (FoA) and formycin B (FoB) are found in a highly rigid *anti* conformation in the solid state,[323–325] but detailed information regarding their three-dimensional conformations in aqueous solutions is more relevant for understanding their carcinostatic properties.

Formycin A **FoA**   Formycin B **FoB**

By looking at the Tables 1–4, the reader may get the definitive conviction that the free-electron doublet on the nitrogen acts by a field effect on both the hydrogens and the carbons of the furanose as previously discussed, thus deshielding most of the hydrogen nuclei and at the same time shielding their attached carbons.[239,242,257,262,311] However, while the *anti* conformation is entirely preserved for the Pys, the Pus slightly move to the *syn* form because of the repulsive electrostatic interaction between this nitrogen electron doublet and the analogue O5′ free electron doublet.[239,257] The magnitude of the extended zig-zag inter-ring coupling constant $^5J_{H5H1'} \approx 0.5$–0.6 Hz in the two Pys, which compares very well with that ones measured for U and dU (0.4–0.5 Hz) whereas no analogous coupling is observed for the *syn* orotidine O, is an additional evidence for the preferred *anti* form.[326] Finally, this strong destabilising effect has a direct influence on both the ribose conformation and the exocyclic C4′–C5′ bond, by diminishing the S, C2' *endo* pucker state and the gg rotamer and thus reinforcing the envelope ^{3}E, C3' *endo*, N state in the Pys, the less common twist $^2T_1$, C2' *endo*, C1' *exo* in the Pus and the gt populations for both the Pys and the Pus aza-nucleosides (*vide infra*).[325–327] Thionucleosides may also be intriguing compounds since they often get a distinct geometry in the solid state as compared to their water solution, and more generally in their solution state. For example, 4-thiouridine (4-SU) crystallises in the monoclinic space group C2 with some water molecules in the *syn* conformation,[328] whereas numerous NMR experiments in water conclude a preferred *anti* form[93,105,118,242,257,311] as it is easily shown from chemical shifts trends in Tables 3 and 4 (*vide supra*). 2-Thiothymidine (2-ST) and 2-thiodeoxythymidine[92] are also *anti* in water. In contrast, 2-thiocytidine (2-SC) crystallises as dihydrated triclinic crystals[329] (space group P1), and crystals of 6-thioguanosine (6-SG) monohydrate are orthorombic[330] (space group C222). In both solids, as well as in their aqueous solutions, the glycosidic torsional angle is *anti*.[184,331–333] This substitution of the carbonyl group by a thiocarbonyl group appears as a way to increase the size of a hydrogen bond by a small increment equal to about 0.45 Å. Surprisingly, it was observed that DNA polymerase I prefers analogues that are larger than natural DNA bases and it is found that 2-thio and 4-thio thymidine can lead to increased efficiency and selectivity in pairing and replication.[334] In the same manner, 2-thiouridine seems to give more important rigidity and thermostabilty of thermophile tRNAs.[92] Such an increase in size by a sulphur in place of an oxygen is also without any sensitive effect on the 2-thio-orotidine which remains in the *syn* form because the more bulky carboxylate is forced to occupy the *anti* side with respect to the ribose.[335] It is well known that the sulphur atom is inferior to the oxygen as a proton acceptor for hydrogen bonding, thus resulting in an increase in size in complexes with adenosine and a profound change in the interaction specificty.[336,337] This aspect is extremely important in view of a difference in the strength of hydration of the sulphur and the oxygen, respectively, in water or aqueous solution and a possible sensitive influence of the dynamics of the nucleosides in these media (*vide infra*). These differences in conformation between the solid state and the solution should be accounted for by these differences in hydrogen bonding. In solid 4-thiouridine, the base residues are stacked along a hydrophobic channel, while two ribose moieties are connected by hydrogen bonding between one water molecule with the O5' oxygen atoms which are arranged in parallel planes along a hydrophilic channel.[328] This arrangement, as proposed early by Saenger and Scheit,[328] seems to be in line with later observations which conclude on a predominance of dipole-induced dipole interactions and a large increase in stacking affinity in thio-nucleosides.[330,338] In water, according to our previous results on uridine[191], we are rather inclined to propose a hydration scheme in which one water molecule is bonded to the sulphur atom S4 and another one is bonded to the oxygen atom O2, each one in a cyclic manner with the labile amide hydrogen H3. In solid 2-thiocytidine, the sulphur atom is hydrogen bonded both to a water molecule and to the amino group of an adjacent molecule and this latter hydrogen bond is the only inter-base hydrogen bond.[329] In water, we may anticipate a completely distinct hydrogen bonding structure, that is, one hydrogen bond between sulphur and water and one or two hydrogen bonds involving the amino group and the solvent, but in an *anti* conformation.[320]

To conclude this section, we must add some words concerning aqueous solutions. As can be seen, most of the NMR experiments have been run in (dry) DMSO-d_6 or in other aprotic organic solvents, mainly because of the very low solubility of most nucleosides in other media. In water, as in liquid ammonia, the solvent participates in hydrogen bonding equally well either as a donor or as an acceptor molecule and this is of special importance if there exists an intra-molecular hydrogen bond like in anhydro cyclic nucleosides or in some amino-substituted ones or, sometimes, in the solid state. In this case, the strongest inter-molecular hydrogen bonds with the solvent can be expected, thus breaking the intra-molecular ones which have produced a noticeable minimum of energy in non-protic solvents. This change naturally induces a change in the solvation and a new more pronounced minimum of energy. For each of the the Pys, this implies a strong modification in the conformation, the *anti* form becoming preferred for a better access of water to the amid group where hydrogen bonding takes place.[190] The same situation certainly prevails also for X and isoguanosine (iG) with their amide groups also pointing toward the opposite direction with respect to the ribose. This *anti* conformation can be anticipated to be also more favourable for better availability of the free electron pair of all the nitrogen atoms of other Pus, given that the nitrogen N1 is the preferred site of protonation in adenosine and the second preferred one in G after the nitrogen N7 atom.[80] These considerations upon solvation may generally explain the distinct conformations that have been observed for some nucleosides in the solid state as compared to their structure in the liquid state.

5.1.1.3. Inter-ring coupling constants   This third method for approaching the rotation about the glycosidic bond and the *syn* ⇌ *anti* conformer equilibrium of base with respect to the sugar ring in nucleosides is, from the author's personal view, one of the most appealing. It is naturally based on Karplus-type relationship (Equation (2)) relating scalar coupling between one nucleus on the base to the other on the ribose with the torsional angle χ. The pioneering work by Lemieux and his colleagues[339] on selectively labelled C2 uridine derivatives can be summarised in terms of Equation (2) with $A = 6.7, B = -1.3$ and $C = 0$. Although their data and graph were based only on ribonucleosides, they can also safely used for deoxynucleosides because the substitution of H for OH on the C2' carbon has only a small effect on the J values and do not alter the overall qualitative conclusions.[115,340] The method has been reviewed in two subsequent papers on the conformational analysis of nucleosides, nucleotides and nucleic acids,[81,305] and is now in current use with the routine development of carbon-13 Fourier transform NMR in the proton-undecoupled or gated mode. However, Davies and co-workers[341,342] have questioned the reliability of this approximate Karplus-type dependence which, by only using the vicinal $^3J_{C2H1'}$ coupling corresponding to the *anti* form, gives four possible values for the angle χ, namely two *syn* and two *anti* conformations.[176,343] Accordingly, they proposed to measure both $^3J_{C2H1'}$ and $^3J_{C6H1'}$ in Pys and both $^3J_{C4H1'}$ and $^3J_{C8H1'}$ in Pus, respectively. Addition and subtraction of the values for each series can lead to the *syn* ⇌ *anti* conformer equilibrium, despite some contradictory claims.[109] An unexpected and surprising feature of this trick[342] is that the magnitude of the sum also varies with the population p_{anti} of the glycosidic bond conformation, thus requiring some extrapolation of the fit.

This approach has been of special interest in adding other independent evidence to the overwhelming body of data indicating that Pys and Pus and other similar compounds exist predominantly or exclusively in the *anti* conformation in solution.[93,98,110,112,115,121,138,153,242,255,272] Naturally, the method also allows the detection of the privileged *syn* form as resulting from special specific or non-specific constraints (*vide supra*)[93,110,115,176] and even seems to indicate a preference for the *syn* conformers in Pus.[138] Its valuable ability to probe all the glycosidic torsion angles χ along the conformation circle (Figure 4) has been given in a noteworthy paper by Cho and Evans.[184] The authors report very interesting and fruitful correlations between the H2′ proton chemical shift and the $^3J_{C4H1'}$ coupling constant as well as between the H2′ proton chemical shift and the $^3J_{C8H1'}$ coupling constant on a series of substituted guanosines, thus corroborating the continuous dependence of the two NMR observables with the conformation.[184] In addition, this analysis has also given further evidence of an *anti* conformation for the C8-amino substituted guanosines satbilised by an intramolecular hydrogen bond between the amino proton and the O5′ atom in DMSO-d_6 (see Section 5.1.1.2). The importance of this $^3J_{CH}$ coupling in the determination of the structure of nucleic acids has not escaped Bax group's attention who has developed a dedicated band-selective HMBC experiment for quantitative interpretation.[304] Hilbers et al.[344] have improved the method by suppression of H, H couplings from the proton of interest by inclusion of an appropriately shaped 180° proton pulse into the sequence.

In an elegant extension of the above ideas, Kline and Serianni[345] have run a thorough conformational analysis of nucleosides by ^{13}C enrichment at C1′ and C2′ of the ribose ring in water to address the utility of inter-ring $^3J_{CC}$ couplings for determining *N*-glycoside bond conformation. By using two conformationally constrained cyclonucleosides, 2,2′-anhydro-(1-β D-arabinofuranosyl)uracil and 2′,3′-isopropylidene-2-5′-O-cyclouridine, to construct a very crude (only three points!) Karplus curve for the ^{13}C–C–N–^{13}C coupling pathway across the *N*-glycose bond in A, C, G and U, they could test current models describing the *syn* ⇌ *anti* conformer equilibrium. Unfortunately, the magnitude of the couplings are too low (curiously near zero!), thus precluding any quantitative evaluation and the authors could only conclude that in each nucleoside the base is probably oriented approximately orthogonal to the C1′–C2′ bond. It is a pity that they could not obtain any measurable coupling constant here because the approach is very interesting and, usually, we should expect $^3J_{CC}$ coupling constants to have more noticeable magnitudes (often near 3–6 Hz). Other related methods have been proposed, especially by using long-range $^5J_{H1'F5}$ coupling constants in 5-fluoropyrimidines nucleosides in water.[346] The conformation *anti* which places the F5 fluorine and the H1′ proton in a "zig-zag" or "W" configuration is the most suited for this long-range coupling to occur and it also means that the coupled nuclei are the farthest apart. The authors stressed upon two important conclusions in their work: firstly, a through-space mechanism and a σ–π through bond inductive mechanism can both be ruled out; secondly, the coupling is bigger for the β compounds (>1.5 Hz) than for the α anomers (<1.5 Hz) and therefore can be a useful measure of the anomeric configuration of 5-fluoropyrimidine nucleosides of unknown structure.[346] In contrast, in some 2',3'-dideoxy-4'-fluoroalkylnucleosides, Mele and collaborators[293] have detected, but only on α anomers, long-range $^7J_{HF}$ and $^6J_{CF}$ couplings. A through-space mechanism for the transmission of the nuclear spin information is demonstrated, thanks to an attractive interaction between F on the ribose and H on the base resulting in a mediated intramolecular hydrogen bonding. Consistent with this hypothesis is the decrease in the coupling magnitude with an increasing polarity of the solvent, which therefore has an increasing capacity to establish intermolecular hydrogen bonding in competition with the intermolecular one.

Direct scalar hetero-coupling $^1J_{CH}$ between the C1' carbon-13 and its attached proton H1' has also been used as a probe for glycosidic torsion angle.[184,347] This observation relies on two hypotheses: on one hand, a markedly change of the s character of the carbon with the glycosidic torsional angle, and, on the other, an effect of the nitrogen atom p orbital. However, the range of variation remains very small, especially for the change in s character with the change in torsion angle χ, which is within the experimental error.[184] Indeed, this coupling seems to be a more relevant probe for predicting the anomeric configuration of the C1' carbon in nucleososides.[85]

Most authors have remarked that this inter-ring coupling is essentially invariant inside each of the Pu's base moiety and inside each of the Py's base moiety.[81,138,242] This view should be somewhat moderated, according to Uzawa et al.,[287] because an orientational effect of oxygen functional groups can reduce its magnitude in a *trans* relationship. It seems also to be completely independent of the solvent and is notably the same in DMSO as in water.[81,105,109,111,242,249,264] This last point will be discussed with more details in the next section since this aspect is naturally related to the observation time scale. In any case, this cannot exclude any particular hydration or conformational changes, but the dynamics of this solvation is too fast to be detected on the slow time scale of the coupling constants.[280,348]

Very recently, Sklenář and his group[153] have reconsidered the validity of one unique Karplus equation for describing the inter-ring coupling in all the nucleosides. First, they quote some possible pitfalls because the nucleosides do not posses sufficient symmetry to restrict the Karplus curve as is generally done. Second, because the coupling pathway is very complex, with mainly significant differences in π-bonding between the bases, it seems unrealistic, if not wrong, to retain only one equation with the same parameters, whatever the base. Third, they question about the modifications induced by oxygen bridges in cyclonucleosides. So, the group has tackled again the problem in the two complementary directions of experience and theory. To be able to measure very small coupling $^5J_{H1'F}$ less than the natural linewidth, they developed a spin-state selective excitation (S2E) α, β NOESY experiment. To overcome the somewhat empirical choice of the Karplus parameters, they calculated the coupling constants after a large number of optimisation of the geometry and the constraints on the sugar by using DFT techniques under the acronym SOS-DFPT, an algorithm that is known to be very efficient. This is a formidable task and we will make only some comments on the results in the quantum mechanical calculations (Section 5.3), we ourselves not being specialists in this sophisticated theory.

5.1.2. Ribose and deoxyribose puckering

Since the proposal by Sundarlingam and coworkers[76,119] of a pseudo-rotational path of the furanose ring in nucleosides, the utility of this concept in the understanding of the ribose puckering by all the NMR observables (chemical shifts, coupling constants, relaxation data, NOE, etc.) on the cycle has been extensively discussed[77,78,122,136,141–143] and reviewed.[41,81–84,105,349] Special computer programs have also been written like DAERM,[258] PSEUDOROT[350] and HETROT[351] for the conformational analysis from well-designed 2D experiments. Under constant feedback from the experiments, these programs can be considered as a prelude to the use of the very sophisticated machinery of one (or many) of the numerous force-field MD calculations like AMBER,[352] BMS,[353] CHARMM,[354] GROMOS[355] or OPLS-AA.[356] Thus, we will limit in this section to a brief survey of the most important results, with some emphasis of specific conformational features in water, if any. Likewise, the pseudo-rotational pathway appears somewhat complex to use with the required accuracy and so far has been drastically reduced by Sarma et al. to a simple C3′, *endo* (^{3}E, N) $\rightleftharpoons$ C2′, *endo* (^{2}E, S) equilibrium without any substantial loss in qualitative information (Figure 5B).[106,327]

That is clear is that the scalar coupling constants, mainly the vicinal ones, are of most important use to determine the sugar conformation. This heuristic approach is based upon the assumptions that both the $^3J_{H2'H3'}$ coupling and the sum $^3J_{H1'H2'} + {}^3J_{H3'H4'}$ of the two other couplings are essentially invariant and are very close to the means 5.3 ± 0.4 Hz and 9.6 ± 0.2 Hz, respectively, and this is so whatever the sugar, its conformation and the solvent.[76,105,348] This increase in the C3′ *endo*, N state with an increase of $^3J_{H3'4'}$ and a decrease of $^3J_{H1'H2'}$—with naturally the reverse situation in the increasing proportion of C2′ *endo*, S state—has been reported by Hruska, emphasising the validity of the previous hypothesis.[117] The Karplus parameters $A = 10 \pm 0.2$ and $B = -1.0 \pm 0.2$ (together with $C = 0$) are then derived for use of Equation (2) with the mean phase angles $^NP = 18°$ (^{3}E) and $^SP = 162$ (^{2}E) and a mean pucker amplitude $\tau_m = 40°$. Alternatively, the method leads to the equilibrium constant $K_{eq} = x_S/x_N = {}^3J_{H1'H2'}/{}^3J_{H3'H4'}$ between the N and S states.[348] As an application of the above rules, the population p_N of the C3′, *endo*, N state is calculated[105,257,348] by the simple expression given by Equation (3):

$$p_N \approx {}^3J_{H3'H4'}/\left({}^3J_{H1'H2'} + {}^3J_{H3'H4'}\right) \tag{3}$$

Another empirical relation[178,241] between the interplay of vicinal coupling constants $^3J_{HH}(S)$ in the S state and $^3J_{HH}(N)$ in the N state resulting from the validity of the Karplus law inside each nucleoside family is summarised in Equations (4a)–(4c):

$$^3J_{H1'H2'}(S) \approx {^3J_{H3'H4'}}(N) \approx {^3J_{H2''H3'}}(N) \approx 9.1\,\text{Hz} \tag{4a}$$

$$^3J_{H1'H2''}(S) \approx {^3J_{H2'H3'}}(S) \approx {^3J_{H1'H3'}}(N) \approx 5.8\,\text{Hz} \tag{4b}$$

$$^3J_{H2'''H3'}(S) \approx {^3J_{H3'H4'}}(S) \approx {^3J_{H2'H3''}}(N) \approx {^3J_{H1'H2'}}(N) \approx 0.0\,\text{Hz} \tag{4c}$$

As a general trend, it is generally observed that the ribonucleosides are in the *anti* conformation, which is in good correlation with the N pucker state. Nevertheless, it appears that this behaviour is less marked with some Pus for which the *syn* arrangement cannot be neglected, but always with an N state for the sugar. When the *syn* form is preferred and/or significantly populated, as in cyclo, ara and bulky substituted compounds (see Section 5.1.1 above), the sugar generally adopt the S state. It seems also to be a common feature that in all the deoxynucleosides the contributions *syn* type for the base and S state for the sugar are increased and sometimes are in a large majority. In this way, these changes are especially marked in non-polar solvents and it has been remarked that they look like those resulting from a removal of the 5′-OH hydroxyl (about 15%).[115] There is no further stabilisation of the *syn* conformation by an intramolecular hydrogen bonding 5′-OH···N3 for all the purine nucleosides in water and in DMSO.

Some common characteristic features of all deoxyribonucleosides, particularly in water, should be mentioned here. Firstly, in all the Pys, the H2″ chemical shifts—which naturally correspond to the protons under the ribose—are at lower field as compared to their counterparts H2′—which are still the protons above the ribose—but the reverse stands for the Pus.[258,274] A rigorous explanation of this difference remains difficult and is demanding for a full QM-DFT calculations, but some qualitative arguments can be advanced. The exposure of each proton to the base ring current is certainly not the same in each type of nucleoside. It has been demonstrated by Giessner-Prettre and Pullman[357] that the diamagnetic anisotropy of this effect is more pronounced in purine than in pyrimidine, so that the H2″ protons in the Pus which are closer to the base are more shielded. Other kinds of effects are also important like the proper diamagnetic anisotropy of the carbonyl bond and of the double bond or the electric fields by the oxygen and the nitrogen atoms.[219] Secondly, an interesting comparison of the two isomers 6-MedU (*syn*) and dT (*anti*), in D_2O and DMSO-d_6, has revealed other important and proper features of the sugar in the deoxy series.[274] There is no sensitive solvent effects from water to DMSO. In both solvents, the magnitudes of the *cis* vicinal couplings $^3J_{H1'H2''}$ (8.0–8.5 Hz) and $^3J_{H2'H3''}$ (8.1–8.5 Hz) are larger in 6-MeU by about 1.5 Hz than the corresponding couplings in the dT series. This means that the dihedral angles about the C1′–C2′ bond and about the C2′–C3′ bond are biased in 6-MeU relative to dT towards the eclipsed ($v \approx v_2 \approx 0°$) situation. According to the Altona–Sundaralingam approach, this can be interpreted as a flattening of the ring, that is, a reduction in the pucker amplitude τ_m, and this reduction should lead to a decrease in the sum of the *trans* vicinal couplings $^3J_{H1'H2'} + {^3J_{H3'H4'}}$ as well as to an increase in the *cis* vicinal couplings $^3J_{H1'H2''}$ and $^3J_{H2'H3'}$. It is seen that these sums, in aqueous solution, lie in the narrow range from 10.8 to 11.4 Hz, with no clear difference between the dT and 6-MedU isomers. The larger *cis* vicinal couplings $^3J_{H1'H2''}$ and $^3J_{H2'H3'}$ couplings in the 6-MedU compound, coupled with the similarity in the *trans* sum $^3J_{H1'H2'} + {}^3J_{H3'H4'}$ for the two derivatives, can be taken as evidence that the sugar ring in the 6-MedU isomer has "pseudo-rotated" away from the pure C3' *endo*, N state and C2' *endo*, S state. This is clearly shown by the application of Equation (3), which gives, in D_2O, $p_N = 38\%$ for dT and $p_N = 56\%$ for 6-MedU, a value not usual for *syn* deoxyribo-sides in water! Faced with this body of evidence, the only valuable conclusion is a flattening of the sugar ring in the 6-MedU, an observation also made by Lee and Sarma[114] in a series of 3',5'-cyclo deoxyribonucleosides. Also, in deoxy series, the near equalities, $^3J_{H2''H3'} \approx {}^3J_{H3'H4'}$ and $^3J_{H1'H2''} \approx {}^3J_{H2'H3'}$ seem always satisfied.[274] Finally, variable substituents at the 5-position of U, 5-XU, do not modify the *anti* character of the glycosidic bond, but have drastic influence on the N1'–C1' and C1'–O4' bond lengths in opposite ways[118]: electron-withdrawing X substituents lengthen the N1'–C1' bond, whereas electron-donating groups shorten this bond with reverse effects on the C1'–O4' bond. From NMR results in water, the authors predict that these slight structural changes, especially on the sugar, induce large consequences on its N, S states: electron-withdrawing groups favour the C3' *endo*, N state, whereas a C2' *endo*, S state is preferred with electro-donating X substituents. Simple MO-SCF calculations (see below) are in line with these experimental data.[118]

The above general rules are naturally more and less respected according to the substituents present on the furanose ring and important modulations can be perceived along these lines. As an example, Gushlbauer and Jankowski[105] have reported a strong influence of the electronegativity of the 2' substiuent in a series of 2'-substituted dU in water: the greater this electronegativity, the greater the H2' chemical shift, but the smaller the $^3J_{H1'H2'}$ coupling and the greatest is the N state population of the deoxyribose. In a previous study of the 2' fluorine effect on the sugar conformation, the same group has even arrived at the conclusion that the sugar conformation of the 2'-fluoro-ribosides is more similar to ribose (N-form) than to the usual deoxyribose (S-form).[105] Uesugi et al. have also observed the same increase of the N state population in correlation with a decrease in the H1' chemical shift with respect to an electronegativity increase of the 2' substtiuent.[107] A plausible explanation is that the electron donation of the O1' oxygen of the ribose, both by a through-bond effect and by through-space interaction, is rein-forced in the N state as was suggested by old SCF and extended Hückel (EH) calculations in the 1980s.[118] Other subtle changes have been observed by Lee and Sarma[114] on fused β-ribonucleosides in water where the sugar ring display a ^2E, C2' *endo*, S pucker state with a glycosidic torsion angle $\chi = 290°$, that is midway between the traditionally accepted values for *syn* and *anti* conformations. In contrast, when the ribose and the exocyclic bond are frozen, the compounds exhibit an exacerbated 3E, C3' *endo*, N form.

Deslauriers and Smith[249] have compared the proton spectra of U, dihydrour-idine DU and β-ΨU in DMSO-d$_6$ and in D_2O and concluded that the conforma-tions of the ribose rings are very similar in the two solvents. Agris and collaborators[93] have shown that the 2-thiourudine 2-SU is more hydrophobic than its analogue U and thus gives a ratio S/N lower than for U.

5.1.3. Free rotation about the exocyclic C4′–C5′ bond

Many people have also tackled with this problem for a long time and the scientific literature is rich with a large number of results from NMR approaches. So, we wish here to stress on two particular and related points: How to safely assign each of the two diastereotopic methylene protons? Is there a specific interaction between water and the 5′-hydroxylic group water which can preferentially accommodate one conformation among the three generally assumed as probable (Figure 6).

The first tentative answer is probably due to Davies and Danyluk[358] who hoped to measure some significant differences in the coupling constants $^3J_{H5'OH5'}$ and $^3J_{H5''OH5'}$ in an extremely dry—if not completely dry—mixture of C_6D_6 and DMSO-d_6. At room temperature, they observed a surprising constancy of this coupling (5.0 ± 0.1 Hz) for all the usual nucleosides (U, C, A, G, dT, dC, dA and dG) they examined, a value in completely in agreement with that expected for a freely rotating group in alcohols.[358] Naturally, they reasonably concluded that the 5′-hydroxyl group of nucleosides is rotating freely about the exocyclic C5′–O5 bond at least on the NMR time scale of nuclear spin–spin couplings. In contrast, the authors quote the great dependence of the other 4′-OH, 3′-OH and 2′-OH coupling constants with their respective vicinal H4′, H3′ and H2′ protons inside the different nucleosides, with an average value less than that of the freely rotating group. In the AB system obtained by spin decoupling from the OH5′ hydroxyl proton, in all the nucleosides the low-field signal (3.61–3.75 ppm) was thus arbitrary assigned to the H5′ proton while the high-field resonance (3.56–3.73 ppm) was named H5″ proton—without any more indications!—but according to a convention largely adopted nowadays (see Tables 1–3).[348] Therefore, some more or less empirical methods for discriminating these hydrogen have appeared.

In 1971, Hruska et al.[117] have observed direct relationships between any perturbation on the sugar pucker and the vicinal couplings $^3J_{H4'H5'}$, $^3J_{H3'H4'}$ and $^3J_{H1'H2'}$. The Karplus equation (2) is once more the cornerstone for interpreting the vicinal coupling constant $^3J_{H4'H5'}$ and $^3J_{H4'H5''}$ based on the three distinct populations p_{gg}, p_{gt} and p_{tg} of three more abundant rotamers (see Figure 6). Clearly, in the gg rotamer in which the 5′-hydroxyl group lies above the sugar ring, the magnitudes of the vicinal coupling constants $^3J_{H4'H5'B}$ and $^3J_{H4'H5'C}$ are predicted by to be small (ca. 2 Hz), since each of the relevant coupled nucleus (H5′B or H5″C) is in a gauche conformation. However, if the 5′ hydroxyl lies off the ring in either the tg or gt rotamer, the 4′ hydrogen is trans to one methylene hydrogen and the corresponding vicinal $^3J_{H4'H5'}$ is predicted to be large (ca. 10 Hz). Thus, any sugar puckering change resulting in a decrease in the proportion of the gg rotamer around the C4′–C5′ exocyclic bond should be manifest *de facto* in an increase in the observed sum $\Sigma = {}^3J_{H4'H5'B} + {}^3J_{H4'H5'C}$. As a consequence, good correlations are observed: (i) between the sum Σ and $^3J_{H3'H4'}$ with a positive slope and (ii) between the sum Σ and $^3J_{H1'2'}$ with a negative slope, since the sum $^3J_{H3'H4'} + {}^3J_{H1'H2'}$ is an invariant in a pyrimidine nucleosides series, whatever the sugar puckering (*vide supra*).[117] One year later, Remin and Shugar[359] tried to solve the assignment dilemma of the H5′ and H5″ proton chemical shifts by using a similar, but somewhat empirical, approach in a large series of pyrimidine nucleosides and nucleotides in water. The inclusion of the influence of the phosphate of nucleotide on the chemical shifts of the methylene protons seem very crucial—and therefore subject to some criticism—in their search for a consistent choice among the only two possible combinations. Another disputable point is the important modification in the ratio p_{gg}/p_{gt} (by a factor of 5) with a small change in the Karplus parameters. Nevertheless, the authors observe two remarkable regularities: (i) $\delta_{H5'}$ > $\delta_{H5''}$ and (ii) $^3J_{H4''H5'} < ^3J_{H4''H5''}$ in all the series. On this basis, they concluded that the proton they have labelled $H_{5'C}$, which, in the right terminology of stereochemistry is the proton H5' prochiral-R, corresponds exactly to the one also named H5'' by Davies and Danyluk and resonates upfield as compared to the proton H5' which is pro-S. In the same manner, by a direct of observation of hydrogen bonding between the 5'-hydroxy and 2-keto groups of N(3) methyl-2',3'-iPU in CDCl$_3$ and CCl$_4$, Davies and Rabczenko[360] have further confirmed this assignment of the methylene protons. They also report, in agreement with a similar previous work by Hruska et al.,[361] a good correlation between the sum of the coupling $\Sigma = ^3J_{H4'H5'} + ^3J_{H4'H5''}$ and the difference $\delta_{H5'} - \delta_{H5''}$ between the two diastereotopic protons and extend the results to the Pus. Finally, they show that this general agreement in the assignment allows prediction of the various population of the three staggered rotamers in the order $p_{gg} > p_{gt} > p_{tg}$ in D$_2$O as in the solid state.[360] More exactly, the observed average couplings $^3J_{H4'H5'}$ and $^3J_{H4'H5''}$ are only a function of the three rotamer populations and by assuming with the Karplus law (Equation (2)) that the relevant couplings in each rotamer, say $J_t \approx$ 10 Hz and $J_g \approx 2$ Hz, are only dependent on the dihedral angle, we can write the two Equations (5a) and (5b):

$$^3J_{H4'H5'} = p_{gg} \times {}^3J_g + p_{gt} \times {}^3J_g + p_{tg} \times {}^3J_t \approx 2.0 \times (p_{gg} + p_{gt}) + 10p_{tg} \quad (5a)$$

$$^3J_{H4'H5''} = p_{gg} \times {}^3J_g + p_{gt} \times {}^3J_t + p_{tg} \times {}^3J_g \approx 2.0 \times (p_{gg} + p_{tg}) + 10p_{gt} \quad (5b)$$

Simple algebraic manipulation of these two equations gives Equation (6), which is reasonably valuable for all the nucleosides rotamers populations that are nearly independent of the solvent, often including water[249]:

$$p_{gg} \approx 0.53, \quad p_{gt} \approx 0.32 \quad \text{and} \quad p_{tg} \approx 0.15 \quad (6)$$

Allowing for an oxygen–oxygen repulsion for all the rotamers by enlarging the dihedral angles by something like 5–15° as observed in X-ray studies on crystalline nucleosides does not really improve the results and modify the conclusions.[362] One of the most surprising feature is that the gg conformer in which the 5'-OH lies above the ribose ring is still preferred for the U derivatives in aqueous solutions.[249,264] There is nevertheless a small decrease in this population as indicated by the two $^3J_{H4'H5'}$ and $^3J_{H4'H5''}$ couplings which are more alike in DMSO-d$_6$ than in D$_2$O.[249] In contrast, an important increase from 50% in DMSO to 70% is observed for adenosine A in water at pH 7.0.[264] This is to mean that interaction of this hydroxyl with water, either by chemical exchange between the protons or by hydration of this group, is not a determining factor in the stability of the exocyclic chain and *per se* of the sugar moiety.

The near equality in the proportion of the rotamers gg and tg for dU in water can be explained by some release in the interaction 3'-OH, 2"H in the C2' *endo*, S pucker state which increases the tg weight at the expense of the gg population.[249] When the base is *syn* like in 6-MedU, the situation is even worse as compared to that of dT in water. The repulsive interaction between the carbonyl and the OH5' leads not only to a strong destabilising effect on the gg rotamer and resulting from the repulsive interaction between the 2-keto oxygen of the base and the OH5' hydroxyl, as demonstrated by a large amplified repulsion when 5'OH is replaced by a phosphate group.[274] In addition, it is pertinent to note that this destabilisation is similar in water and in DMSO ($p_{gg} \approx 0.31$, $p_{gt} \approx 0.49$, two values largely distinct from that above in Equation (6)) not indicating therefore any significant hydrogen-bonding effect with water.[274]

There are many other subtle variations when nitrogen or sulphur atoms are present on the base ring as in aza-nucleosides where the electronic interaction between the electron doublets on the nitrogen and on the oxygen destabilises the gg rotamer.[257,327] For example, in water at room temperature, the Pus are still 70% in gg conformation, but this proportion falls to about 0.53 for the 8-aza Pus and in the meanwhile the tg conformer ($p_{tg} \approx 0.32$) dominates over the gt conformer.[239] In the same manner,[92] the C3' *endo*, gg form (80%) of the sugar moiety is remarkably stabilised in water on modification of T to 2-Thio thymidine 2-ST, but not on modification of dT to 2-SdT (only 59% against 54% in dT)). The steric effects between the 2-thiocarbonyl group and the 2'-hydroxyl group may certainly cause the rigidity of the C3"-*endo*-gg form of 2-ST. Such rigidity of 2-ST probably contributes to the thermostability of 2-thiopyrimidine polyribonucleotides and extreme thermophile tRNAs. Agris and collaborators[93] have even shown that 2'thiouridine 2-SU is more hydrophobic than its analog U, so that in water the percentage of gg conformer reaches an amount equal to 94%!

Naturally, conformational constraints on the sugar should induce important effects on the orientation of this exocyclic bond. Likely, in 2',3'-O-isopropylidene nucleosides, the ribose often tends toward to the C2' *endo*, S pucker, an effect is severely amplified by two other factors, namely, a *syn* glycosidic torsion and the existence of an intramolecular hydrogen bond. Such a situation is found with N(6)-dimethyl-2',3'-isopropylidene adenosine in CDCl3 where the exocyclic bond adopts the gg conformation in order to bring out the 5' OH in the most favourable position for a hydrogen-bonding donation to the nitrogen N3 of the purine (90% S and 90% gg).[241]

It is impossible to end this section without making some comments about what we believe to be the most reliable method for the chemical shifts assignment of the two exocyclic protons on nucleosides. This success is a part of the elegant work by the Serianni–Kline group in their continuing and coherent effort to solve this reminiscent problem.[363–366] In this way, they have undertaken an extensive study of stereoselective deuterium exchange of methylene protons in methyl pentofuranosides. The cornerstone of their nice method is the selective deuteration of the pro-S methylene proton at the C4 position of methyl β-D erythrofuranoside. They were then able to establish the stereospecificity of the reaction by all the battery of proton and carbon-13 NMR, including isotopic ^{13}C enrichment,

INADEQUATE experiments and ^{13}C longitudinal relaxation measurements. Empirical rules have been developed that predict ^{13}C chemical shifts in aldopyranose rings, and, correlatively, the effect of pyranose structure on these shifts has been examined. By using carbon-13 chemical shits and ^{3}J$_{C-H}$ coupling constants across the furanose ring, the group has also definitively shown that the pro-S H$_4$ proton, which by definition is *cis* to OH$_3$, is definitively the more shielded of the two methylene diastereotopic protons in tetrafuranosyl rings.[363] With all this material in their hands, Kline and Serianni[364] have recorded proton spectra of the four usual nucleosides A, C, G and U in deuterated water together with the parent methyl β-D ribofuranoside for comparison. They hopefully confirmed that the previous empirical assignment of the proton named H5″ or H5′C, according to the two different nomenclatures (*vide supra*), is right; more exactly, it corresponds to the proton H5′ pro-R which resonates at higher field with respect to the other diastereotopic proton H5′ which is pro-S and deshielded. They also calculated the three rotamer populations p_{gg}, p_{gt} and p_{tg} in good agreement with results obtained in Equation (6), but show that substitution of the base by the methyl group in the parent ribofuranoside derivative changes drastically the proportions of the gg and gt rotamers by about 0.40 and 0.50, respectively. In their next paper,[365] these authors extended the analysis to the analogous deoxynucleosides dA, dC, dG, dU plus dT and checked that the order of the resonances H5′ pro-R and H5′ pro-S is unchanged as compared to the parent ribonuclcosides, but the shieldings are much closer. Interestingly, the vicinal couplings ^{3}JH4′H5′ are reported and the coupling of H4′ with H5′ pro-R always appears greater ($\approx$ 1 Hz) than that one between H4′ and H5′ pro-S. The gg rotamer is also less abundant (by about 0.10) at the profit of each of the gt and tg conformers which grow up in the same extent. They further ran a thorough conformational analysis of ^{13}C enriched at C$_{1'}$ of the furanose ring of many erythronucleosides and ribonucleosides in water to address the utility of ^{3}J$_{HH}$, ^{3}J$_{CH}$ and other long-range couplings for determining for determining the preferred North and South conformers and their puckering amplitudes.[366] We have already shown how this group by this blend of methods has confirmed the preferential *anti* conformation for Pys and to a lesser extent for Pus in water by using ^{4}J$_{CCNC}$ carbon-13, carbon-13 couplings as well as as ^{1}J$_{CH}$ couplings.[145,345]

5.1.4. α, Arabino, cyclo and dideoxy nucleosides

In this short section, we wish to describe some particular properties of other families of nucleosides in water with emphasis on the differences, if any, with their parents. We have deliberately limited our discussion to what we think to be the most important class of compound and this is our personal choice. But, remember that the list is very long and an exhaustive approach not only would need pages and pages but is also simply not realistic.

The first compounds of interest in this way are naturally the α isomers which are epimers on the carbon C1′ of the β nucleosides (Figures 2 and 3). They are not found in nucleic acids, but do occur as constituents of smaller molecules in living cells, for example in vitamin B$_{12}$, and can exert a biological activity sometimes equal or even exceeding that of the β-anomers. There are no general rules relating the properties of the α-ribonucleosides to their β-anomers, and the "inverse" configuration of the H2' and H2" protons has been invoked in the deoxyano-mers.[103,117] At first sight, steric hindrance is much larger for α than for β ribonu-cleosides because of the interactions between the OH2' and the base, so that one may expect significant changes in the glycosidic torsion angle. The following conclusions have been derived by Gushlbauer et al.[367] from proton NMR mea-surements in water, including some longitudinal relaxation:

- In general, the protons of α anomers are at lower field than those of the β-anomers, the largset difference is naturally for the anomeric protons.
- α-C is likely in the same conformation as β-C with an *anti* form, a most preponderant C3' *endo*, an N pucker state and the populations p_{gg}, p_{gt} and p_{tg} which are identical to those in its anomer.
- α-ΨU is in rapid equilibrium between the *syn* and *anti* forms, with a large predominance of the N state (79%) and the same conformational distribution of rotamers about the exocyclic bond as in β-ΨU. This first point is clearly in full disagreement with observations by Grey, Smith and Hruska,[265] who on the basis of chemical shifts on the base and on the ribose support an *anti* glycosidic torsion angle. Indeed, the two groups have measured the same range of coupling constants on the ribose and the important fact, in our view decisive, is the amplitude of the long-distance inter-ring coupling between H5 and H1' measured by the Canadian researchers in the α-ΨU (1.3 Hz) which certainly is in favour of the *anti* form.

Ever since 1961, Lemieux has established beyond any reasonable doubt that NMR spectral parameters like H1', H2' and H2" chemical shifts allow the assign-ment of the configuration of each of the α and β-anomers of dT in water because the compounds posses "considerable conformational purity".[368] This approach has then been extended by many others by using proton,[104,114] carbon-13 chemical shifts[369], 1D or 2D NOE spectroscopy[109,261,281,288,292] and relaxation data.[367,370] Note that, in this deoxy class, the "inverse" configuration is interpreted as fol-lows: on one hand, *cis* H2' β-anomer and *cis* H2" in α-anomer have nearly similar shielding and, on the other hand, *trans* H2" in β-anomer and *trans* H2' in α-anomer have also similar shielding, and this only for a given glycosidic conformation, that is *anti* or *syn*.[103,117] A comparison within a large set of C5-substituted dUs does not reveal any important modifications to the base torsion, the sugar puckering and the *exo* cyclic bond from the β-series to the α-series.[103]

The second family of interest is the β-arabinose series which is the other epimer on C2' carbon of the β-nucleosides (Figures 2 and 3) Although it seems admitted that they also adopt generally in water the *anti* form, they have been the object of controversial views about the existence of an O2'–O5' hydrogen bond in which OH5' serves as the donor. While the Shugar' group[127,132] invokes this possibility of this bonding in a basic medium (pH $\geq$ 11) where the sugar hydroxyls dissociate, Huska and collaborators[264] argue instead in favour of a repulsive O2'–O5' inter-action, thus opposing this bonding in aqueous solution. The first view is sup-ported by old SCF-MO calculations[128] and crystal structure, but the absence of any overwhelming bias from DMSO to water is against such a bonding, the preference for a C2′ *endo*, S state and a gg rotamer not being a sufficient argument. What is clear is that the same trends as in β-nucleosides are also observed in this series, viz. correlations between the vicinal couplings, chemical shifts on the base and ribose C3′ *endo*, N state, exocyclic bond rotation and specific effect by bulky substituents on the base.[114,127,182] In this last case, it seems that a 8-*n*-butyl substituent in the ara-A[371] leads to a surprising *anti* glycosidic torsion as speculated by the low-field OH5′ resonance which, however, is known to be a poor probe!

Cyclic nucleosides do not manifest special properties, as compared with the previous families, except that they are naturally and by constraint in the *syn* glycosidic angle.[110,114,119,182,267] The most important changes appear to be on the ribose puckering which often exhibits a very unusual state, like a C4′ *exo* state[120] or a C4′ *endo* state[267] and even an O4′ *exo* state for the 5′,2-O6-cyclo,2′,3′-O-isopropylydene uridine,cyclo-iPU.[372] More generally, we often find again the C2′ *endo*, S state[110,267] and the exocyclic rotamer distribution is very different in the various compunds.[267] Mention is made here to the in-depth proton spectral analysis of a cyclo ara-U in D_2O at 80 °C by Hruska et al.[373] who were able to observe fine long-range four-bond and five-bond couplings between the furanose hydrogens; especially, the ^5J coupling implying the anomeric proton H1′ and the exocyclic protons H5′ is reported for the first time and is certainly another proof of a unique and favourable geometry for the sugar ring.

AZT

The last family is that of the dideoxy nucleosides, like AZT and their analogues which have emerged as widely used drugs as potent HIV-inhibitors and it was claimed that this crucial therapeutic property should be the result, *inter alia*, of a particular ribose pucker.[96,259,374–376] This last point is still unclear and is always disputed because the presence of the polar azido group should have a considerable functional role.[98,311,377] In their proton and carbon-13 NMR report on AZT in water and in DMSO, Kunwar et al.[98] have determined an *anti* glycosidic torsion angle, a sugar pucker equilibrium between the usual C2′ *endo*, S and C3′ *endo*, N geometries and a predominant gg conformation around the exocyclic C4′–C5′ bond, a result that does not differ much from those derived for most of nucleosides (see above). Their results, in liquid state, are clearly at variance with those obtained in the solid state by X-ray diffraction where two distinct molecules A and B are observed. It has been conjectured that the very unusual B structure (in the extreme limit of the *anti* form), with a C3' *exo*, C4' *endo* sugar pucker and a gt rotamer) may represent the biologically active form of AZT,[374,376] but this assertion cannot be supported by this study in water, a medium much close to the physiological one than the solid state! In a next paper,[98] the same group has also studied in water several 2',3'-dideoxynucleosides, ddN, which are assumed to be the alternative of AZT in the acquired immune deficiency syndrome (AIDS). With the only exception of a 3'-fluoro derivative of the dideoxythymidine, FddT, which exists almost totally in the S pucker state, all the other compounds exhibit the same characteristics as the majority of other nucleosides in solution. Here also, the results differ quite largely from solid-state observations and the authors are inclined to think that the presence of a polar group at the 3' position increases the S-type population with a correlated improvement in the anti-HIV activity. The well-known "gauche effect" which is described as "a tendency to adopt that structure which has the maximum number of gauche interaction between the adjacent pairs and/or polar bonds"[378] may play an important role in this stabilisation of the S pucker state. Similar conclusions have been arrived at by others either on 2',3' dideoxy nucleosides,[375] including a "particular" iso-ddA related to the usual one by transposition of the base moiety from the C1' position to the C2' position while maintaining its *cis* relationship with the exocyclic bond.[259] In this way, 4' azidothymidine, which is also in an *anti* geometry but with an extremely rare 4' envelope conformation for the ribose, appears as an exception, probably because of the special position of the azido substituent which can interact with the exocyclic hydroxymethyl group. In summary, it is possible, as been stated elsewhere, that the *syn* conformation is probably deleterious to HIV inhibition.[375]

5.2. Investigations by other nuclei

From a simple glance at the chemical structure of the nucleosides, any chemist immediately sees the prime importance of the nitrogen and oxygen atoms with their free electron doublets at the molecular periphery. They are incontestably the best probes for studying intermolecular interactions with the solvent, especially for hydrogen bonding since they can act both as a proton acceptor and as a proton donor. The major problems are naturally those inherent, on one hand, at the low natural abundance of the two NMR active spins (^{15}N and ^{17}O) and, on the other hand, to their relative small magnetogyric ratio (moreover negative and even a quadrupolar spin $I = 5/2$ for ^{17}O). [221,222] Nevertheless, they have probably, more than many other nuclei, and this mainly for nitrogen-15, benefited from the continuous advances of NMR technology. In addition, because of the constant lowering of isotope enrichment costs, they are more and more currently investigated, especially once more for nitrogen-15.

We wish also to quote the abundant nitrogen-14 nucleus, although its considerable and harmful quadrupolar effects on the linewidths in high-resolution NMR. But, in the early days of nuclear quadrupole resonance spectroscopy (NQR), it was considered as very promising for the study of amino acids, peptides, nucleosides and nucleotides, polynucleotides and all nitrogen-containing compounds, which are all structures of interest in biology. The reasons are quite clear: the quadrupolar effect (quadrupole coupling constants and/or quadrupole frequency resonance shifts in NQR) can be related via the electric field gradient (efg) to the electron density around the nitrogen by using the Townes–Dailey theory.[379] In this way, several very rare experiments on deuterium NQR can be also pointed out as they should give some valuable information on the efg along the C–D bond. Note also that oxygen-17 is also (in principle) a good candidate for NQR experiments on nucleosides, but at this time we are not aware of any reports on such studies.[270]

Finally, this reviewer is a bit surprised to not find many fluorine NMR data since this nucleus is among the most sensitive one and incorporation of fluorine on nucleosides has revealed a noticeable anti-cancer therapy. Until now, mainly indirect fluorine effects on proton and carbon-13 have been reported. In view of its van der Waals radius being relatively close to that of hydrogen (1.47 Å against 1.20 Å), we may expect that this should consist in a very reliable probe for NMR analysis without any drastic modification of the nucleoside molecular geometry. We will comment further about this point.

5.2.1. NQR experiments

NQR is typically a solid experiment because the random motion of the quadrupolar interaction in liquid (a rank two tensor like the dipolar interaction) averages to zero all the energy level splitting. Nevertheless, we will quickly summarise here the few results available from the literature because interaction with water which can lead to stable hydrates should be investigated in the future by this method. Some of these aspects, including hydrogen bonding with water, have been outlined in the past.

^{14}N NQR and ^{2}H experiments on the nucleosides U, C, A, I and X have been reported by Edmonds and collaborators at 77 K.[380] Unfortunately, the resolution and the sensitivity were poor so that some interesting transitions are missing. In our view, the two important conclusions that can be deduced from their work are (i) an increase in the efg at the N1 and N3 nitrogen atoms of the pyrimidines by introduction of the amino group in C and (ii) a systematic increase in the efg from I to X and then to A in the Pus. It also seems that the same trends are observed with deuterium. Other deuterium studies of these compounds undertaken by the Volds[381] and by Day et al.[381] have underlined the possibility to observe weakly hydrogen bonded hydroxyl deuterons on the ribose and some water effects in the solid state. The usefulness of ^{35}Cl NQR spectroscopy,[382] combied with DFT calculations, has also been assessed for comparison of the electron distribution in chloro-A and chloro-dA, which suggests a transfer of electron density of about 0.341 electron from the base to the ribose.

The difficulties inherent to NQR spectroscopy are its very low sensitivity in conjunction with the very poor resolution. Also, the experiments are generally run by using the Hahn technique of double resonance in the laboratory frame (DRLF) or in the rotating frame (DRRF), which notably increases their signal-to-noise ratio, but at the expense of very delicate adjustements.[383] Remember also that all the experiments are run at 77 K, the nitrogen liquid temperature, confining all the information to the solid state. In this way, the very recent development by the

Canet's group[384] in Nancy of a cutting-edge technology using pulsed NQR coupled with all the instrumentation of modern NMR spectroscopy should herald a renewal in this technique. We think—and hope—that it should open up very promising new horizons for the future studies in the field of nitrogen-compounds of biological interest.

5.2.2. Nitrogen-15, Oxygen-17 and Fluorine-19 data

5.2.2.1. Nitrogen-15   By looking at Table 6, it is readily observed that Pu's nitrogen nuclei are more deshielded than their analogues in the Pys. As we have already shown, this difference can be accounted for by the more important diamagnetic current on the purine ring as compared to the one on the pyrimidine ring.[357] A second characteristic and related feature is that the Pu's N1 shifts are always in the limited range 165–175 ppm whereas the analoguous Py's N1 stand in another narrow range centred at about 150 ppm. Unequivocal assignment of the chemical shifts has been ensured by specific labelling, by direct or long-range couplings with the protons or even by 2D correlation spectra between nitrogen and proton in the direct or reverse modes (HSQC, HMQC and HMBC experiments). When data in DMSO and water are both available, it is also evident that there is no sensitive solvent effect on the nitrogen screenings. This means that at the nitrogen chemical shift range, which is notably greater then that of proton and carbon-13 nuclei and thus more favourable than the other two, one cannot detect any differential solvation or hydration on the respective sites. The lifetime of such an interaction remains too short to be accessible from the chemical shifts.

Another power of nitrogen-15 NMR here holds in giving information about the possibility to compare the nucleophilities and the protonation potentialities of all the sites when varying the pH of the aqueous solutions. Such knowledge is naturally an important key for the study of nitrogen site selectivity in binding with diamagnetic metal ions. For example, it has been shown that the N3 site of C is the more basic[121,385] and even more basic than the N7 site of G, which is itself the more basic site of this nucleoside.[386,387] For A, the N1 nitrogen is the preferred site of protonation,[121,246,388] but the N7 nitrogen seems to indicate a reduced paramagnetic contribution due to partial protonation.[389] In the case of 8-hydroxy adenosine, 8-OHA, a dramatic change on the imidazole ring takes place which takes a keto form at C8 as evidenced by the appearance of a well-resolved one-bond N7–H7 coupling ($^1J_{N7H7} = -98.7$ Hz).[246] 1-Methyladenosine, 1-MeA, and 7-methylguanosine, 7-MeG, deserve special mention[121,390] because they are fully recognised as the products of methylation of DNA by carcinogenic agents such as dimethyl sulfate. During protonation of 1-MeA, the N3 nitrogen of the purine is shifted downfield by more than 20 ppm and an upfield shift of about 90 ppm for the NH2 is observed. The N1 nitrogen is practically unaffected and the results suggest an important redistribution of the positive charge inside the purine base, including the imidazole ring.[121] Protonation of 7-MeG has been extensively examined by Barbarella, Bertoluzzi and Tugnoli[390] in DMSO-d$_6$. By a thorough comparison with the 1-methyl guanosin, 1-MeG, the authors have obtained two important results. Firstly, the protonated structure of 7-MeG looks like that of protonated G with a hydrogen attached to the N2 nitrogen and the positive charge localised at the methylated nitrogen N7 (structure A, left below). Secondly, this unusual structure then can associate one to one with the C to give a very stable canonical Watson–Crick base pair (structure B, centre below) whereas no such pairing can be formed between 1-MeG and C in DMSO (structure C, right below).

A

B

=Ribose

C

Formycine A, FoA, is also very exemplary in that protonation at N4 in water results in a transfer of a pyrazolo ring hydrogen from N1 to N2.[250] Unfortunately, this assertion has been made on the basis of carbon-13 spectra and it is a pity that nitriogen-15 NMR has not been used here as it is well known to be an elegant method to approach tautomerism.[221,388] I is mainly protonated at the N7 site[340] which is generally the more basic one in Pus with the remarkable exception of A, probably because of a possible unfavourable steric strain between the amino group and the adjacent H7 proton.[246]

These are, among many others, some indications that can help in understanding preferential hydration sites. In addition to the mechanism of direct protonation, abstraction or transfer of a proton, the exchange of the labile protons on the nucleoside with the water may be mediated through the surrounding nitrogens of the base.[391] This aspect has been investigated by postulating the formation of a complex between a hydronium ion, the NH_2 amino of C, A and G and a ring nitrogen, respectively N3, N1 (or possibly N7) and N1. By acting as an auxiliary base, the predominant role of the ring nitrogen may explain the relatively high exchange rate that is observed at neutral pH. If such a hypothesis seems very plausible, complementary data will be needed such as from a full exchange study by varying the temperature and pH. Unfortunately, to the best of our knowledge, these studies are, to date, missing. Another aspect that is of much concern is that the proposition of the authors is severely restricted to amino-group-bearing compounds.[391] The question is thus two-fold: Is this mechanism really limited to the amino nucleosides? Is there an interplay with the hydronium in all the other bases? If no, what are then the ring nitrogens implied instead? These key aspects should find some partial answers with the help of dynamic investigations that allow scanning intermolecular interactions at a much small time scale.

5.2.2.2. Oxygen-17 By its localisation at the extreme border of the molecular surface of nucleosides, oxygen—and mainly carbonyl oxygen—constitutes a privileged probe to look at intermolecular interactions with the solvent. The major problem, as we have quickly surveyed previously, is the sensitivity and linewidths of oxygen-17 nucleus. Nevertheless, some interesting papers deal with the effect of solvation of carbonyl compounds on oxygen-17 screenings.[222,392] Naturally, all experiments require an unavoidable labelling at the site under examination and the considerable cost of such an operation has clearly limited the number of studies. One of the first papers dealing with oxygen-17 on thymidine (T or dT?) and some pyrimidine bases has appeared in 1982 and is strictly limited to the enriched O4 of thymidine in neutral and anioinc forms whose shift indicates a preferential protonation at the N3 site.[393] In the field of nucleic acid dynamics, longitudinal and transverse relaxation of a sample of polyadenylic acid enriched at the phosphoryl group was investigated.[394] We do not further comment on the results as this work is clearly outside the scope of this review.

The elegant series of three key papers on the structure and hydration of nucleosides by Schwartz, Mc Coss an Danyluk[395–397] is, on the contrary, of direct concern with this review and we will now discuss them in detail. The first rapid communication,[395] reports about the important hydration effect on oxygen O4 and O2 of U, iPU and 3-MeU, specifically oxygen-17 enriched at these two positions. The shifts on the methylated compound 3-MeU are used to ascertain the expected diketo form of the three compounds in water and acetonitrile. Linewidths changes (800–500 Hz) within the temperature covered, viz., 30–75°C, appear relatively important whereas no variation in both carbonyl chemical shifts is detected. In contrast, going from the aprotic solvent CH_3CN to the protic one (D_2O) causes a 32 ppm upfield shift of O4 oxygen but only a very small upfield shift for the O4 oxygen; The authors note some contradiction with the carbon-13 data which indicate more involvement of C2 carbonyl in hydrogen bonding and self-association with adenine derivatives,[398] but the situation is completely different here as we will see later. The important downfield shifts of O4 nuclei as compared O2 in water (more than 50 ppm) are interpreted as a decrease in the π bond order for the C2=O2 carbonyl bond. In our opinion, other effects may apply, such as some differential modifications in the electron density at each oxygen. In the second paper,[396] the authors determine the different equilibrium constants implied in the process of hydration at the two carbonyls by varying water concentration and temperature. The presence of two nucleoside–water bonding equilibria is then clearly shown at room temperature. A one-to-one complex ($K_1 = 0.28$, $\Delta H = -5.2$ kcal mol^{-1}) is first formed and a second one to two complex ($K_2 = 0.065$, $\Delta H = -11.2$ kcal mol^{-1}) appears then in a water excess. The detailed analysis is very complex with many hypotheses on the limits of ^{17}O shifts of the unbonded partners in the equilibria and the distinction of two types of hydrogen bond in water, one named L for the lone electron pair on the oxygen and the other called P for the water proton. Surely, some of these *a posteriori* assumptions can be disputed. The merit of the paper is a search for a full consistency in the approach by a detailed study centred on the sole iPU compound, selectively labelled at O2 or O4, and then to extrapolate the conclusions to U which do not markedly differ apart from a more *anti* conformation in U as compared to iPU known to be more *syn*! As a conclusion, at high concentration in water, one water molecule acts as a P hydrogen bond type to each of the O4 lone pairs, so that two water molecules are hydrogen-bonded in the P type to the carbonyl O4. The carbonyl O2 seems to "participate relatively weakly in H_2O hydrogen bonding and may be weakly intramolecularly hydrogen-bonded to the 5' OH". Naturally, this last affirmation should be accepted with caution for mainly two reasons. First, the *syn* conformation of iPU is highly favourable for this kind of intramolecular hydrogen bond, but renders the comparison with the anti-U, in which such a bonding cannot exist, at least is very risky. The second observation is that the authors never take into account to the possibility of an L-type hydrogen bond of the water with the "acid" H3 of the base. These two points will be underlined in the Dynamics section (*vide infra*). The last paper[397] is more traditional in that it tries to observe the influence of C5 substituent in U and T on the oxygen-17 chemical shifts, including some arabino and cyclo compounds. Data are discussed here in terms of π electronic density at the oxygen, π bond order at the carbonyl bond and inverse energy of the lowest lying n→π* transition energy[399]. The previous trends in hydrogen bonding by water at the O4 oxygen are found again, and it seems that the substituents do not severely affect it. In the competition between hydrogen bonding and substituent effects at the O4 carbonyl oxygen, it is clear that the substituents play only a minor role and do not necessarily change hydrogen-bonding ability. It appears also that the largest effect of hydrogen bonding to a given carbonyl is an increase in the π electronic charge on the carbonyl oxygen, so that O2 with a higher π electronic charge is shifted upfield relative to O4 for the non-hydrogen-bonded nucleosides.[400] Insofar as a hydrogen bond at the site is created, an upfield shift should be observed in agreement with the 32 ppm shift reported by the authors in their first paper. This means that the effect of a change in the π bond order is minimal and must be treated with extreme caution. A indirect conclusion derived from this study is that the poor ability of O2 in hydrogen bonding with water is not the result of π bond order or π electronic charge, but simply steric hindrance at O2 as has been addressed by Scheiner.[400] Because of this steric effect, one, and only one, molecule can bind to O2 by using only one of the two lone pairs on the oxygen, contrary to O4 which is fully exposed to the solvent. This has been corroborated by our personal investigations by intermolecular relaxation (see below, Section 6)

5.2.2.3. Fluorine-19   Most of the rather rare works on ^{19}F NMR on nucleosides mainly concern 2' fluoro,2' deoxy nucleosides.[94,105,378,401,402] This is an interesting aspect since the fluorine is generally more sensitive to the environment than the proton and then acts as an excellent probe for potential structural modifications. No major changes in the glycosidic torsion angle, in the sugar puckering and in rotamer populations of the exocyclic are observed in β ribonucleosides.

pH titration curves show that the chemical shift of the fluorine is influenced by the protonation of the base at low pH.[402] As a result, it seems reasonable to think that the major change induced by the strong electron-withdrawing fluorine is in the pK_a of the nucleosides.

The fluoro arabino series has received much attention because of the known antiviral activity of these compounds which are inhibitors of the herpes virus.[94,378,401] Some conformational modifications have been observed here as compared to the canonical series. In water, the 3'-deoxy 3'-fluoro ara-U remains in the *anti* form, but the ribose pucker in a twisted conformation with O4' *endo* and C1' *exo*, like in a perfect pentane, exhibits a drastic modification not observed for the 2'FdU. Fluorine, hydrogen couplings are in good agreement with this structure and help determine the new pucker. Good correlations of usual vicinal proton, proton $^3J_{HH}$ and Huggin's electronegativity values E_R of the substituent are in line with dihedral angles. A plot of the difference in the proton shifts $\delta_{H1'} - \delta_{H2'}$ versus the same Huggin's electronegativity value E_R is explained by an angular dependence of the shifts in relation with the dihedral angles and, *per se*, with the inductive effect generated by the substituent.[94] On a study on another antiviral compound, 2'-fluoro,5-iodo ara-C, FIAC, no sensitive effect is observed on the pucker sugar, but "the CH_2OH side chain is disordered" and the usual populations undergo the "gauche" effect so that the gg and gt rotamers are in nearly equal proportions.[378] It is remarkable that Lipnick and Fissekis arrive to the same conclusions in a study of 2'-deoxy,2'-fluoro ara-C in water, calculating values equal to 41% gg, 41%gt and 18%tg for the three rotamer populations.[401] This distribution is quite different from that of dA, but similar to that of other ara-C derivatives such as the above FIAC in which the iodine substituent seems to be without any sensitive effect. By introducing two fluorine atoms at the positions 2' and 3' of U, ara-U and 6-aza-U, and by computing furanose conformation by the PSEUDOROT program, Barchi et al.[158] have given a more quantitative insight in the sugar puckering induced by the two fluorines. In the three compounds, the gg and gt rotamers, jointly with the gauche effect, are operative to maintain a pseudo-diaxial arrangement of the C2' and C3' vicinal fluorine atoms where their interactions are minimal. But the influences on the ribose pucker are more complex with either a preferred S state (U) or N state (ara-U and aza-U). As expected, fluorine chemical shifts are good probes of sugar conformations since the "*endo*" fluorines are shifted upfield whereas the "*exo*" ones are shifted downfield ($\Delta\delta \approx$ 14 ppm). This value has been checked by DFT calculations. In this work, long-range proton fluorine has been of great help in the structure assignment.

The paper by Schweitzer and Kool[291] describing the design, synthesis and structures of three non-polar nucleoside isosteres of normal Pys and Pus is interesting in view of the chemical and biological implications of such an introduction of fluorine at distinct sites, either on the base or on the ribose. Here fluorine acts as the isostere for oxygen since they have practically the same van der Waals radius (1.47 and 1.52 Å, respectively) and C–H replaces N–H and–CH_3 replaces –NH_2. It is demonstrated that the isosteres are very close mimics of their natural counterparts, mainly in the steric aspect. They have no covalent bonding characteristics and no hydrogen-bonding possibilities and thus can be very useful for studying all other interactions, and by a way of consequence, hydrogen-bonding ability. They are naturally more hydrophobic and more lipophilic than their natural analogues. Indeed, a number of studies have shown that introduction of the fluorine atom in the nucleoside backbone has no measurable conformational effect on the sugar puckering and glycosidic torsion angle; only a minor but perceptible effect is visible on the exocyclic bond conformation.[105,291,378,401,402]

This is to say that structural analysis should gain from fluorine NMR, whereas structural chemistry has nothing to lose in this operation. For example, Cushley et al.[346] have shown that the magnitude of the long-range inter-ring constant $^5J_{H1'F5}$ is a good indication of the stereochemistry α or β of the nucleoside (vide supra). Why not measure directly the shifts on the fluorine or carbon-13 fluorine couplings to get other similar invaluable information about conformation? Another point of interest concerns the possibility of the through-space coupling interaction with the fluorine which has been indirectly postulated as a possible specific effect in α anomers.[293] In our opinion, a direct fluorine NMR analysis can help to elucidate this way, mainly by 2D fluorine, fluorine COSY and NOESY experiments as well as ^1H, ^{19}F or ^{19}F,^{13}C HOESY experiments, the fluorine detection or carbon-13 detection modes, respectively. The last, but not the smallest problem with these fluorinated compounds, should be their potential low solubility in water, but this is not an unsolvable issue.

5.3. Quantum mechanical calculations

Before dealing with the modern NMR computational methods, which are today a part of the toolbox of all spectroscopists, chemists and biochemists, we wish to say a few words about the old semi-empirical or minimal basis set ab initio methods in use at a time where the computational power available was very limited. It remains that they have played an important role in the understanding of structural chemistry and spectroscopy and, especially, in the analysis of chemical shift trends in a large variety of aromatic compounds. As we have noted at the beginning of this section, the driving force was, inter alia, the use of the Karplus-Pople Equation (1), which was very soon after followed by ab initio calculations of screening constants and spin–spin coupling constants.

The group of Grant[273] has obtained a very good correlation between all the observed chemical shifts of the Py and Pu carbons of the naturally occurring nucleosides against the electronic π-charge estimated by very crude MO approximations derived either from the Hückel approximation or from Pariser–Par parameterisation. Furthermore, the slope in all cases varies as 160 ppm/electron as generally assumed.[220,403] Some minor improvements can be obtained with the CNDO/2 MO method or by introducing Mulliken gross populations, but main difficulties persist, probably at the level of the carbon–nitrogen bond polarisation and of the neglect of orbital overlap. The role of the C–N bond polarisation has been further confirmed in more sophisticated ab initio HF calculations on a Pu model in which the glycosidic torsion angle χ has been varied in steps of 30° in order to quantify this effect on the mimic of carbon C8 of Pus.[404] In the 360° range of variation, the amplitude of the torsion effect is about 5 ppm from the upfield side (*anti* form) to the downfield side (*syn* form) with a maximum of about 12 ppm near $\chi \approx 60°$. The N1–C2 and C2–N3 bonds contribute for about 7 and 2.5 ppm, respectively, to this full variation. This effect of torsion of the glycosidic angle upon the proton chemical shifts of ribose in Pys and Pus has been investigated many years ago by Giessner-Prettre and Pullman[405] by calculating the diamagnetic effect in the framework of the SCF-MO Pariser–Parr–Pople procedure and by adding an electric effect as estimated by Buckingham[312] and Musher[313]. Ring current is negligible in Pys, but more important in Pus. The contribution of the C=O group is important not only by its magnetic anisotropy but also by its polarisation effect which can affect distinctly the various protons. In addition, as we have previously noted, the free electron doublet on the nitrogen gives a local magnetic anisotropy effect which is important mainly in the Pus.

Others have also observed the correlation between chemical shifts and electronic charge. Alderfer, Loomis and Zielinski have reported such a dependence for carbon-13 chemical shifts in A, C, U and their fluorinated analogues against CNDO/2 charge densities, except—once more!—when a nitrogen atom is situated β to the fluorine atom.[135] Shugar et al.[132] have also reported that the charge effect by a methyl upon substitution of 2'-hydroxyl proton in an ara-C is only significant on the C2' carbon and the O4' oxygen but not on the other carbon and protons atoms and so the effect seems to be negligible on the chemical shifts (except C2' and O4'). This assertion should be disputed in view of the possible charge transfer from the oxygen via σ^* localised orbitals and also via a through-space interaction resulting from an overlap between some π and π^* of the C5=C6 pyrimidine double bond.[118] In contrast, in a study of the effect of methylation at the various N positions on 22 nucleosides in water and DMSO,[406] it has been postulated that the trend in carbon-13 chemical shifts could reflect a change either of the bond order or of the excitation energy in the Ramsey theory of screening constants.[217]

Few *ab initio* SCF-MO calculations of nuclear screenings and spin–spin coupling constants have appeared also during this period (1975–1995), probably because of a lack of new data (mainly carbon-13 and nitrogen-15) at the time when Fourier transform spectroscopy began to be popular. To the best of our knowledge, only a full calculation, limited to cytosine as the base, has been done.[407] The authors have used gauge invariant atomic orbitals (GIAO) in their perturbation approach which is known to be advantageous in giving shielding constants that are completely independent of the coordinates reference frame as are each of the diamagnetic and paramagnetic contributions as well. All the shielding constants of all the nuclei of cytosine (H,C,N,O) have been obtained and they reproduced correctly the experimental trends, including the experimental shielding order $\sigma_{N3} < \sigma_{N1} < \sigma_{NH2}$ and the good position for oxygen-17. Faced with the difficulties with the application of Karplus Equation (2) for investigating sugar pseudo-rotation for *cisoîdal* couplings, Jaworski et al. have run perturbative INDO-SCF-MO to calculate this coupling on isomeric β-arabino and β-xylofuranosyl nucleosides.[133]

Quantum chemical calculations of nucleoside structures have significantly progressed in the past years with highly developed software based on *ab initio* post

Hartree-Fock (HF) at the MP2 (Möller–Plesset) level of electron correlation[33–37,122,149] and/or even more sophisticated DFT performed by means of the hybrid exchange and correlation functional (B3LYP) with an extended 6-31G* basis set.[38,151,154,155,231,232] This last DFT approach can be regarded a cost-effective alternative to the more time-consuming MP2 second-order perturbation for ground-state analysis of nucleosides. It has been used, by employing the GIAO-DFT method, for estimating fluorine chemical shifts in 2′,3′-difluorinated nucleosides with good success to predict the separation between *endo* and *exo* fluorine atoms, the first being about 14.1 ppm upfield from the second.[158] At the HF-MP2 level and by using the classical Fermi-contact perturbation,[227,230,231] Serianni et al.[151] have observed a great dependence of all carbon-13 proton spin–spin couplings ($^1J_{CH}$, $^2J_{CH}$ and $^3J_{CH}$) with the ribose geometry. For example, larger $^1J_{CH}$ values were found when a C–H bond is quasi-equatorial where the bond is shorter than in quasi-axial orientation, a result which suggests a role for this direct coupling as a complement of all other $^3J_{HH}$ and $^3J_{CH}$ couplings in the conformational analysis of furanose rings.

NMR calculations require an energy expression for the system in presence of an infinitesimal perturbing magnetic field, thus needing a good approximation for the exchange-correlation functional for the density and thereby becoming very time consuming.[153,231,232,408] Today, one of the best models, going beyond the uncoupled DFT, uses the finite perturbation theory (FTP) approach with sum-over-states (SOS) approximations combined with density functional methodology, named under the acronym SOS-DFPT.[408] The method, which is "rotationally invariant", is an improvement as compared to the response theory of the exchange-correlation potential to an external magnetic field and handles cases where the HF-MP2 approach works very poorly and with some gain in computing time.[231,408] In addition, the essential introduction of water either implicitly by models like the polarized continuum model (PCM) and Onsager model[150] or mimicking its dielectric properties[157] or explicitly by introducing many water molecules in the "tool box"[157] has a strong influence on the most stable conformation obtained. We do not wish to pursue this matter which is clearly outside the field of this review, but we will only add an example to underline these difficulties.

In their recent paper, Barbe and Le Bret[157] have studied the effect of water on the sugar puckering of U, dU and 2′-OMeU. Without water or with its implicit mimic, South puckering is found for U and dU while 2′OMe is North, the first result being in complete disagreement with NMR data in water (see above). When water is explicitly introduced, but at specific positions, that is between the base O2 and the sugar OH2′, the N puckering is obtained for U but no changes are observed for the two other compounds dU and 2′-OMeU. In addition, some very questionable hydrogen bonding implying hydroxyl groups with water and with the base proton H6 are deduced, in contrast to our own work (see below Section 6.6). At nearly the same time, and without any explicit or implicit consideration of water, Ghomi et al.[38] have calculated by DFT that the C3′ *endo*, N state is the slightly more preferred conformation to the C2′ *endo*, state and have found other hydrogen bonds inside the sugar. If it is agreed now that that theoretical chemistry is a full partner with experiment, the way remains long to obtain fully reliable comparison of theoretical results, first among themselves and then with experimental measurements.

By using DFT and individual gauge for localised orbital (IGLO), Dejaegere and Case[409] have obtained what is certainly one of the most interesting results. Probably due to complexity, calculations were limited to ribose and deoxyribose rings, but they agree very well with experimental data. As an example, they show the deshielding in C3' *exo*, S as compared to C3' *endo*, N, the important low-field shift of the C1' while the C2' is correctly as the high-field carbon. The same is true for the range of proton chemical shifts, and in both cases the slope of calculated shifts versus experimental data is fairly good, being very close to one for the deoxyribose model. The influence of the thymine base is also well reproduced, especially the inversion of H2' and H2" shifts, which means that H2" is deshielded as compared to H2' in dT (see above).[258,274] In addition, some new information about the chemical shift anisotropy (CSA) is given which can be of potential help in the nucleic acids structure determination process.

Carbon-13 solid-state NMR is an invaluable tool for nucleoside structural studies since it provides the means to acquire spectra that correspond to single conformation and can give information about CSA, as opposed to ^{13}C solution NMR methods. These distinct effects of sugar puckering on the ribose carbons C2', *endo* (S type) and C3' *endo* (N type), dependence of the C1' and C54 resonances on the glycosidic and C4'–C5' exocyclic torsion angles, respectively, have been explored by Harbison and collaborators.[410] The authors then construct, from data analysis and discriminant analysis, two canonical coordinates—linear combinations of chemical shifts which give a desired statistically optimal conformation from the NMR. The first one concerns the S and N sugar puckers and is only a function of the C1', C4' and C5' chemical shifts. The second one is intended to define the effect of the gg and gt orientations (but not tg?) about the exocyclic torsion angle by using the C2', C3', C4' and C5' chemical shifts. Two types of calculations are then run by the authors, one by using the SCF-GIAO *ab initio* approach and the second by using the DFT-GIAO method and a model compound where the base is restricted to a simple pyrimidine substrate. Either optimisation of geometry from crystallographic data was done or solid-state coordinates were used as they stood. The GIAO-DFT method systematically overestimates the shifts by about 5 ppm, whereas the GIAO-SCF underestimates them by the same amount. In each case, the errors appear to be systematic and the slope of the plot of the calculated values versus experimental shifts is always very close to unity. Apart some important deviations with the C2' and C3' resonances, the agreement between experiment and theory is excellent, thus validating *a posteriori* the protocol retained by the authors. Finally, the importance of electron correlation in the calculation is duly pointed out. A very similar, but more limited, approach has also been published for analysing the C3', C4' and C5' chemical shifts of ribose ring on 31 nucleosides and nucleotides.[152]

As we have seen above, the group of Sklenář[153] has also addressed many questions concerning the use of Karplus relation (Equation (2)) for elucidating nucleoside conformation from the inter-ring long-range couplings $^nJ_{HH}$ or $^nJ_{CH}$ or $^nJ_{HF}$ and, mainly, the imperfections of its parameterisation because of a lack of experimental data covering all the (uncertain) range of glycosidic torsion angle. First, the authors observe that the restriction of the Karplus curve to the extremes $\chi = 60°$ and $\chi = 240°$ cannot be legitimate in view of the insufficient symmetry of all the nucleosides. Second, there is a large diversity of coupling pathways, so that significant differences in π-bond order in unsaturated bonds can exist between the various bases and one needs separate A, B and C coefficients. Third, the oxygen bridges formed in cycle nucleosides may also influence distinctly the various couplings, for example $^3J_{C2H1'}$ or $^3J_{C4H1'}$ and $^3J_{C6H1'}$ or $^3J_{C8H1}$. So far, they have tackled with a theoretical study of the Karplus equation first on dA, dG, dC, dT and dU[153] and, more recently, $^5J_{H1'F}$ coupling on 5-fluoropyrimidine-substituted RNA.[153] This last case is particularly stimulating because the coupling is smaller than the natural linewith and requires, to be accurately measured, sophisticated, dedicated 2D experiments. In order to overcome this difficulty, the authors have developed a $^1H,^{19}F$ spin-state selective excitation (S^2E) pulse sequence combining a 1H, 1H NOESY mixing scheme and water suppression by excitation sculpting. Some standard gradient-selected HMBC experiments were run to exploit the $^5J_{H1'F}$ in order to facilitate the chemical shift assignment and also as a check; 30 compounds derived from the HIV virus and labelled with either 5F-uracil or 5F-cytosine have been investigated in water. Geometries have been optimised at the DFT-B3LYP level. Spin–spin coupling constants were calculated by the functional perturbation theory SOS-DFPT as described above and by neglecting the spin–dipolar contribution, but taking only account of the Fermi-contact term as implemented in the deMon NMR code.[408] These studies have shed new light on the relationship between the glycosidic angle torsion and the couplings, sometimes at variance with some previous assessments. A clear distinction between the Pys and the Pus appears, in contrast with experiment, the results suggesting the surprising order $^3J_{C6H1'(cis)}$ > $^3J_{C6H1'(trans)}$ for the Pys and this in full agreement with the data retained. However, the authors recognise that their parameterisation has been obtained for S nucleoside conformer and that the effect of sugar pucker on the spin–spin coupling is extremely important. There is also a non-negligible shift in the syn $\rightleftarrows$ anti conformer equilibrium (about 10°) which is attributed to the non-symmetrical substitutions of C1′ carbon by atoms with different electronegativities. Generally, the glycosidic torsion angle χ largely depends upon the sugar pucker and upon the exocyclic bond rotamer. For a change in the state of the sugar from S to N, all $^3J_{CH1'}$ couplings decrease for the syn form and increase for the anti form and the $^1J_{CH}$ couplings have a reverse relation. The exocyclic bond conformation influences all the couplings in the same way. The sugar conformation affects the $^1J_{C1'H1'}$ by the relative orientation of the $n_{O4'}$, $\sigma^*_{C1'—H1'}$ orbital with a direct effect involving a $\cos \chi$ dependence and an indirect effect involving a $\cos^2 \chi$ dependence.[173] For the long-range $^5J_{F5H1'}$ couplings, the phenomenon is well rationalised on the classical W zig-zag configuration of the five intervening bonds in the preferred anti conformer, with only a negligible effect of the sugar pucker in this case.[173]

6. DYNAMICS IN WATER

This section is intended to describe time-dependent phenomena and includes a survey on the magnitude of the various barriers to rotation, their heights and the potential influence of water on them. This concern the $syn \rightleftharpoons anti$ conformer equilibrium, the sugar puckering, the rotation around the exocyclic bond and, possibly, the restricted rotation around the C–N bond in amino-substituted compounds. A thorough discussion on chemical exchange of acidic protons with water protons will follow, as this is an important point in biology. Transverse and longitudinal relaxation data will then be examined, and we will end this section by addressing some issues on the very promising diffusion studies. Along these lines, we wish to cover a very large time scale of dynamic processes, from a few seconds as deduced from classical NMR exchange to several picoseconds (or less) for rotational correlation times as deduced from relaxation measurements and translation times as obtained with diffusion studies.

6.1. Rotation barrier about the glycosidic bond

There is such a scatter in the data reported—from 4 to 16 kcal mol^{-1}—that it seems very difficult to get a good idea of the dynamics of the conformational equilibrium. The lowest extreme value has been reported for G in water by Petersen and Led,[411] whereas the highest extreme is derived from an *ab initio* SCF-HF/6-31 G calculation on a flexi-guanosine.[188] Naturally, it is evident that the structure of the base and the presence of substituents, either on the ribose or on the base, have large effects on this amplitude. For these reasons, the highest value should be taken with care because, in the flexible structure, the imidazole ring is only linked to the pyrimidine part by a "single" bond. As a result, the electronic delocalisation is severely restricted between these two rings with respect to its natural counterpart G where they are fused. By a way of consequence, the electronic interaction between the imidazole group and the ribose moiety, and especially with O4' oxygen, is increased, so that the C1'–N bond order is certainly also increased. The right value is certainly in the middle of the range $\Delta H^{\neq} \sim 6$–12 kcal mol^{-1}. Indeed, in 1971, by using ultrasonic relaxation in water, Rhodes and Schimmel[412] were the first to determine an activation energy $\Delta H^{\neq} \approx$ 6.2 kcal mol^{-1} for A in water, after a careful examination and a rigorous removal of all the other superposing effects like sugar pucker. This corresponds to a relaxation time of 4.0 ns for A, which is greater by a factor 2 than that obtained for dA (2 ns) because the steric interaction between the 2'OH and N3 of the purine in the C2' *endo*, S state of the ribose in A notably increases the height of the glycosidic barrier as compared to dA. The same order of magnitude is observed for Fo, G and I. The method has been refined some years ago to obtain the same qualitative results with an equilibrium constant in water $K_{eq} \approx 4.0$ and the following lifetime for each conformer $\tau_{syn} \approx 5$ ns and $\tau_{anti} \approx 20$ ns.[413] By using SCF–INDO computations in a purine nucleoside model, Boerth and Bhowmik[136] have obtained an energy barrier $\Delta H^{\neq} \sim 10$ kcal mol^{-1} which, of course, corresponds to the gas phase. Interestingly, extensive electronic delocalisation on the purine cycle in case of cation nucleoside indicates a considerable fall of this barrier, up to 5 kcal mol^{-1}. However, it is well known that there are considerable solvent effects on the magnitude of this barrier, mainly in water which can act as well as a donor or an acceptor of hydrogen bonding.[37,138–140,145,147–150] Very recently, by using MD and CPMD calculations, Murugan and Hugosson[138] have clearly corroborated this view and shown that the inter-conversion process has a larger barrier in chloroform as a solvent, whereas the barrier height is brought down significantly in water as the solvent due to the stabilisation of transition state by intermolecular hydrogen bonding with water. An experimental value $\Delta H^{\neq} = 10.7$ kcal mol^{-1} ($\Delta G^{\neq} = 13.7$ kcal mol^{-1} and $\Delta S^{\neq} = -11.1$ cal mol^{-1} K^{-1}) has been measured for Tph, and this barrier seems even higher when a S atom replaces the O6 so as to preclude the rotation around the glycosidic bond so that only the *anti* conformer is present.[254]

We are not aware of any reported experimental value for the Py's barrier, but in the absence of any ultrasonic relaxation in water for C and U,[412] we presume that this value is higher by something like 2–4 kcal mol^{-1}. This opinion is corroborated by the small ring current in the base as compared to the Pus,[357] so that the electronic interaction between the base and the ribose is strengthened as well as by a through-space influence on the O4' free doublet or by a through C1'–N through bond effect, decreasing the ability to free rotation. Another proof is given by the temperature-independence of the ribose shifts and coupling constants in the range 30–70 °C for the Pys[249,265,414] as opposed to a small change of these parameters for the Pus.[252] The small change of the H6 resonance with temperature that has been sometimes reported for the U derivatives (about 0.05 ppm upfield) should be taken with extreme care and is undoubtedly an ambiguous monitor of the conformation about the glycosidic linkage.[267] Other influences like field effects should be operative in this case. The strong preference for an *anti* form for the Pys, as we have seen above, is in line with this point of view.

Methylation at the ribose O2' oxygen of ribonucleosides has no effect upon the three-dimensional structure of a nucleoside in water and no temperature effect is observed in the interval from 30 to 60 °C, again suggesting some rigidity about the glycosidic bond.[414] Introduction of a hetero-atom like N in place of a CH group, or S in place of O, here also changes the magnitude of the barrier. One can think that in aza-derivatives the glycosidic rotation is more free than in normal Pus because the steric hindrance in the transition state is somewhere removed.[325] 2-SU has a glycosidic barrier greater than about 5 kcal mol^{-1} as compared to U because of the large van der Waals volume of S ($r_{vdw} \approx 1.80$ Å) compared to that of O ($r_{vdw} \approx 1.52$ Å).[415] Phillips and Lee[335] have demonstrated that introduction of S at the 2-position in place of O raises the barrier and has a reverse effect as compared to that of replacement of O4 by S in O.

α-Nucleosides and arabinonucleosides are also other particular cases for which data are completely missing. Nevertheless, empirical PF calculations indicate that the barrier is higher than in β-nucleosides, especially for α-ribonucleosides (by more than 3 kcal mol^{-1}) because of the specific, strong steric interactions between the base and the 2' hydroxyl group and a very high energy transition state in which the sugar adopts the O4' *endo*, C4' *exo* conformation.[145] In addition, the gauche effect with the O4' oxygen also contributes to the raising of the barrier.

6.2. Sugar pseudo-rotation barrier

There is a good agrement[138,347,351] between the different authors here to adopt a pseudo-rotation barrier in the range 2–5 kcal mol^{-1} as a result of a number of theoretical approaches (MD, CPMD, BD, HF, DFT)[37,137,143,149–151,157,409] or semi-empirical calculations (MM, PF).[77,78,122,142–148] To our knowledge, the only one measurement has been provided by a very detailed analysis of carbon-13 longitudinal relaxation times of A, G, I and X in liquid ND$_3$.[416] An average activation energy of 4.7 kcal mol^{-1} was found for the four Pus, just near the higher boundary of the range.

This value should be considered only as a rough approximation. It is well known that in nucleosides, like in nucleic acids, the furanose ring is asymmetrically substituted, and therefore the different conformations are weighed unevenly around the pseudo-rotation cycle and span all the thermodynamically accessible regions of this space. Traditionally, the pseudo-rotation wheel is divided into four equally sized quadrants centred around pseudo-rotation phases P equal to 0°, 90°, 180° and 270° and are termed the northern (N), eastern (E), southern (S) and western (W) quadrants, respectively[41] (Figure 5A). S states and N states represent energy minima of comparable depth, whereas eastern states E and western states W form energy barriers. The eastern barrier (2–4 kcal mol^{-1}) is always lower than the western one (4–5 kcal mol^{-1}).[37,143] Generally, the interconversion between the N and S states proceeds via this east barrier and involves a decrease in the amplitude of furanose during the puckering pathway.[37,78] It should be noted that both barriers are lower than N/S exchange via the P form, providing evidence that pseudo-rotation is more preferred than inversion. For ribose and deoxyribose, Dejaegere and Case[409] have calculated an eastern barrier equal to 3.7 kcal mol^{-1} and a western barrier equal to 4.4 kcal mol^{-1}, in full agreement with the above data and HF-MP2 results by Serianni et al.[151]

They are some subtle changes in these figures as it is observed that the barrier is higher in ribose than in deoxyribose,[77] but is lower in Pys than in Pus,[78] in the first case because of the gauche effect. On similar lines, the ribose in 2-SU has a barrier equal to 2.9 kcal mol^{-1} while it is 2.1 kcal mol^{-1} in U.[143] This barrier is increased in α-nucleosides and arabinonucleosides as compared to their β-homologues.[145] It also seems that it is raised in AZT by about 1.7 kcal mol^{-1} compared to that of deoxyribose because of the rare occurrence of the sugar ring pucker.[374] These five examples are again a demonstration of the very subtle importance of steric hindrance in the transition state as it is well recognised that the major contribution to the barrier in ribose is the van der Waals repulsion between the 2' and 3' eclipsed hydroxyl groups. In addition, DFT calculations with the introduction of explicit water by Barbe and Le Bret[157] have pointed out that in water the barrier is lowered by about 0.6 kcal mol^{-1}, in rough agreement with Olson's predictions.[77]

The easiness of the S ⇌ N Interconversion by crossing the pseudo-rotation barrier is clearly demonstrated by the very small lifetimes of each conformer along the pathway, probably less than 1 ns.[347] Such a value has been exactly measured in a recent carbon-13 longitudinal and transversal relaxation times study with NOE measurements in D_2O, combined with an MD approach on the ribose moiety of dC from a double-stranded DNA.[417] The authors find that repuckering is a likely motional model for the ribose moiety which occurs with a time constant of around 100 ps. This coherence in the two datasets is very interesting from the dynamic NMR view point, confirming the postulated time scale of the interconversion but ruling out any observation at the very low time scale of chemical shifts in water.

6.3. Rotation barrier about the exocyclic hydroxymethyl bond

There is a big lack of data here, probably because it is very low and certainly much less that the 3.5–4.0 kcal mol^{-1} claimed elsewhere,[252] with a more realistic value ranging certainly in the interval 2.5–3.5 kcal mol^{-1}. The few available experimental results strongly substantiate this view. For example, the good correlation between the difference of the H5′ and H5″ chemical shifts and the sum of their respective coupling with H4′ has been observed in a large range of temperatures and for a large population of nucleosides (both Pys and Pus).[361] The negative slope is a good indication of a drastic change in the rotamer population in favour of p_{gt} and p_{tg}. For cyclo-U in water,[267] p_{gg} decreases from 0.50 at 27 °C to 0.33 at 88 °C. The temperature effect appears less pronounced in water as compared to DMSO because of potential hydrogen bonding in the first solvent.[178,267]

Due to this smallness and the presence of the two diastereotopic protons H5′ and H5″, always chemically distinct whatever the temperature, it is not surprising that the barrier cannot be measured within the chemical shift time scale. Nevertheless, it should be kept in mind that this barrier is highly sensitive to solvent effects and, naturally, is slightly higher in water (certainly by something like 2 kcal mol^{-1}) because of intermolecular hydrogen bonding and chemical exchange with water. Let us add a final word about what we think as one of the best methods to attain this parameter. It consists in measuring deuterium, proton and carbon-13 longitudinal relaxation at variable temperature of the asymmetrical top. By using the classical formalism of relaxation, mainly by estimating the quadrupolar effect and by separating the intra dipole–dipole contribution from the spin–rotation in the different terms, it is then possible to extract all the relevant correlation times τ_Q, τ_{eff} and τ_{SR} of the rotational motion.[58,418] The dependence of these three correlation times with temperature leads directly to the activation energy of the process. This reviewer has not been aware of such an analysis so far, and deeply hopes to find it in the very near future in the literature!

6.4. Rotation barrier about the exocyclic amino bond

Exocyclic amino groups of adenine, guanine and cytosine are key sites in the formation of hydrogen bonding for maintaining canonical Watson–Crick pairing of the bases in DNA and RNA. Their exposure in the grooves of double-helical structures makes them privileged targets for binding with drugs as well for specific interaction with water. The non-planarity of the amino groups with the base plane has also been recognised for a long time and water can influence both its non-planarity and the rotation around the C–N bonds. It is then of special interest to know the activation energy of this restricted rotation.

There is large agreement for the activation energy of the Pus, where all experimental results,[316–319,419,420] generally obtained in water and by the classical line shape analysis,[421] are in the range $\Delta H^{\neq} = 11$–18 kcal mol^{-1}. It is noticeable that this interval nicely encompasses the one which is predicted by semi-empirical SCF-MO-CNDO/2 and EHT theoretical calculations (12–16 kcal mol^{-1}).[422] The only divergence seems to appear with the Pus for which some reported data are largely outside this range, being sometimes greater by a factor of 2.[315] An indication to this disagreement should be found in the result given by Jordan[320] for 4-aminopyrimidine in water at pH 1.56 and 12 °C and for which a value $\Delta H^{\neq} = 14.6$ kcal mol^{-1} is reported. This last one is clearly in the calculated range and does not differ much from that data reported by Shoup et al.[315] for a series of cytosine derivatives with methyl-substituted amino groups which are also in the previous range ($\Delta H^{\neq} = 15$–18 kcal mol^{-1}). It is very difficult to propose a plausible explanation for this divergence, but we are inclined to think that the barrier is certainly greater than that deduced by Martin and Reese.[319] Our feeling is founded on the idea that this barrier should not be smaller in Pys than in Pus in view of our previous discussion about the distinct ring current effects in the two kinds of compounds (see above). In their careful measurements on the N6 amino group of A by ^{15}N-edited magnetisation transfer experiments, Michalczyk and Russu[420] have seriously stressed upon the spurious contribution from the amino proton exchange with water. Even if this process is generally low (less than 2.1 s^{-1} in the present case) as compared to the hindered amino rotation (about 80–90 s^{-1} here), it should be taken into proper account in the line shape analysis. Otherwise, the rate of the amino rotation k_r will be systematically overestimated and the activation energy E_a will be, by way of consequence, systematically underestimated. This is certainly one of the reasons why NMR spectra always give too low activation energies, even in organic solvents which are never free from some traces of water.

The important point to keep in mind is that this barrier is largely greater than all the other barriers; especially it is about twice that about the glycosidic bond. As a result, it should be expected that the presence of an amino group on the base may impart some supplementary rigidity to the nucleoside. A proof of this assertion may gained from NMR relaxation works at the non-substituted amino nitrogen as well as at mono- and disubstituted amino groups on the base in order to be rid of the exchange with water. At this time, it does not seem that works in this direction has been reported, but they are highly welcome. According to its large chemical shift range, its long relaxation times and the availability of isotope-enriched compounds today at relatively affordable price, nitrogen-15 NMR at the amino site, at the base ring and above all at the glycosidic linkage should be of great advantage for evaluating all these processes.[221,423]

6.5. Chemical exchange with water

We have already underlined the great importance of this aspect in biological processes where water mediates proton transfer between the different bases in DNA and RNA and therefore influences the structure and function of the biological molecules. The overall mechanism of such a transfer is very complex and it supposes a good knowledge of all the distinct exchange between the partners, that is to say at a very rough level, exchange of the nucleoside labile protons with water protons, base-pair opening, transfer from water to base and then base-pair closing. Generally, the kinetics of all these steps is very slow at the biochemical scale, that is, in the range from millisecond to second or even more. Detailed examination of the mechanisms implies a careful description of all the parameters like pH, catalysis by acids and bases, temperature, concentration, etc., and is clearly outside the scope of this review. Among many others, Guéron and collaborators[39] have given many fruitful suggestions on all these key points and have clarified many discrepancies concerning base-pair lifetimes; the interested reader can consult their papers for useful guidelines. Moreover, a number of sophisticated and well-designed specific sequences, often for only one measurement, have been implemented by different authors to tackle this kinetics in water. It is also out of the question to describe them here and the interested reader is referred to periodical reviews dedicated to this field. In this section, we will limit ourselves to giving the prominent points about the nucleoside proton exchange with water from a chemist's point of view.

Some earlier reports[174,233–238,424] have addressed a large variation of the H8 proton in Pus, mainly in A, with temperature, solvent and/or pH along with along with a facile base-catalysed deuterium exchange in D_2O. This suggests that this proton possesses some acidic character and may form hydrogen bonds under certain conditions. The H-8 proton is shifted to low field in proton-acceptor solvents and the magnitude of the shift change is roughly correlated with the relative proton-acceptor strengths of the solvent. These results have been explained on the basis of a hydrogen-bonding interaction between H8 and the proton acceptor group of the solvent.

There is severe scarcity of data about the direct exchange of the imino protons with water in Pus and Pys and these data are generally obtained indirectly within some postulated and often controversial mechanism of exchange protons in base pair inside a short oligonucleotide duplex. As a consequence, the figures reported here should be taken with caution and are only intended to give some trend for valuable comparison of the lability of the base protons. They all concern pseudo first-order rate constant k_{exch}, and the reciprocal of k_{exch} is assumed to be the proton lifetime τ_{exch} on the imido nitrogen atom at neutral pH.

The first set of data was reported for the imino H3 proton of U by Mandal et al.[425] in 1979, who have used a spectral difference method on stopped-flow spectrophotometry ($k_{exch} \sim 130\,s^{-1}$) and by Ts'o et al.[426] 2 years later by only using a crude analysis of the line shape broadening ($k_{exch} \sim 85$–$160\,s^{-1}$), both at room temperature. The activation energy (about 6 kcal mol^{-1}) is very sensitive to the pH and seems to decease with increasing pH, as expected. There is a remarkable agreement between the authors in spite of the largely distinct approaches. By performing many diffusion measurements in a mixture of DMSO and water, we have obtained by ourselves[191] values slightly smaller than these two previous ones, for example, $k_{exch} \sim 14 \text{ s}^{-1}$, which roughly corresponds to $k_{exch} \sim 45 \text{ s}^{-1}$ in neat water at room temperature (see below). Before concluding with isolated nucleosides in water, we wish to mention the brief report of Guéron et al. in *Nature*[39] where they point out, as a manuscript in preparation, a τ_{exch} value close to 8 ms for T in water at room temperature and pH 6 which decreases to 4 ms at pH 5. We failed to the final paper relating to this work which is of direct concern here and we risk anticipating a value of 12 ms at pH 7 which would be in very good concordance with our results on U.

Unfortunately, we have not been aware of other experiments describing pure exchange of Pus and Pys in water. For these reasons, we now give only short comments on the results extracted from proton exchange either from nucleotides or from oligonucleotides duplexes as mediated by water with all the risks of misinterpretations as outlined above. In this field, data are clearly more abundant because it was, in the 1980s, a privileged area of research for a large number of NMR scientists. Experimental results can be classified in two categories depending on whether they were obtained by the (old) crude linewidth broadening analysis or, more recently, by the most reliable approach using measurements of longitudinal relaxation with modified Bloch–Solomon equations for many-site exchange[51] as given by Equation (7).

$$\frac{d\mathbf{M}(k,t)}{dt} = \mathbf{L}\,\mathbf{M}(k,t) \quad \text{with} \quad \mathbf{L} = -\begin{bmatrix} R_{1II} + k_I & R_{1IS} - k_S \\ R_{1SI} - k_S & R_{1SS} + k_I \end{bmatrix} \tag{7}$$

In this equation, R_{1II} and R_{1SS} are the direct longitudinal relaxation rates and R_{1IS} and R_{1SI} are the cross-relaxation rates in the two states during the chemical exchange. k_I and k_S are, respectively, the rate for proton leaving the environment I (imino site on the base) and entering in the environment S (water), and vice versa for k_S according to the Scheme 1.

The magnetisation vector $\mathbf{M}(k,t)$ is thus a column matrix consisting of the two longitudinal magnetisation $\mathbf{M}_I(t)$ and $\mathbf{M}_S(t)$ of the exchanging sites, while the vector $\mathbf{k}$ includes the two chemical exchange rates k_I and k_S. In this simple two-site exchange, integration of the system (7) is straightforward and analytical solutions are readily availble[51] (see Equations (18a) and (18b)).The resulting magnetisation intensities $\mathbf{M}_I(t)$ and $\mathbf{M}_S(t)$ depend upon the initial conditions, according to whether the first applied inversion impulsion is selective, bi-selective or non-selective.

In this first category, we only found a study on a ribonucleic acid mini-duplex.[427] Large differences are observed for the imino proton between the

$$I-H \underset{k_S}{\overset{k_I}{\rightleftharpoons}} S-H$$

Scheme 1 Two-site exchange model.

internal pair A–U (151 s^{-1}) and the external pair A–U (518 s^{-1}) indicating an easier access of water in the last case, while the central pair G–C gives an even slower exchange (104 s^{-1}). These three overall rate k_{exch} are deduced from an underlying three-state model of hydrogen exchange, proton transfer mechanism and exchange catalysis with simplifying hypotheses.

In the second category, Bendel[428] has obtained k_{exch} (T) $\sim 0.11 \text{ s}^{-1}$ and k_{exch} (G) $\sim 0.24 \text{ s}^{-1}$ for the imino H3 proton of T in the pair A–T and for the imino H1 proton of G in the pair C–G, respectvely, at room temperature and pH 6.2 on calf-thymus DNA. Hypothesis has been made by the authors of a two-site exchange with a water concentration in large excess so that its magnetisation is not affected during the course of the experiment. The same method, but accounting for the water magnetisation, applied on a self-complementary oligonucleotide strand of DNA at pH 8.0 leads to $\tau_{exch} \approx 150 \text{ ms}$ water for the imino proton H3 of T.[429] Results by Mirau and Bovey[430] for the imino proton H3 of U in PolyA–PolyU in water at pH 7 again gives another distinct value, k_{exch} (U) $\sim 22 \text{ s}^{-1}$. By using both 1D and 2D EXSY quantitative methods as a check, the group of Johnston[431] has measured the proton exchange between 9-ethyladenine and 1-cyclohexyluracil as a test model of two nucleosides in DMSO-d_6. Although the results are internally consistent, they are systematically smaller by an order of magnitude with respect to previously published data. For the exchange of the imino proton H3 of U with water, they find k_{exch} (U) ~ 3.4–5.8 s^{-1} and for the amino proton NH_2 of A the exchange is quasi null (k_{exch} (A) ~ 0–0.33 s^{-1}), but the figures are in good order.

A last comment is to observe that the above data are not as inconsistent as we might think at the first glance. For example, for U and T (which are very similar), all the experimental conditions differ from author to author, mainly sample and pH, and these greatly influence an already complex exchange in the base pair. This is to say that the value k_{exch} (U) $\sim 22 \text{ s}^{-1}$ reported by Mirau and Bovey[430] is clearly in full agreement with our own findings on U in water.[191] Finally, exchange with water is always slower with imino protons in Pus than with their analogues in Pys.[39]

McConnell and collaborators[427] have run the first measurements of the exchange with water of the exocyclic amino hydrogen in A and G in 5′-AMP and 5′-GMP, respectively, always using the linewidth broadening technique. The two values they have obtained, 69 and 38 s^{-1}, respectively, compare very well between them, in view of the very approximate method they have used. By a more sophisticated stopped-flow ultraviolet spectroscopy,[432] this exchange has been investigated in A at a neutral pH, which lead to a pseudo first-order exchange, $k_{exch} \sim 4.5$–6.5 s^{-1} with an activation energy $E_a \sim 10$–11 kcal mol^{-1}. These figures are in line with an expected exchange slower than that of the imino proton on the base, as it is well known that the amino protons are less acidic than the imino ones. The kinetics is also slower than the NH1 exchange with water in the protonated form AH^+ at the N1 nitrogen for which k_{app} is $\sim 160 \pm 20 \text{ s}^{-1}$ within the proposed mechanism. This is also in line with the predictions. More recently, Kettani et al.[433] have reported for the amino proton of C and by using magnetisation transfer experiments[434] a lifetime $\tau_{exch} \approx 100 \text{ ms}$ for the exchange with water at pH 7. The authors have particularly pointed out the difficulties in separating this exchange from the rotation of the group and so far have limited their discussion to the two limiting cases where the rotation is either fast or slower than the exchange and get evidence that exchange with water does not trigger rotation in protonated C. Incidentally, we can note than in cAMP and cGMP this time is largely increased to reach a value near 1 s, also in line with predictions.

6.6. Relaxation time measurements

One of the best ways to tackle the dynamic features of nucleosides in water is certainly to measure the relaxation times of the many distinct nuclei that constitute the chemical structure of the compounds (proton and deuterium, carbon-13, nitrogen-15 and oxygen-17). Nitrogen-14 which is largely quadrupolar, especially in the base ring where its quadrupolar constant may reach many megahertz, remains completely inaccessible by NMR and is more relevant of the future development of NQR (Section 5.2.1).[384] Because of its specific location in the three-dimensional chemical structure, each nuclear relaxation is a unique probe of this dynamics and makes up with all the other ones the overall motional fingerprint. As this is clearly dependent upon the specific or non-specific solvent interaction at each of the sites, the individual relaxation rates are the best sensors of these influences. Longitudinal relaxation times and transverse relaxation times are the two important parameters for the dynamic investigations in solution. NOEs (conventional longitudinal Overhauser effects) and ROEs (transverse Overhauser effects) are naturally the other essential parts of these studies since they carry information about intramolecular flexibility as well as intermolecular interactions. Extraction of all the spectral densities from experimental data should be a formidable task, but concerning nucleosides in solution at moderate magnetic field, the usual well-known approximation of "extreme narrowing conditions" $[(\omega\tau_c)^2 \ll 1]$ is an invaluable tool for the chemist.[51–56,58–63] There are many distinct correlation times and each characterises a given motional reorientation (a dipole–dipole vector, a rotation about an axis, a quadrupolar axis or a chemical exchange, e.g.) and, as such, probe the relevant solvent interaction on this motion. Faced with the physical complexity of the interactions, chemists often use many simplifications to overcome these difficulties while keeping the most prominent features of solvent influence at the observed nucleus. One of them is to define an effective correlation time τ_{eff} from the true relevant correlation τ_R adding the solvent contribution as described in Equation (8) where τ_M is the lifetime of the solvent in the environment of the nucleus on consideration (inverse sum).[63]

$$\frac{1}{\tau_{eff}} = \frac{1}{\tau_R} + \frac{1}{\tau_M} \tag{8}$$

In addition, major instrumental difficulties for measuring potential long transverse relaxations times have severely hampered using these data in microdynamics, so that they are generally limited to the determination of moderate to slow chemical exchanges.[435] Therefore, and hereafter, relaxation rate refers only to longitudinal relaxation, unless otherwise indicated.

The three major modes of internal mobility of nucleosides that we have ana-lysed above by chemical shifts and coupling constants (Section 5) are clearly also prone to an investigation by relaxation. This has been done at variable tempera-tures by Lüdemann et al.[271] for protons of the two amino-substituted (normal) 2'-dA and 3'-dA in ND_3 at variable temperature. The authors have based their simple approach on the assumption that the three internal motions are much slower than the overall reorientation of the molecule which is assumed to be isotropic. Note that Equation (8) applies very well to this situation if we replace the lifetime of the solvent τ_M by the correlation time for the internal rotation τ_i and if we assume, as the authors, $\tau_I \gg \tau_R$. Combined with homonuclear NOE, the German group found again that the 3'-amino 3'-dA is mainly in the overall conformation a*nti*, N, gg, while the 2''amino 2'-dA prefers the form *syn*, S, gt. Carbon-13, proton relaxa-tion and NOE on A, G, I, X in ND3 also confirm that ribose N state is correlated with *anti* form while the S state has a strong preference for the *syn* position.[268,271] Interestingly, variations of the carbon-13 rates with respect to temperature seem to corroborate the previous assumption of an isotropic motion.[271] These conclu-sions are at serious divergence with those of Guschlbauer et al.[367] from measure-ments on α-C, β-C and $\beta\Psi U$ in water, who claim that there is no correlation N, *anti* and S, *syn*, but they also defend the (wrong) *syn* conformation for β-ΨU, the same conclusions which unfortunately Guéron and collaborators[285] have arrived at by proton relaxation and NOE measurements (Section 5.1.1.1).

Carbon-13 relaxation of the tertiary carbons of ribose is very instructive in that it is reasonable to accept it as the only dipole–dipole relaxation mechanism with the directly attached proton. In doing so, a good probe for pseudo-rotation of the sugar is at the experimentalist's disposal. Wherever internal motion is not observ-able, all relaxation rates and there temperature dependences are identical. Indeed, these were just the features observed by the German group in another elegant work on A, G, I and X, always in ND_3 as the solvent.[416] Introducing rotation of the relaxing vector C–H, firstly according to a simple small-step diffusion model[436] and secondly according to the Woessner's model of equal jumps among three sites,[437] the authors observed that for both models the limiting cases (fast and slow) are identical and that the intermediate cases are not too distinct. A good estimation of an energy activation unique for the ribose puckering in all the Pus in then deduced (4.7 kcal mol^{-1}) in good agreement with previous published figures (*vide supra*). Moreover, they observe a smaller rotation for the exocyclic carbon C5', which is another proof of the free rotation of the hydroxymethyl group about the C4'–C5' bond (rotation of an asymmetrical top). Unfortunately, their results on iPU and C are very fuzzy. Other studies run on I and its molecular complex Inosiplex by using proton and carbon-13 relaxation suggest a distribution of several conformations in DMSO, but with a predominance of the *syn* form.[269]

Another, also extremely instructive, example is given by the detailed study of the microdynamics of C in DMSO-d_6.[192] It is shown that all carbon-13 relaxation rates are directly proportional to the number of attached protons, thus corrobor-ating the strong predominance of the carbon-13–proton dipole–dipole interaction as the main relaxation mechanism. In addition and based upon this observation, it can also be concluded that the random reorientation of all the C–H vectors is well described by a single correlation time τ which is the overall tumbling time of C and is found to be approximately $\tau_R \approx 150$ ps. This value fairly agrees with the expected one given by the well-known Stokes–Einstein–Debye Equation (9) in which v is the dynamic

$$\tau_R = \frac{4\pi v r^3}{3 k_B T} \qquad (9)$$

viscosity, r is the radius of the molecule, assumed to be spherical, and the other symbols have their usual meanings. A radius $r = 3.3$ Å is then obtained in good agreement with the value predicted from a molecular model. The final point reported by the authors is the correlation time τ_I of the free internal rotation of the amino group of C. It is obtained by measuring the relaxation times of the two amino protons which give two well-separated resonances in DMSO and lead to an effective correlation time $\tau_{eff} \approx 35$ ps, assuming that they relax only by their intra dipole–dipole mechanism. When inserted in Equation (8) together with $\tau_R \approx 150$ ps, this value gives $\tau_I \approx 46$ ps as the correlation time for this internal rotation. Note that the above two figures, which differ by more by a factor 3, are well in line with usual predictions.

Based on a thorough analysis of NOE and relaxation times in nucleotides, but which could also be of interest for nucleosides, Guéron et al.[169,294] have proposed a simple expression for estimating the probability of the *syn* orientation where the NOE enhancements on the H2', H3' and H8 for Pus during RF irradiation of H1' are taken into account. This equation has been later slightly modified by Chachaty and collaborators[295] for extending it to the Pys, and the authors have interpreted the experimental results as a weighting of relaxation rates by both the *syn* and *anti* forms. Nevertheless, it seems that the conclusions are always in favour of a preponderant *syn* form in water. The question is then to know if there is not some preferential influence by the phosphate group! Indeed, more recent results on an extensive relaxation study of non-exchangeable protons of adenosine 5'-phosphate (5'-AMP) seems to indicate a major *anti* conformation in D$_2$O with also about 20–40% for the *syn* form.[438] Some enlightening experiments in this way should be pointed out. By using the full magnetisation modes formalism,[439] Konrat et al.[440] have compared the relaxation of C, U and their monophosphates CMP and UMP in DMSO-d$_6$ and/or water, depending upon their respective solubilities. They have found that all the nucleosides and nucleotides have similar rotation diffusion constants for their ribose units, whereas the diffusion of the base in U derivatives exhibits a significant internal mobility. It seems that the phosphate group in UMP imparts a great internal mobility which solely affects the base unit. This aspect has been also reported 10 years ago in dinucleoside monophosphate in which internal motions are described by an appropriate model of free jumps between sites.[441] As a result, the transition rate in the *syn* ⇌ *anti* equilibrium appears to be strongly increased.[298]

An interesting observation can been made concerning the different stereo-chemistry of the α- and β-epimers. In the first compounds, the H1' proton is *cis* as compared to the H2' proton and then relatively close to it (~ 2.2 Å), while in the second case the two protons are in a *trans* position and more distant from each other (~ 3.1 Å). Because of the main dipole–dipole mechanism in the H1' relaxation, one may expect a shorter relaxation time for the α epimer H1′ proton as compared to its analogue in the β epimer in identical experimental conditions. This has been effectively experimentally substantiated on some isomers of FoA. However, as they have very poor, but distinct, solubilities in D_2O and DMSO-d_6, both the solvent and the temperature were changed from one set to the other in such a way to keep the viscosity constant as has been checked by a constant overall correlation time (100 ± 5 ps).[263] This identification of the two epimers is greatly helped by cross-correlated relaxation[442,443] between chemical shift anisotropy of the proton and H1′, H2′ dipolar relaxation. It is easily shown by a differential relaxation time of each of the H1′ doublet lines that such a mechanism is only active in the α-nucleosides where the proton H1′ and H2′ are close together, but not in the β-nucleosides.[444] Caution has been recently suggested for using this approach which is claimed to be limited because of the existence of a modulation of the cross-correlated relaxation by the sugar pucker.[445] In the same manner, modulation of J couplings by cross-relaxation of the sugar when the correlation time is longer—but greater than 2 ns—has been invoked, and attention has been drawn then in such a case that the use of the simple Karplus relation (Equation (2)) for determining conformations can lead to severe errors.[446]

Deuterium relaxation has been used in different ways to get information about the dynamics of nucleosides in water. The first one consists to selectively deuterate the molecule, for example the proton H8 in Pus, and compare the relaxation of the other proton with that obtained in the normal non-deuterated molecule. This method, dubbed DESERT[447] for deuterium substitution effect on relaxation times, has proven to get useful inter-proton distances in molecule with a better accuracy than NOE and some advantages over it like a higher sensitivity and better resolution for nearly chemically equivalent protons. The second way is to selectively deuterate successively all the positions in nucleosides as, for example, the two diastereotopic exocyclic H5′ and H5″ in order to elucidate all the relaxation mechanisms in the relevant nuclei.[448,449] Thus a significant contribution of chemical shift anisotropy (CAS) on the base carbon C2, C4 and C5 of U and T derivatives has been observed. Measurements of transverse relaxation times T_2 of carbons—which here are of same magnitude as their longitudinal relaxation times T_1 and never greater then 5 s—together with deuterium quadrupolar relaxation are naturally of help for these investigations. Cross-correlation between dipolar relaxation and quadrupolar relaxation in two methyne hydrogens have been observed in a particular case of one 1′,2′-disubstituted iP derivative. Finally, the same technique of deuterium labelling has allowed the conclusion that the pseudorotation of the sugar moiety in Pys occurs at the same time scale as the overall tumbling.[449] Quadrupolar effective rotational correlation times τ_D for deuterium ranges from 30 to 45 ps as compared to an overall correlation time $\tau_R \approx 55$ ps. The apparent activation energy of the deuterium relaxation is in the range 4.8–5.5 kcal mol^{-1}, which is nearly the same as that of ribose puckering (Section 6.2).

Previously we have stressed upon the key interest in nuclear quadrupole relaxation of nitrogen-14, a particularly important nucleus from the biochemical and biophysical points of view with all the inherent difficulties—not to say the impossibilities—to observe it by NMR (Section 5.2.1). Indeed, there exists an elegant method for overcoming these problems: that is, by recourse to a measure of the relaxation time of the nucleus in the rotating frame, generally known as $T_{1\rho}$.[450,451] This parameter is often assumed to have a value identical of that T_2 for small molecules in solution, but in well-designed situations it carries in itself other useful information than T_2. As an example, by providing the *a priori* knowledge of the scalar coupling between nitrogen-14 and proton J_{NH} or carbon-13, it is possible to indirectly access to the relaxation time of the quadrupolar nucleus via the relaxation time of the coupled nucleus, here proton or carbon-13, and *vice versa*.[451] Moseley and Stilbs[452] have taken advantage of this method for determining several lifetimes of the imino and amino protons of A, C, G and T in DMSO-d_6 or in the natural base pairs A–T and C–G. Nevertheless, it seems that the figures they have obtained are largely overestimated.

For a long time, most chemists have neglected intermolecular interactions between the solute and the solvent as a means of relaxation, as they were primarily interested in the structure. However, the theoretical aspect of an intermolecular dipole–dipole relaxation has been predicted by physicists ever since the early days of NMR by Bloembergen, Purcell and Pound.[453] It is not the goal of this review to discuss it, and the interested reader is referred to Abragam's text book[450] and other recent reviews and books on this topic.[58–63] The very complex phenomenon of solvation includes weak non-covalent interactions, hydrogen-bonding formation, electrostatic and dispersion effects, charge-transfer interactions, and, more generally, formation of solute–solvent complexes that are long lived relative to the time associated with diffusive encounters of the solute and the solvent is the heart of chemistry. It is evident that one can gain fruitful interpretation from this intermolecular relaxation. Naturally this means that the solute as well as the solvent carries with it NMR-sensitive nuclei and indeed this is practically always the case with the modern technology. Biologists have recently been aware of the great importance of this interaction with water as the solvent and have therefore developed, from the usual NOESY and ROESY techniques, a whole lot of well-designed sequences to get the most pertinent view of hydration of biomolecules.[7–20] To the best of our knowledge, studies relating to intermolecular homonuclear NOE between water and nucleosides are very limited and we can only point out the works by two groups. Newby and Greenbaum[454] have depicted water molecules bridging NH1 group of β-ΨU and nearby phosphate oxygen atoms in RNA. The experiment was duly run by using the CLEANEX-ROESY experiment in order to cancel any intramolecular NOE or ROE and exchange-relayed effects.[455] Two important papers by Berger and his group[456] report on the preferential hydration site of A by using a DPFGSE-NOE sequence and water suppression for improving detection of the small signal enhancements observed.[457] NOEs and ROEs have also been reported for A in DMSO-d_6 containing 2 mol% HOD and have allowed observation of strong interactions with 2′ and 3′ hydroxyl. Separation of dipole–dipole interaction of the hydroxyl group with water from their exchange with water is hardly obtained and the 5′ OH group of A seems to be unaffected by water.[458] This is in complete contradiction with our own findings on U in DMSO-d_6 where this resonance is easily identified. In addition, in water it is in such a fast exchange that if cannot be detected (see below).

Despite their great interest and their high sensitivities, NOESY and ROESY experiments suffer from a major insufficiency, in our opinion. The observed nucleus, hydrogen, is by nature at the very molecular surface and reflects only partial interactions. It should be often more interesting to probe the extent of the interaction more inside the solute heart to get a better understanding of its overall effect. In our opinion, this has become possible with the development of hetero-nuclear Overhauser spectroscop (HOESY) in the 1980s, allowing probing any Overhauser effect between any solute nucleus and any solvent nucleus, an idea published independently and at the same time in the same issue of the *Journal of the American Chemical Society* by Rinaldi[459] and Yu and Levy[460] as two urgent notes! In this way, it is noteworthy that Yu and Levy,[460] in their check of the new proposed sequence on ATP in water have the following comments about their results. "All three phosphates in ATP show intense cross peaks with the HOD signal. This means there are strong dipolar interactions between solvent protons and the phosphorus nuclei. The γ terminal phosphorus, apparently most accessible to the water, has the strongest cross peak and hence intermolecular dipolar interactions. The α phosphorus shows reduced intermolecular dipolar interactions and also intramolecular dipolar interactions with the sugar 5'-hydrogens. The β phosphorus shows only low-level dipolar interaction with solvent protons." Some digressions about the author's vocabulary are important such as the choice of the words "accessible, reduced and low-level" as an emphasis on the distinct magnitude on the ATP folded phosphate chain. Finally, the authors conclude their first communication: "This HOESY experiment should prove useful for evaluation of molecular conformation and also for studies of solvent–solute interactions." It is true that for investigations of intramolecular interactions much better and much sensitive NMR methods are available, but, to be honest, this last phrase was the stimulus of our personal works in this field of interactions of biological bases[189,461] or of their models[462,463] or of nucleosides[190,191] with water.

Probably the first qualitative observation of this intermolecular NOE of nucleo-sides with water was by Mantsch and Smith.[276] When they ran the carbon-13 spectrum of U in D_2O, the carbonyl carbons C2 and C4 were "barely distinguishable" whereas in H_2O there was a clear increase in their resonance magnitudes, and in DMSO-d_6, where the proton H3 of the uracil cannot exchange with the solvent, the signals were even much more intense (by a factor of about 4). This clearly indicates some proximity of water with the two carbonyl groups.

In our opinion, the best method to get the best possible quantitative aspect about the intermolecular interaction with water of these two carbons is to run two distinct and complementary measurements of carbon-13 relaxation times with proton broadband decoupling. First, in fully enriched deuterated "heavy" water D_2O the two relaxation rates $R_{D_2O}(C2)$ and $R_{D_2O}(C4)$ will be obtained completely free of any intermolecular proton interaction. A second experiment in normal "light" water H_2O will give $R_{H_2O}(C2)$ and $R_{H_2O}(C4)$ which now includes the intermolecular contribution by water. Finally, the ^{13}C, ^1H dipole–dipole relaxation contribution $R_1^{dd} = 1/T_1^{dd}$ to the relaxation rate $R_1 = 1/T_1$ is given in the extreme narrowing conditions[26] by Equation (10) in which η is the NOE factor.

This last value is obtained in two successive experiments by gating the broadband decoupling (direct and inverse gated decoupling).[26]

$$R_1^{dd} = 1/T_1^{dd} = \frac{\eta}{1.989 T_1} \tag{10}$$

The difference between the two dipole–dipole contributions in H_2O and D_2O $R_1^{dd}(H_2O) - R_1^{dd}(D_2O)$ leads to the pure intermolecular dipole–dipole interaction of the two carbonyl carbons C2 and C4 with the water protons allowing approach to the selective hydration at each carbonyl carbons. The bases of this interpretation are founded upon an expected, completely negligible dipole–dipole interaction of the carbon atoms from intermolecular deuterium isotopes in D_2O as compared with that from proton nuclei in H_2O. This is clearly indicated by the theoretical ratio $R_1^{dd}(H_2O)/R_1^{dd}(D_2O)$ of their respective contributions as given by Equation (11).[52–56,58–63,447]

$$\frac{T_1^{dd}(H_2O)}{T_1^{dd}(D_2O)} = \frac{R_1^{dd}(D_2O)}{R_1^{dd}(H_2O)} = \left(\frac{\gamma_D}{\gamma_H}\right)^2 \times \frac{I_D(I_D+1)}{I_H(I_H+1)} \approx 6.3 \times 10^{-2} \tag{11}$$

In Equation (11), γ_D and γ_H are the magnetogyric ratio of the deuterium and the proton and I_D (=1) and I_H (=1/2) their spin quantum numbers, respectively. Experimental results for U are shown in Table 8, where in place of neat water H_2O a mixture 50/50 of H_2O/D_2O has been used for locking the magnetic field. Carbon-13 effective correlation times τ_{eff} have been estimated in the extreme narrowing conditions from the well-known expression (Equation (12)) in which all the symbols have their usual meanings.[191,418]

$$R_1^{dd} = 1/T_1^{dd} = \left\{\frac{\mu_0}{4\pi}\right\}^2 \gamma_H^2 \gamma_C^2 \hbar^2 \tau_{eff} \sum_{i=1}^{n} \langle r_i^{-3} \rangle^2 \tag{12}$$

Because of the very low sensitivity of the ^{13}C, ^1H 2D HOESY experiment, the crucial step for observing this intermolecular effect is the optimisation of the mixing time t_m between the two magnetisations of the water protons and the detected carbon. Theoretical aspect of this point has been discussed by Macura and Ernst in their landmark paper.[464] Indeed, as it is very difficult to know exactly all the relevant relaxation rates parameters to get accurate expressions of both the cross relaxation rate R_C and the leakage relaxation rate R_L, other ways are searched. The best one is certainly to draw the proper NOE build-up curve for each observed site as a function of time.[59,60] Naturally, this first alternative is very time consuming and in case of ^{13}C, 1H HOESY it is generally prohibited unless carbon-13 enriched samples are available. In addition, the curves have generally a very smooth maximum for intermolecular interaction[59–63,190] and, jointly with the inherent in line integration errors, they might be sometimes badly defined. This is also another difference with homonuclear NOEs where the maxima are rather sharp. The other way implies a pertinent analysis of the interplay of all the parameters affecting this mixing time in order to get the best approximate experimental value.[465,466] It is this method we have always chosen in all our studies and the details are given elsewhere.[189–191,461,463] For example, with the NOE factors

TABLE 9 Carbon-13 longitudinal relaxation times T_1, NOE factors η, dipole–dipole relaxation rates R_1^{dd} and correlation times τ_{eff} for uridine in water (≈ 1 mol l^{-1}) at 300 K (dynamic viscosity $\nu \approx 1.0$ cP)[64]

Carbon number	C2	C4	C5	C6	C1'	C2'	C3'	C4'	C5'
			Uridine in $H_2O + D_2O$ (50–50)						
T_1 (s)	14.2	14.1	1.11	0.97	1.07	1.16	1.28	1.08	0.72
η	0.72	0.84	1.30	1.18	1.23	1.29	1.40	1.26	1.00
τ_{eff} (ps)	25.9	27.7	25.5	26.5	24.9	23.9	23.4	24.7	17.1
			Uridine in pure D_2O						
T_1 (s)	19.7	19.3	1.14	1.06	1.22	1.13	1.31	1.16	0.80
η	0.59	0.76	1.42	1.39	1.46	1.33	1.54	1.43	1.16
τ_{eff} (ps)	27.6	25.9	27.2	28.6	25.4	25.2	25.1	26.1	17.9
			$R_1^{dd}(H_2O) - R_1^{dd}(D_2O)$						
$\times 10^4$ (s^{-1})	104	101	≈ 0	≈ 0	≈ 0	≈ 0	≈ 0	≈ 0	≈ 0

given in Table 9 and the T_1 relaxation time of water measured to be 1.5 s in the mixture H_2O/D_2O (50/50 v/v), the optimum mixing time t_m is in the range 6.5–7.5 s.

Figure 8A represents the 2D HOESY $^{13}C,^{1}H$ plot of U in water recorded with $t_m = 7$ s and with pumping NOE in order to increase the sensitivity by a factor 2.[467] Note in the spectrum the complete absence of any direct correlation between a carbon and its attached proton. The 1D trace extracted from the 2D map along the water chemical shift clearly shows the same extent of the water interaction at the C2 and C4 carbons, respectively. For comparison, the same 1D trace from a 2D correlation recorded under the same conditions, but for only the base uracil u in water, exhibits at the C2 carbon an effect twice that observed at the carbon C4. This difference, which points out the prominent role of the labile H1 and H3 hydrogens in the hydration scheme, has been corroborated by NOESY and 15 N, 1H HOESY experiments on an nitrogen-15 enriched sample of uracil u.[189] With the help of some long-range interactions between selected carbons and hydrogens for which the separation distance is accurately known, a semi-quantitative interpretation of the hydration scheme of U and its base uracil has been proposed (see Scheme 1). Due to the linkage of the ribose and thus to the loss of its labile proton H1, the base in U is only dihydrated while it is trihydrated in uracil (Figure 8B–D) Streric hindrance by the ribose should also hamper the close approach of water near the amide group. Although they bear hydroxyl groups, the C2', C3' and C5' are never affected by water, whatever the mixing time used. The C4' carbon is no more affected by water. This is clearly evident in Table 9, which does not show any intermolecular dipole–dipole relaxation for all the ribose carbons in water. It seems that all stand as if water is maintained relatively far from the ribose or if its residence time near the sugar is too short to lead to an interaction. NOESY spectra of U in water also confirm the absence of any correlation peak of water

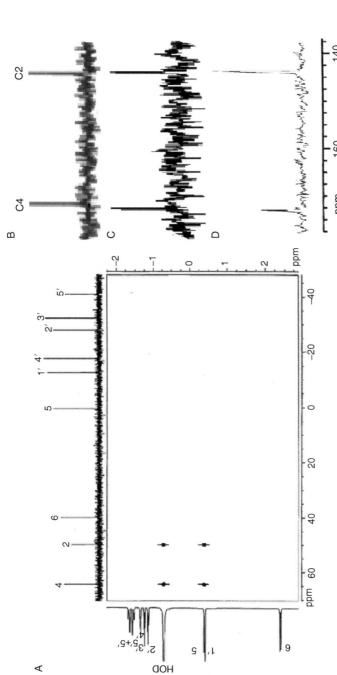

Figure 8 2D ^{13}C, ^{1}H HOESY spectrum of uridine U in neat water recorded at 75.4 MHz at room temperature on a Bruker Avance spectrometer (mixing time $t_m = 7$ s). The entire 2D map is shown with the 1D ^{13}C, {^{1}H} and 1D ^{1}H spectra represented along the horizontal and vertical axes, respectively (A). 1D trace extracted from the carbonyl region along the water chemical shift in the 2D map (B). Same trace for uridine U in DMSO-d_6/H_2O (83/17 v/v) (C). The same trace along the water for uracil u in DMSO-d_6/H_2O (83/17 v/v) and recorded in the same conditions is shown with the same scale for comparison in (D). Note that the ratio of intensities C2/C4 is 2/1 for uracil u, but only 1/1 for uridine U in the two solvents (from Reference 64).

with the hydroxyl groups which are in fast exchange with water and, as expected, give exchange peaks with the (bad) positive phase. To be right, some very small correlations of H1', H2', H3' and H4' with water can be guessed, but they are too low (and a bit random!) to conclude about any specific hydration in the environment of C2' and C3', contrary to what has been claimed elsewhere for adenosine A.[458]

6.7. Diffusion experiments

These experiments are certainly as old as NMR itself, probably because at that time people were intrigued by the broadening of lines resulting from the very large static and fluctuating gradients in the poor homogeneity magnetic field at their disposal and the drastic effect on the resonance lines and spectra. After some decades during which it went into oblivion, the technique is now in full renewal mode since the 1990s with faster development in physics, chemistry, biology and medicine. Since that time, the number of reviews and textbooks dealing with it has been steadily increasing. The interested reader will find in this series *Annual Reports in NMR Spectroscopy* an excellent and recent review article by F. Stallmach and P. Galvosas[468] with many references in all the modern trends. For our direct concern, we will simply note that this is coincident with the commercial availability of reliable hardware like gradient coils, suitable power sources and interfaces to the conventional NMR spectrometers and well-designed software. At the same time, the chemists have become more and more interested in the microscopic features of all types of intermolecular interactions and have discovered that diffusion is an essential complement of relaxation studies. Making and breaking bonds in a complex is for example typically relevant for diffusion which gives insight of the overall lifetime, of the kinetics of formation and dissociation and of the cooperative motion of each of the partners during all these steps.[61–63,200–203,468,469] As for chemical shifts and coupling constants, chemical exchange and relaxation rates, in principle, we can distinguish three regimes for water bound to a biomolecule:

In the slow one, the two water resonances δ_W (bound) and δ_W (free) are well separated and the bound water diffusion D_W (bound) can be distinguished from the free water diffusion D_W (free). It is sufficient to measure each diffusion constant in the proper relevant experiment. This corresponds naturally to a long lifetime of the water in the bound state. In the intermediate and fast regimes (short lifetime of the bound water as compared to the slow NMR observation time scale), only a weighted water resonance signal δ_W (obs) is detected and, as a consequence, only a weighted water diffusion D_W(obs) is observed according to Equation (13) (p_W(bound) + p_W(free) = 1).

$$D_W(\text{obs}) = p_W(\text{bound}) \times D_W(\text{bound}) + p_W(\text{free}) \times D_W(\text{free}) \qquad (13)$$

The other point concerns the relation of the measured diffusion constant with the size and shape of the diffusing species, which, in theory, can be obtained from the Stokes–Einstein Equation (14) in which r_H is the so-called hydrodynamic radius of the diffusing species[470] and the other symbols have their usual meanings.

$$D = \frac{k_B T}{6\pi v r_H} \tag{14}$$

This identification analysis may involve the correlation of the experimental diffusion coefficients with calculated values to their molecular weight. For a simple application of Equation (14), let us to suppose an inverse cubic root of the molecular weight. In reality, it has been claimed[471–473] that the ratio of the diffusion coefficients (D_i/D_j) for two different molecular species i and j is rather inversely proportional to the square root or to the cube root of the ratio of their molecular weight for rod-like and spherical molecules according to Equation (15).

$$\sqrt[3]{\frac{M_j}{M_{ij}}} \leq \frac{D_i}{D_j} \leq \sqrt{\frac{M_j}{M_i}} \tag{15}$$

This discrepancy results from the complex factors involved in translational diffusion like the shape, the microfriction of the solute with the solvent (stick or slip hydrodynamic) and density and many others.[470] For these reasons, the best method for finding the molecular weights remains completely empirical and is based upon experimental correlation between diffusion and molecular weight on similar and well-known compounds in the given solvent.

The literature is very rich in diffusion experiments, but remains relatively poor concerning interaction of nucleosides in water.[203] However, understanding all the details of hydration of biomolecules, their other interaction and the identification of all the solutes according to the two above-mentioned points are especially relevant to this technique. As an example, Cohen et al.[474] have approached the hydration of crown ether and they were able to measure two distinct diffusion constants for water in two cases and to determine the number of water bounded to the crown ether or its complex with potassium iodide by using Equation (14). They suggest different modes of solute–water interaction as well as for the free crown ether and for its complex with potassium iodide. In another work, the same group[473] has characterised all the hydrogen-bonded assemblies of calix[4]arene (rosettes) with the help of Equation (15).

Nearly related to the scope of this review, we can quote some studies of interactions of C and U and nucleotides in water with divalent ions added to study their aggregation,[204] cation-template self-assembly of guanosine[214] or iso-guanosine,[209] interactions of UMP with several monovalent cations,[475] complexation of sugars[210] and complexes formed by copper with the nucleoside antibiotic sinefungin.[216] Diffusion coefficients of nucleic acids in DNA have been measured in order to approach the changes in apparent hydrodynamic parameters near the strands[206] and compared with theoretical models of diffusing spheres, a cylindrical rod and a prolate ellipse.[476]

In the field of dynamics in water, diffusion occupies a special place, as chemical exchange or binding process with the solvent or any solute (ligand) involves modulation of the translational diffusion coefficient by an order of magnitude or more. Before introduction to DOSY experiments, relatively simple sequences, but always using pulsed gradients and stimulated echo,[477] have been developed 10 years ago and have been included in conventional 2D NOESY, ROESY and EXSY correlations. We will only report two distinct examples in which the rate of exchange of an amide proton with water has been obtained and compared with that measured by conventional methods. In its continuing interest for gradient enhanced spectroscopy, the group of Van Zijl and Moonen[478] has proposed a very simple extension of the EXSY sequence with a pulsed encode gradient at the end of the evolution period and a decode gradient at the beginning of the acquisition period. By varying the diffusion time, they found for the amide proton of N-aceylaspartate in water a lifetime $\tau_{exch} \approx 93-122$ ms, the longest diffusion time giving the smallest value.[478] Andrec and Prestegard[479] introduce very simply a short spin–echo diffusion filter in the exchange period of a selective-inversion exchange ^{15}N, ^1H HMQC sequence. Experimental results for exchange of each of the four amino acids that constitute the acyl carrier protein are fully quantitatively consistent with the exchange rates measured using a selective-inversion exchange experiment, particularly when the most intense gradients, which however remain very low (29 G/cm), are used in the filter. Rates in the range 15– 67 s^{-1} are determined, depending on the amino acid observed. The authors have also examined the more complex three-site exchange. Recently, in very simple DOSY experiments on sucrose in a mixture DMSO-d$_6$ + H$_2$O, Cabrita and Berger[480] have noted the effect of "the diffusion-sensitive experimental time" (i.e. the diffusion time) in the apparent diffusion- coefficient-measured chemical exchange for hydroxyl protons in fast or slow exchange (Equation (13)).

In continuation of our work on intermolecular NOE and for completeness, we were also interested by DOSY experiments for measuring the exchange rates. This has been done of the amide proton of uracil[461], thymine[461] and U[191] in water and/ or DMSO, keeping in mind the two complementary imperatives, that is instrumental reliability and user-friendly data processing.

The number of sequences now available for measuring the translation diffusion D has been considerably growing in the last 10 years according to the proper motivation and interest of each author.[200–203] The main property of a good sequence is to be as simple as possible with a minimum number of RF pulses and gradients to avoid instrumental and software complications. At the same time, the inherent instrumental artefacts resulting from application of gradient pulses, mainly background gradients, eddy currents, convection effects and finite duration of pulses, should be minimised. The sequence should be also as free as possible from the unavoidable NMR artefacts resulting from RF pulses, mainly chemical shift encoding, J-modulation and zero quantum coherence (ZQC). To this end, a very convenient sequence that satisfies these imperatives and now largely used by chemists is the bipolar pulse pairs longitudinal eddy current delay (or longitudinal encode decode) sequence, proposed in 1995 by Wu, Chen and Johnson and shortly known as the BPP-LED sequence.[481] It was used in this work and it is well described in the original paper and by Johnson in his review.[201,481] Nevertheless, it has two major drawbacks: (i) the requirement of a considerable phase cycling (minimum 16) for the best efficiency and (ii) its total duration time which precludes any use in case of short transverse relaxation time T_2. These two points are not generally a problem, even with the proton, because the great dilution of the solute in the solvent (U in water or DMSO-d_6 in our case) results in a low signal-to-noise ratio and thus needs signal averaging. For a signal of initial amplitude $S(0)$ at the beginning of the diffusion period, application of the BPP-LED sequence gives at the time of detection a signal $S(\Delta + \delta + 2\tau)$ which is attenuated, neglecting transverse relaxation, according to Equation (16) in which $q = \gamma g \tau$ is the gradient pulse area, also called spatial wave vector[477] and the other symbols have their usual meanings.

$$S(\Delta + d + 2\tau) = S(0) \exp[-Dq^2(\Delta - \delta d/3 - \tau/2)] \qquad (16)$$

In theory, the Inverse Laplace transform (ILT) leads to the diffusion constant D. Second, in practice, the ILT is well known to be an ill-conditioned problem because of the inherent noise in the data and to overcome this difficulty, many algorithms and processing data have been developed, each with its own advantages and drawbacks.[482] In most methods, the DOSY processing begins as usually by a Fourier transformation in all but the diffusion dimension and then proper computing scheme follows to tackle the ILL. On the contrary, the IRRT (for inverse regularised resolvent transform) as proposed by V. Mandelshtam[483] does not proceeds sequentially, but it is a multi-scheme processing which handles the whole dataset at once. The Fourier and Laplace transforms are replaced by a non-linear method which is able to process indistinguishably both the frequency and diffusion dimensions. It is fast and simple because it only requires one adjustable parameter (the regularisation parameter) and gives good resolution and sensitivity in both dimensions.

The effects of chemical exchange of the proton amide H3 of U with water can therefore be estimated qualitatively by varying the diffusion interval Δ in a series of DOSY experiments (Figure 9). The proton H3 exhibits during the diffusion an apparent diffusion coefficient which is distinct from that of the other protons remaining attached to the molecule U. These latter ones represent the proper diffusion of U in water. In fact, during the diffusion interval, the observed diffusion coefficient D_{obs} for the H3 proton lies between the diffusion coefficients of U and water, being modulated by the exchange rate k_I and k_S and weighted according to Equation (13) by its relative lifetimes τ_U in U and τ_W in water, respectively. This is especially clear for the "intermediate" diffusion time $\Delta = 100$ ms (Figure 9B). For 50 ms (Figure 9A), the observed diffusion coefficient of the H3 amide proton and the other U protons are very similar because exchange is slow, but as the diffusion interval Δ increases, the amide proton H3 and water diffusion coefficients tend to the same value because full exchange has taken place (Figure 9C). In Figure 10, we present also the DOSY spectrum of U in "dry" DMSO-d_6 where the exchange of the OH2', OH3' and 5OH5' hydroxyl protons with the residual water is clearly visible as a slow process in this solvent. Naturally, in DMSO/water (83/17 v/v), the exchange of the hydroxyl protons with water is too fast to be observed as compared to the much slower exchange of the H3 amide proton on the uracil moiety (compare to Figure 9).

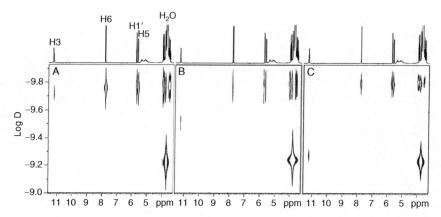

Figure 9   DOSY Map of Uridine U in DMSO-d_6/H_2O (83/17 v /v) recorded on a Bruker Avance spectrometer at 400 MHz and room temperature. The BPP-LED sequence was used and three different diffusion times Δ are shown: $\Delta = 10$ ms (A); $\Delta = 200$ ms (B); $\Delta = 1000$ ms (C). Note the overlap between the ribose protons and the water. The data have been processed by the iRRT method (from Reference 482).

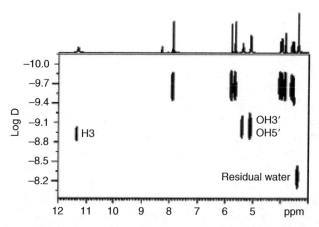

Figure 10   DOSY map of the Uridine U in "dry" DMSO as obtained with the BPP-LED sequence for a diffusion time $\Delta = 100$ ms. The Levenberg–Marquardt algorithm has been used here for processing the data. Note that the exchange of the hydroxyl protons on the ribose leads to practically the same apparent diffusion as that of the amide proton H3. The 1D spectrum on the frequency axis indicates a relative sharpness of the hydroxyl protons resonance as a proof of a very slow exchange (compare to Figure 9) (from Reference 482).

In order to get a quantitative estimation of the exchange rates of the labile proton from U with water, we need to determine the expression describing the longitudinal magnetisation in the two-site exchange approximation (Scheme 1). Direct longitudinal relaxation R_{II} and cross relaxation R_{IS} are negligibly small or nearly identical in the two states during the chemical exchange, so that the diffusion-modified Bloch equations[201,484–486] may be written as a pair of coupled differential equations (Equation (17)) analogous to Equation (7). It suffices to replace in Equation (7) R_{1II} by $D_I q^2$ and R_{1SS} by $D_S q^2$ and to suppress R_{1IS} and R_{1SI}.

$$\frac{dM(q,t)}{dt} = L M(q,t) \quad \text{with} \quad \mathbf{L} = -\begin{bmatrix} D_I q^2 + k_I & -k_S \\ -k_I & D_S q^2 + k_S \end{bmatrix} \tag{17}$$

We give here the solutions, Equations (18a) and (18b)) which are readily available for this simple case of the two-site exchange.[201,479–481]

$$M_I(t) = \left[\frac{M_{I0}}{2} - \frac{(\mu M_{I0} - k_S M_{S0})}{2\lambda}\right] \exp[(-\sigma + \lambda)t] + \left[\frac{M_{I0}}{2} + \frac{(\mu M_{I0} - k_S M_{S0})}{2\lambda}\right] \exp[(-\sigma - \lambda)t] \tag{18a}$$

$$M_S(t) = \left[\frac{M_{S0}}{2} + \frac{(\mu M_{S0} + k_I M_{I0})}{2\lambda}\right] \exp[(-\sigma + \lambda)t] + \left[\frac{M_{S0}}{2} - \frac{(\mu M_{S0} + k_I M_{I0})}{2\lambda}\right] \exp[(-\sigma - \lambda)t]$$

$$\text{with} \quad \sigma = \frac{1}{2}[k_S + k_I + (\gamma \delta g)^2 (D_I + D_S)],$$

$$\mu = \frac{1}{2}[k_I - k_S + (\gamma \delta g)^2 (D_I - D_S)]$$

$$\text{and} \quad \lambda = \sqrt{\mu^2 + k_I k_S} \tag{18b}$$

In order to calculate the exchange rates, the intensity of the H3 resonance was plotted as a function of the square of the gradient pulse area $q^2 = (\gamma g \tau)^2$ for the 16 experimental points ranging generally from $g = 5\%\ G_{max}$ to $g = 90\%\ G_{max}$ where $G_{max} = 56\ \text{G cm}^{-1}$ is the maximum strength of the gradient. The experimental decay was fitted by a three-variable ($k_I = k_U$, $k_S = k_W$, $M_{S0} = M_{W0}$) non-linear least-squares procedure based on Equation (18a).[191] For the fitting, the diffusion coefficient $D_U = D_I$ of U was obtained by analysing the ethylenic proton H6 of U and averaging the values obtained from all experiments performed with the all the diffusion times Δ. The water diffusion $D_W = D_S$ was obtained by analysing the water resonance in an independent experiment recorded in the same experimental conditions but without U. These calculations were performed for each NMR diffusion experiment and led to a k_U value equal to 14 s^{-1} for the H3 proton with a fitting error of about $\pm 3\%$, and a statistical error of about $\pm 6\%$ within all the measurements. This result compares very well the exchange constant of 18 s^{-1} that was obtained in the case of uracil for the H3 proton in the same experimental conditions.[461] This indicates that the ribose moiety does not induce any significant (electronic and/or steric) perturbation on the H3 proton lability. It should be noted that this value has been obtained in neat DMSO-d_6 containing very little water. Comparison with the literature values in water implies a correction which leads at about $k_{exch} \approx 45\ \text{s}^{-1}$, which is in the range reported by others[39,425,426] (see Section 6.5).

6.8. Microdynamics in water: A tentative interpretation

For concluding this section—and this review—we wish to give some personal views and tentative assumptions about the most intimate process of hydration, that is, at the atomic level and at the picosecond time scale. The assumed microdynamics is described hereafter on the examples of uridine U and its base uracil u for comparison and for which it has been shown that a dihydrate and a trihydrate, respectively, are formed in water according to Scheme 2.

We think that this analysis can be also extended to other Pys, providing they do not have any exocyclic exchanging proton, like C with its NH2 protons for example. For a complete comparison with u for which all data have been recorded in a mixture DMSO-d_6/H_2O (83/17 v/v) because of the insolubility of u in neat water, all the data concerning U in water have been also extrapolated to the same mixture DMSO-d_6/H_2O (83/17 v/v), called simply DW in the following. The whole microdynamics data postulated for U and u are then summarised in Table 10 according to the discussion and assumptions below.

We have previously shown that the microdynamics of both U and u in DMSO can be approximated to a nearly symmetrical tumbling spherical top. This means that for the U microdynamics in the solvent, we can neglect the internal reorientation of the ribose ring puckering, the free rotation both around the glycosidic bond and around the C4'–C5' bond. Moreover, it has also been demonstrated that the hydrates have very short lifetimes and that the reorienting species are always the unsolvated molecules which never carry with them the water molecules in their rotational motion in the solvent.[189–191]. Thus, the quasi-spherical U with a radius $r_W \approx 3.59$ Å reorients by an overall rotation time $\tau_R \approx 25.7$ ps in water, that is about 50 ps in the solvent DW according to the correction from the water dynamic viscosity ($v = 1.04$ cP) to the DW dynamic viscosity ($v = 2.01$ cP) (Equation (9)). For u, the overall correlation time has been obtained directly. By assuming a complete loss of correlation for a translation of U by a distance equal to twice its radius, the usual Einstein–Smoluchowski Equation (19)

$$\tau_T = \frac{4r_W^2}{6D_{U(DW)}} \tag{19}$$

gives the translational correlation time τ_T of U in DW where $D_U = 1.63 \times 10^{-10}$ m^2 s^{-1} is the measured diffusion constant D_U of U in DW and the factor 6,

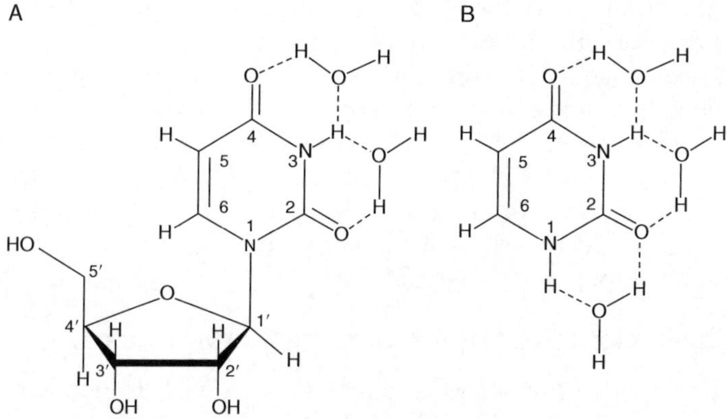

Scheme 2   Molecular structure of the dihydrated uridine U (A) and of the trihydrated uracil u (B).

TABLE 10   Solvation microdynamics and chemical exchange dynamics in DMSO/water (83/17 v/v) as compared for uridine U and uracil u

		Uridine U	Uracil u
$10^{10}\ D\ (m^2s^{-1})$		1.63	2.60
$V_W\ (\mathring{A}3)$		194.4	90.4
$r_W\ (\mathring{A})$		3.59	2.78
$\tau_R\ (ps)$		≈ 50	≈ 24
$\tau_T\ (ps)^a$		≈ 530	≈ 50
$\tau_{res}\ (ps)^b$		≈ 820	≈ 130
$\tau_{exch}\ (ms)$	H1		125
	H3	71	56

^{a} $\tau_T = \frac{4r_W^2}{6D}$ and $\tau_{res} = \frac{\langle d_{NOE}^2 \rangle}{(D_i + D_w)}$.
^{b} Calculated by Equation (19) by replacing the factor 6 by 2 (see text). D_{water} (DMSO) $= 6.56 \times 10^{-10}\ m^2\ s^{-1}$ (measured). $\tau_{HB} \approx 2\text{--}10$ ps (literature).

not 2, is included because of the three-dimensional random walk of U in DW.[12,17,450,487] A value $\tau_T \approx 530$ ps is thus obtained, which is about 10 times the rotational correlation times τ_R in the same solvent.

Now, the first question is: what about the hydrogen bond with water during the rotational and translational motions? The answer is provided by recent *ab initio* molecular dynamics simulations[12,488] as well as by the tremendous advance in ultrafast femtosecond spectroscopy,[12,489] both of which indicate that the lifetime of a hydrogen bond with water is typically in the range from 2 to 10 ps. In our opinion, the important point is that this hydrogen-bonding lifetime has a lower limit, in the range of the rotational reorientation time of pure water, which is now very well documented as 2 ps at room temperature[490] and probably in the range 3–5 ps in DW according to Equation (9). In fact, it seems very reasonable to think of hydrogen bond breaking and hydrogen bond making as the time-limiting processes for the overall tumbling of water. So far, during the U tumbling by a typical angle of about 1 radian and in which U does not carry with it any water molecule, there are numerous processes for hydrogen bond breaking and hydrogen bond making, say, something like 5–25. Hydrogen bond breaking occurs via librations or hindered rotations by very small angles, which occur at relatively high frequencies, and play a dominant role in the initial role of solvation. In the same manner, during the free diffusion translation of U in water, hydrogen bond breaking and hydrogen bond making occur via the intermolecular vibration (hindered translation) with water, also at high frequency, and this process repeats a large number of times (between 50 and 270).

The second question is then: can we estimate the residence time of the water in the hydration shell? By assuming that there is no more intermolecular Overhauser effect for hydrogen and carbon for more than 3.5 Å and that water has left the first hydration shell at this distance, we can use the approximation recently developed for estimating the residence time of water in proteins.[12,17] The residence time is then obtained from an expression analogous to Equation (19) but in which the diffusion constant D is replaced by the mean value of the diffusion coefficient D_U of U and that of water D_W in DW ($6.56 \times 10^{-10} \, \mathrm{m^2 \, s^{-1}}$). In Equation (19), the factor 2 is reintroduced in place of 6 because of the unidirectional character of the process, that is, approximately perpendicular to the C–N bond of the amide group. This gives a value of about 42 ps, which is in reasonable agreement with that generally agreed for proteins.[12,17]

At the other side of the time scale ($\tau_{exch} \approx 10-20 \, \mathrm{ms}$), the slow dynamic exchange of the amide proton with bulk water occurs, which is mediated by this fast dynamic hydrogen bonding with water. Naturally, this chemical exchange is crucially dependent on the thermodynamic properties of the solution (temperature, pH, catalysis) and can be, but in rare cases, diffusion-limited.[39]

Comparative results for the assumed microdynamics of uridine U and uracil u in the mixture DW DMSO-d_6/H_2O (83/17 v/v) are summarised in Table 9. The relevant values for uracil u (e.g. τ_T and τ_{res}) have been calculated using the same approximations as above for uridine U and according to the previous reported data.[189,461] It is instructive to observe that the translational correlation times τ_T and the residence times of water τ_{res} in the first hydration shell for U are much greater (by a factor 7–10 than those for u in the same conditions). They are also much closer to the data reported for proteins in water.

7. CONCLUSION

This review has attempted to present a detailed description of the different aspects of the hydration of nucleosides. It is hoped that the breadth of these intimate effects on the NMR time scale has been communicated and that the survey has been the most comprehensive as possible. The literature review has shown that much work has been done in the structure with a large body of papers on chemical shifts and coupling constants. Continuous advances in instrumentation for a higher sensitivity and cutting-edge technology with more and more intense magnetic fields can again be a benefit to probe more and more detailed effects on the electronic environment of nuclei of interest and to detect the smallest J coupling constants. Nevertheless, it remains that much progress for the future is in dynamic studies with diffusion experiments which are only at their beginning in liquid state. The development of high power gradients with short duration times and short recovery times is one of the ways to reach the "Holy Grail" of an improved time scale for NMR observation. In these fundamental intermolecular processes at the heart of chemistry, the NMR specialist should be helped by the spectacular advances in the other fields of spectroscopy like time-resolved Raman spectroscopy, laser femtosecond spectroscopy or neutron diffusion scattering, and many others.

ACKNOWLEDGEMENTS

This work began with HOESY experiments in the 1990s in Paris (H. B. Seba, PhD Thesis, Université Paris Diderot, Paris, 1990). It has further been greatly benefited from diffusion measurements which were run in Spectropole, Marseille, during a sabbatical leave from September 2004 to August 2005

(P. Thureau, PhD Thesis, Université de Provence, Marseille, 2006). I am pleased to thank Professor André Thevand for his kind hospitality and Stéphane Viel for enlightening discussions. I would especially like to warmly thank Professor Pierre Turq who welcomed me to his talented team in the Unversité Pierre et Marie Curie, Paris, and who has made this review possible. Without the constant and patient help with bibliography from Guilhem Pages, this work would never have been possible. Special greetings to him. Latifa Ziani, Werner Kunz and Hanna Sierzputowska-Gracz are acknowledged for proving me with many data. Lastly, I am indebted to Pascal Bernard, Vincent Dahirel and Guillaume Meriguet for fruitful discussions and for their help during the preparation of the manuscript.

REFERENCES

1. W. Sanger, *Annu. Rev. Biophys. Biophys. Chem.*, 1987, **16**, 93.
2. E. Westhof, *Annu. Rev. Biophys. Biophys. Chem.*, 1988, **17**, 125.
3. M. Berman, *Curr. Opin. Struct. Biol.*, 1994, **4**, 345.
4. T. V. Chalikian and K. J. Breslauer, *Biopolymers*, 1998, **48**, 264.
5. G. A. Jeffrey and W. Saenger, *Hydrogen Bonding in Biological Structures*. Springer-Verlag, New York, 1991.
6. S. N. Timasheff, *Biochemistry*, 2002, **41**, 13473.
7. H. Pessen and T. F. Kumosinski, *Methods Enzymol.*, 1985, **117**, 219.
8. S. K. Burley and G. A. Petsko, in: *Advances in Protein Chemistry*, C. B. Anfinsen, J. T. Edsall, F. M. Richards, and D. S. Eisenberg (eds.), Volume 39, Academic Press, New York, 1988, p. 125.
9. R. J. A. McCammon and S. C. Harvey, *Dynamics of Proteins and Nucleic Acids*. Cambridge University Press, Cambridge, 1988.
10. (a) Y. Q. Qian, G. Otting and K. Wüthrich, *J. Am. Chem. Soc.*, 1993, **115**, 1189; (b) M. Billeter, *Prog. NMR Spectrosc.*, 1995, **27**, 635; (c) K. Wuthrich, M. Billeter, P. Guntert, P. Luginbuhl, R. Riek and G. Wider, *Faraday Discuss.*, 1996, **103**, 245; (d) G. Otting, *Prog. NMR Spectrosc.*, 1997, **31**, 259.
11. B. Halle, V. P. Denisov and K. Venu, in: *Biological Magnetic Resonance*, N. R. Krishna and L. J. Berliner (eds.), **17**, Kluwer Academic Publishers, New York, 1999, p. 419.
12. S. K. Pal, J. Peon, B. Bagchi and A. H. Zewail, *J. Phys. Chem. B*, 2002, **106**, 12376.
13. (a) R. W. Kriwacki, J. R. B. Hill, J. M. Flanagan, J. P. Caradonna and J. H. Prestegard, *J. Am. Chem. Soc.*, 1993, **115**, 8907; (b) V. Dötscht and G. Wider, *J. Am. Chem. Soc.*, 1995, **117**, 6064.
14. B. Bagchi, *Chem. Rev.*, 2005, **105**, 3197.
15. B. Gregory, *Protein–Solvent Interactions*. Dekker, New York, 1995.
16. N. Prabhu and K. Sharp, *Chem. Rev.*, 2006, **106**, 1616.
17. (a) G. Otting and K. Wüthrich, *J. Am. Chem. Soc.*, 1989, **111**, 1871; (b) K. Wuthrich, G. Otting and E. Liepinsh, *Faraday Discuss.*, 1992, **93**, 35; (c) G. Otting and K. Wuthrich, *Acc. Chem. Res.*, 1995, **28**, 171; (d) G. Otting, E. Liepinsh, B. Halle and U. Frey, *Nat. Struct. Biol.*, 1997, **4**, 39.
18. M. G. Kubinec and D. E. Wemmer, *Currr. Opin. Struct. Biol.*, 1992, **2**, 828.
19. J. A. Ernst, R. T. Clubb, H. X. Zhou, A. M. Gronenborn and G. M. Clore, *Science*, 1995, **267**, 1813.
20. (a) B. Halle, T. Andersson, S. Forsén and B. Lindman, *J. Am. Chem. Soc.*, 1981, **103**, 500; (b) V. P. Denisov and B. Halle, *Faraday Discuss.*, 1996, **103**, 227; (c) K. Venu, V. P. Denisov and B. Halle, *J. Am. Chem. Soc.*, 1997, **119**, 3122; (d) V. P. Denisov and B. Halle, *J. Am. Chem. Soc.*, 2002, **124**, 10265; (e) K. Modig, E. Liepinsh, G. Otting and B. Halle, *J. Am. Chem. Soc.*, 2004, **126**, 102; (f) B. Halle, *Philos. Trans. Roy. Soc. Lond. B*, 2004, **359**, 1207.
21. D. P. Kharakoz, *J. Phys. Chem.*, 1991, **95**, 5634.
22. L. R. Olano and S. W. Rick, *J. Am. Chem. Soc.*, 2004, **126**, 7991.
23. J. Fitter, R. E. Lechner and N. A. Dencher, *J. Phys. Chem. B*, 1999, **103**, 8036.
24. (a) D. Russo, G. Hura and T. Head-Gordon, *Biophys. J.*, 2004, **86**, 1852; (b) D. Russo, R. K. Murarka, J. R. D. Copley and T. Head-Gordon, *J. Phys. Chem. B*, 2005, **109**, 12966.
25. I. Michalarias, X. Gao, R. C. Ford and J. Li, *J. Mol. Liq.*, 2005, **117**, 107.
26. Y. Z. Chen and E. W. Prohofsky, *Nucleic Acids Res.*, 1992, **20**, 415.
27. M. G. Kubinec and D. E. Wemmer, *J. Am. Chem. Soc.*, 1992, **114**, 8739.
28. E. Liepinsh, G. Otting and K. Wüthrich, *Nucleic Acids Res.*, 1992, **20**, 6549.

29. M. Kochoyan and J. L. Leroy, *Curr. Opin. Struct. Biol.*, 1995, **5**, 329.
30. B. Schneider and H. M. Berman, *Biophys. J.*, 1995, **69**, 2661.
31. M. Egli, S. Portmann and N. Usman, *Biochemistry*, 1996, **35**, 8489.
32. (a) G. N. J. Port and A. Pullman, *FEBS Lett.*, 1973, **31**, 70; (b) A. Goldblum, D. Perahia and A. Pullman, *FEBS Lett.*, 1978, **91**, 213; (c) B. Pullman, S. Miertus and D. Perahia, *Theor. Chim. Acta*, 1979, **50**, 317.
33. J. Leszczynski (ed.) (1999). *Computational Molecular Biology*. Elsevier, Amsterdam, 1999.
34. (a) V. I. Poltev, A. V. Teplukhin and G. G. Malenkov, *Int. J. Quant. Chem.*, 1992, **42**, 1499; (b) E. González, A. Deriabina, A. Teplukhin, A. Hernández and V. I. Poltev, *Theor. Chem. Acc.*, 2003, **110**, 460.
35. (a) P. Hobza and J. Šponer, *Chem. Rev.*, 1999, **99**, 3247; (b) M. Kabeláč and P. Hobza, *Phys. Chem. Chem. Phys.*, 2007, **9**, 903.
36. D. C. Clary, D. M. Benoit and T. van Mourik, *Acc. Chem. Res.*, 2000, **33**, 441.
37. (a) N. Foloppe and A. D. Mc Kerell, *J. Comput. Chem.*, 2000, **21**, 86; (b) N. Foloppe and L. Nilsson, *J. Phys. Chem. B*, 2005, **109**, 9119.
38. (a) N. Leulliot, M. Ghomi, G. Scalmani and G. Berthier, *J. Phys. Chem. A*, 1999, **103**, 8716; (b) N. Leulliot, M. Ghomi, H. Jobic, O. Bouloussa, V. Baumruk and C. Coulombeau, *J. Phys. Chem. B*, 1999, **103**, 10934; (c) A. Hocquet, N. Leulliot and M. Ghomi, *J. Phys. Chem. B*, 1999, **104**, 4560; (d) Y. P. Yurenko, R. O. Zhurakivsky, M. Ghomi, S. P. Samijlenko and D. M. Hovorun, *J. Phys. Chem. B*, 2007, **111**, 9655; (See also 2008, Volume 112, p. 1240); (e) A. Hocquet and M. Ghomi, *Phys. Chem. Chem. Phys.*, 2000, **2**, 5351.
39. (a) M. Guéron, M. Kochoyan and J.-L. Leroy, *Nature*, 1987, **328**, 89; (b) M. Kochoyan, J. L. Leroy and M. Guéron, *J. Mol. Biol.*, 1987, **196**, 599; (c) J. L. Leroy, M. Kochoyan, T. Huynh-Dinh and M. Guéron, *J. Mol. Biol.*, 1988, **200**, 223; (d) M. Guéron and J. L. Leroy, *Methods Enzymol.*, 1995, **261**, 283.
40. M. Egli, *Angew. Chem. Int. Ed. Engl.*, 1996, **35**, 1894.
41. W. Saenger, *Principles of Nucleic Acid Structure*. Springer-Verlag, New York, 1984.
42. (a) A. J. Dingley and S. Grzesiek, *J. Am. Chem. Soc.*, 1998, **120**, 8293; (b) A. J. Dingley, J. E. Masse, J. Feigon and S. Grzesiek, *J. Biomol. NMR*, 2000, **16**, 279; (c) A. J. Dingley, F. Cordier and S. Grzesiek, *Concepts Magn. Reson.*, 2001, **13**, 103.
43. G. Cornilescu, J.-S. Hu and A. Bax, *J. Am. Chem. Soc.*, 1999, **121**, 2949.
44. G. Gemmecker, *Angew. Chem. Int. Ed. Engl.*, 2000, **39**, 1224.
45. N. Müller, *J. Solution Chem.*, 1988, **17**, 661; (See also *Acc. Chem. Res.* 1990, **23**, 23).
46. G. Némethy and H. A. Scheraga, *J. Chem. Phys.*, 1962, **36**, 3382.
47. M. Ikeguchi, S. Shimizu, S. Nakamura and K. Shimizu, *J. Phys. Chem. B*, 1998, **102**, 5891.
48. R. M. Levy, L. Y. Zhang, E. Gallicchio and A. K. Felts, *J. Am. Chem. Soc.*, 2003, **125**, 9523.
49. K. Lum, D. Chandler and J. D. Weeks, *J. Phys. Chem. B*, 1999, **103**, 4570.
50. F. H. Stillinger, *J. Solution Chem.*, 1973, **2**, 141.
51. R. R. Ernst, G. Bodenhausen and A. Wokaun, *Principles of Nuclear Magnetic Resonance in One and Two Dimensions*. Clarendon Press, Oxford, 1987.
52. (a) K. Wüthrich, *NMR of Proteins and Nucleic Acids*. Wiley, New York, 1986; (b) J. Cavanagh, W. J. Fairbrother, A. G. Palmer and N. J. Skelton, *Protein NMR Spectroscopy: Principles and Practice*. Academic Press, New York, 1996.
53. G. Wider, *Prog. NMR Spectrosc.*, 1998, **32**, 193.
54. F. J. M. van de Ven, *Multidimensional NMR in Liquids*. VCH Publishers, New York, 1995.
55. M. H. Levitt, *Spin Dynamics: Basics of Nuclear Magnetic Resonance*. Wiley, Chichester, 2001.
56. J. Keeler, *Understanding NMR Spectroscopy*. Wiley, Chichester, 2005.
57. F. Franks (ed.) (1972). Water, A Comprehensive Treatise, Volume 1. Plenum Press, New York, 1972; (See also 1973, Volume 2).
58. J. Kowalewski and L. Maier, *Nuclear Spin Relaxation in Liquids*. Taylor & Francis, Boca Raton, 2006.
59. H. Mo and T. C. Pochapsky, *Prog. NMR Spectrosc.*, 1997, **30**, 1.
60. (a) J. H. Noggle and R. E. Schirmer, *The Nuclear Overhauser Effect*. Academic Press, New York, 1971; (b) D. Neuhaus and M. Williamson, *The Nuclear Overhauser Effect in Structural and Conformational Analysis*. VCH Publishers, Weinheim, 1989.

61. T. Brand, E. J. Cabrita and S. Berger, *Prog. NMR Spectrosc.*, 2005, **46**, 159.
62. A. Bagno, F. Rastrelli and G. Saielli, *Prog. NMR Spectrosc.*, 2005, **47**, 41.
63. J. T. Gerig, *Annu. Rep. NMR Spectrosc.*, 2008, **64**, 21.
64. H. B. Seba, PhD Thesis, Université Paris Diderot, Paris, 1990.
65. P. J. Hore, *Methods Enzymol.*, 1989, **176**, 64.
66. (a) M. Guéron, P. Plateau and M. Decorps, *Prog. NMR Spectrosc.*, 1991, **23**, 135; (b) M. Guéron and P. Plateau, in: *Encyclopedia of Nuclear Magnetic Resonance*, D. M. Grant and R. K. Harris (eds.), Volume 8, Wiley, Chichester, 2002, p. 4931.
67. M. H. Levitt, *Concepts Magn. Reson.*, 1996, **8**, 77.
68. X. A. Mao and C. H. Ye, *Concepts Magn. Reson.*, 1997, **9**, 173.
69. C. Szantay and A. Demeter, *Concepts Magn. Reson.*, 1999, **11**, 121.
70. M. P. Augustine, *Prog. NMR Spectrosc.*, 2002, **40**, 111.
71. J. Jeener, in: *Encyclopedia of Nuclear Magnetic Resonance*, D. M. Grant and R. K. Harris (eds.), Volume 9, Wiley, Chichester, 2002, p. 642.
72. L. B. Townsend (ed.) (1994). *Chemistry of Nucleosides and Nucleotides.* Plenum Press, New York, 1994.
73. IUPAC-IUB Joint Commission on Biochemical Nomenclature, IUPAC-IUB Joint Commission on Biochemical Nomenclature. *Eur. J. Biochem.*, 1982, **131**, 9.
74. J. E. Kilpatrick, K. S. Pitzer and R. Spitzer, *J. Am. Chem. Soc.*, 1947, **69**, 2483.
75. D. Cremer and J. A. Pople, *J. Am. Chem. Soc.*, 1975, **97**, 1354; (See also p. 1358).
76. (a) M. Sundaralingam, *Biopolymers*, 1969, **7**, 821; (b) M. Sundaralingam, in: *Conformation of Biological Molecules and Polymers, The Jerusalem Symposium on Quantum Chemistry and Biochemistry, Jerusalem, 1972*, E. D. Bergmann and B. Pullman (eds.), Volume 5, Academic Press, New York, 1973, p. 417; (c) C. Altona and M. Sundaralingam, *J. Am. Chem. Soc.*, 1972, **94**, 8205; (See also 1973, Volume 95, p. 2333); (d) E. Westhof and M. Sundaralingam, *J. Am. Chem. Soc.*, 1980, **102**, 1493; (e) S. T. Rao, E. Westhof and M. Sundaralingam, *Acta Cryst.*, 1981, **A37**, 421.
77. (a) W. K. Olson and J. L. Sussman, *J. Am. Chem. Soc.*, 1982, **104**, 270; (b) W. K. Olson, *J. Am. Chem. Soc.*, 1982, **104**, 278.
78. (a) S. C. Harvey and M. Prabhakaran, *J. Am. Chem. Soc.*, 1986, **108**, 6128; (b) H. A. Gabb and S. C. Harvey, *J. Am. Chem. Soc.*, 1993, **115**, 4218.
79. (a) C. R. Cantor and P. R. Schimmel, *Biophysical Chemistry.* W. H. Freeman, New York, 1980; (b) D. E. Metzler, *Biochemistry.* Academic Press, New York, 2001; (c) J. M. Berg, J. L. Tymoczko and L. Stryer, *Biochemistry.* W. H. Freeman, New York, 2007.
80. R. Marek and V. Sklenář, *Annu. Rep. NMR Spectrosc.*, 2004, **54**, 201.
81. D. B. Davies, *Prog. NMR Spectrosc.*, 1978, **12**, 135.
82. R. V. Hosur, G. Govil and H. T. Miles, *Magn. Reson. Chem.*, 1988, **26**, 927.
83. S. S. Wijmenga and B. N. M. van Buuren, *Prog. NMR Spectrosc.*, 1998, **32**, 287.
84. P. E. Hansen, *Prog. NMR Spectrosc.*, 1981, **14**, 175.
85. R. Cortese, H. O. Kammen, S. J. Spengler and B. N. Ames, *J. Biol. Chem.*, 1974, **249**, 1103.
86. K. A. Watanabe, in: *Chemistry of Nucleosides and Nucleotides*, (L. B. Townsend ed.), Volume 3. Plenum Press, New York, 1994.
87. E. Goldwasser and R. L. Heinrikson, *Prog. Nucleic Acids Res. Mol. Biol.*, 1966, **5**, 399.
88. (a) C. K. Chu and D. C. Baker (eds.), *Nucleosides and Nucleotides as Antitumor and Antiviral Agents.* Plenum Press, New York, 1993; (b) C. K. Chu, F. M. El-Kabbani and B. B. Thompson, *Nucleosides Nucleotides*, 1984, **3**, 1.
89. G. Cristalli, S. Costanzi, C. Lambertucci, S. Taffi, S. Vittori and R. Volpini, *Il Farmaco*, 2003, **58**, 193.
90. (a) M. S. Poonian, W. W. McComas and M. J. Kramer, *J. Med. Chem.*, 1979, **22**, 958; (b) A. Mikhailopulo, Y. A. Sokolov, J. He, P. Chittepu, H. Rosemeyer and F. Seela, *Nucleosides Nucleotides*, 2005, **24**, 701.
91. A. Goldin, H. B. Wood and R. R. Engle, *Cancer Chemother. Rep.*, 1968, **1**, 1.
92. (a) G. Alonso, M. Fuertes, G. Garcia-Muñoz, R. Madroñero and M. Stud, *J. Med. Chem.*, 1973, **16**, 1056; (b) K. Watanabe, S. Yokoyama, F. Hansske, H. Kasai and T. Miyazawa, *Biochem. Biophys. Res. Commun.*, 1979, **91**, 671; (c) Y. Yamamoto, S. Yokoyama, T. Miyazawa, K. Watanabe and S. Higuchi,

FEBS Lett., 1983, **157**, 95; (d) J. K. Buolamwini and J. J. Barchi, *Nucleosides Nucleotides*, 1997, **16**, 2101; (e) J. Buckingham, J. A. Brazier, J. Fisher and R. Cosstick, *Carbohydr. Res.*, 2007, **342**, 16.

93. (a) H. Sierzputowska-Gracz, E. Scchacka, A. Malkiewicz, K. Kuo, C. Gehrke and P. F. Agris, *J. Am. Chem. Soc.*, 1987, **109**, 7171; (b) P. F. Agris, H. Sierzputowska-Gracz, W. Smith, A. Malkiewicz, E. Sochacka and B. Nawrott, *J. Am. Chem. Soc.*, 1992, **114**, 2652.

94. (a) R. J. Cushley, J. F. Codington and J. J. Fox, *Can J. Chem.*, 1968, **46**, 1131; (b) K. A. Watanabe, U. Reichman, K. Hirota, C. Lopez and J. J. Fox, *J. Med. Chem.*, 1979, **22**, 21; (c) K. A. Watanabe, T. L. Su, R. S. Klein, C. K. Chu, A. Matsuda, M. W. Chun, C. Lopez and J. J. Fox, *J. Med. Chem.*, 1983, **26**, 152; (d) K. A. Watanabe, T.-L. Su, U. Reichman, N. Greenberg, C. Lopez and J. J. Fox, *J. Med. Chem*, 1984, **27**, 91; (e) M. E. Perlman, K. A. Watanabe, R. F. Schinazi and J. J. Fox, *J. Med. Chem.*, 1985, **28**, 741.

95. (a) S. Nesnow, A. M. Mian, T. Oki, D. L. Dexter and C. Heidelberger, *J. Med. Chem.*, 1972, **15**, 676; (b) J. A. Montgomery, A. T. Shortnacy-Fowler, S. D. Clayton, J. M. Riordan and J. A. Secrist, *J. Med. Chem.*, 1992, **35**, 397; (c) G. Gumina, R. F. Schinazi and C. K. Chu, *Org. Lett.*, 2001, **3**, 4177; (d) X. H. Xu, X. L. Qiu, X. Zhang and F. L. Qing, *J. Org. Chem.*, 2006, **71**, 2820.

96. (a) H. Mitsuya and S. Broder, *Proc. Natl. Acad. Sci. USA*, 1986, **83**, 1911; (b) J. Balzarini, G. J. Kang, M. Dalal, P. Herdewijin, E. De Clercq, S. Broder and D. G. Johns, *Mol. Pharmacol.*, 1987, **32**, 162; (c) H. Maag, J. T. Nelson, J. L. Riossteiner and E. J. Prisbe, *J. Med. Chem.*, 1994, **37**, 431.

97. (a) E. De Clercq and J. Balizarini, in: *Design of Anti-AIDS Drugs*, (E. De Clercq ed.), Elsevier, Amsterdam, 1990; (b) L. S. Kucera, N. Iyer, E. Leake, A. Raben, E. J. Modest, L. W. Daniel and C. Piantadosi, *AIDS Res. Human Retroviruses*, 1990, **6**, 491; (c) D. M. Huryn and M. Okabe, *Chem. Rev.*, 1992, **92**, 1745; (d) R. Alvarez, M. L. Jimeno, F. J. Tomás-Gil, M. J. Pérez-Pérez and M. J. Camarasa, *Nucleosides Nucleotides*, 1997, **16**, 1399.

98. (a) G. V. T. Swapna, B. Jagannadh, M. K. Gurjar and A. C. Kunwar, *Biochem. Biophys. Res. Commun.*, 1989, **164**, 1086; (b) B. Jagannadh, D. V. Reddy and A. C. Kunwar, *Biochem. Biophys. Res. Commun.*, 1991, **179**, 386.

99. A. Bloch, *Ann. N.Y. Acad. Sci.*, 1975, **255**, 576.

100. P. Langen, *Antimetabolites of Nucleic Acid Metabolism*. New York, Gordon and Breach, 1975.

101. M. Dračínský, M. Krečmerová and A. Holý, *Bioorg. Med. Chem.*, 2008, **16**, 6778.

102. (a) M. Perbost, T. Hoshiko, F. Morvan, E. Swayze, R. H. Griffey and Y. S. Sanghvi, *J. Org. Chem.*, 1995, **60**, 5150; (b) R. H. Griffey, B. P. Monia, L. L. Cummins, S. Freier, M. J. Greig, C. J. Guinosso, E. Lesnik, S. M. Manalili, V. Mohan, S. Owens, B. R. Ross, H. Sasmor, E. Wancewicz, K. Weiler, P. D. Wheeler and P. D. Cook, *J. Med. Chem.*, 1996, **39**, 5100.

103. J. Poznanski, K. Felczak, M. Bretner, T. Kulikowski and M. Remin, *Biochem. Biophys. Res. Commun.*, 2001, **283**, 1142.

104. H. Follmann and G. Gremels, *Eur. J. Biochem.*, 1974, **47**, 187.

105. (a) M. Blandin, S. Tran-Dinh, J. Catlin and W. Guschlbauer, *Biochim. Biophys. Acta*, 1974, **361**, 249; (b) W. Guschlbauer and K. Jankowski, *Nucleic Acids Res.*, 1980, **8**, 1421; (c) W. Guschlbauer, *Biochim. Biophys. Acta*, 1980, **610**, 47.

106. H. Singh, M. H. Herbut, C. H. Lee and R. H. Sarma, *Biopolymers*, 1976, **15**, 2167.

107. (a) M. Ikehara, S. Uesugi and K. Yoshida, *Biochemistry*, 1972, **11**, 830; (b) S. Uesugi and M. Ikehara, *J. Am. Chem. Soc.*, 1977, **99**, 3250; (c) S. Uesugi, H. Miki, M. Ikehara, H. Iwahashi and Y. Kyogoku, *Tetrahedron Lett.*, 1979, **42**, 4073.

108. C. Thibaudeau, J. Plavec, N. Carg, A. Papchikhin and J. Chattopadhyaya, *J. Am. Chem. Soc.*, 1994, **116**, 4038.

109. H. Rosemeyer, G. Toth, B. Golankiewicz, T. Kazimierczuk, W. Bourgeois, U. Kretschmer, H. P. Muth and F. Seela, *J. Org. Chem.*, 1990, **55**, 5784.

110. (a) R. Stolarski, L. Dudycz and D. Shugar, *Eur. J. Biochem.*, 1980, **108**, 111; (b) R. Stolarski, C. E. Hagberg and D. Shugar, *Eur. J. Biochem.*, 1984, **138**, 187.

111. P. A. Hart and J. P. Davis, *J. Am. Chem. Soc.*, 1972, **94**, 2572.

112. M. P. Schweizer and G. P. Kreishman, *J. Magn. Reson.*, 1973, **9**, 334.

113. D. Suck and W. Saenger, *J. Am. Chem. Soc.*, 1972, **94**, 6520.

114. (a) R. H. Sarma, C. H. Lee, F. E. Evans, N. Yathindra and M. Sundaralingam, *J. Am. Chem. Soc.*, 1974, **96**, 7337; (b) C. H. Lee and R. H. Sarma, *J. Am. Chem. Soc.*, 1976, **98**, 3541.

115. (a) R. Pless, L. Dudycz, R. Stolarski and D. Shugar, *Z. Naturforsch. C*, 1979, **33**, 902; (b) L. Dudycz, R. Stolarski, R. Pless and D. Shugar, *Z. Naturforsch. C*, 1979, **34**, 359; (c) R. Stolarski, A. Pohorille, L. Dudycz and D. Shugar, *Biochim. Biophys. Acta*, 1980, **610**, 1.
116. (a) M. L. Post, G. I. Birnbaum, C. P. Huber and D. Shugar, *Biochim. Biophys. Acta*, 1977, **479**, 133; (c) G. I. Birnbaum and D. Shugar, *Biochim. Biophys. Acta*, 1978, **517**, 500.
117. (a) F. E. Hruska, A. A. Smith and J. G. Dalton, *J. Am. Chem. Soc.*, 1971, **93**, 4334; (b) F. E. Hruska, in: *Conformation of Biological Molecules and Polymers, The Jerusalem Symposium on Quantum Chemistry and Biochemistry, Jerusalem, 1972*, E. D. Bergmann and B. Pullman (eds.), Volume 5, Academic Press, New York, 1973, p. 345; (c) J. Cadet, C. Taïeb, M. Remin, W. P. Niemczura and F. E. Hruska, *Biochim. Biophys. Acta*, 1980, **608**, 435; (d) F. E. Hruska and J. P. Blonski, *Can. J. Chem.*, 1982, **60**, 3026; (e) G. I. Birnbaum, W. J. P. Blonski and F. E. Hruska, *Can. J. Chem.*, 1983, **61**, 2299.
118. E. Egert, H. J. Lindner, W. Hillen and M. C. Buhm, *J. Am. Chem. Soc.*, 1980, **102**, 3707.
119. (a) D. C. Rohrer and M. Sundaralingam, *J. Am. Chem. Soc.*, 1970, **92**, 4956; (b) S. T. Rao and M. Sundaralingam, *J. Am. Chem. Soc.*, 1970, **92**, 4963; (c) T. Brennan and M. Sundaralingam, *Biochim. Biophys. Res. Commun.*, 1973, **52**, 1348; (d) R. G. Brennan, G. G. Privé, W. J. P. Blonski, F. E. Hruska and M. Sundaralingam, *J. Am. Chem. Soc.*, 1983, **105**, 7737.
120. (a) M. Remin, E. Darżynkiewicz, A. Dworak and D. Shugar, *J. Am. Chem. Soc.*, 1976, **98**, 367; (b) G. I. Birnbaum, J. Giziewicz, C. P. Huber and D. Shugar, *J. Am. Chem. Soc.*, 1976, **98**, 4640.
121. (a) L. Kozerski, H. Sierzputowska-Gracz, W. Krzyzosiak, M. Bratek-Wiewiórowska, M. Jaskólski and M. Wiewiórowski, *Nucleic Acids*, 1984, **12**, 2205; (b) H. Sierzputowska-Gracz, H. D. Gopal and P. F. Agris, *Nucleic Acids Res.*, 1986, **14**, 7783.
122. S. Raić, M. Mintas, A. Danilovski, M. Vinković, M. Pongračić, J. Plavec and D. Vikić-Topić, *J. Mol. Struct.*, 1997, **410–411**, 31.
123. G. A. Jeffrey, H. Maluszynska and J. Mitra, *Int. J. Biol. Macromol.*, 1985, **7**, 336.
124. A. Gelbin, B. Schneider, L. Clowney, S. H. Hsieh, W. K. Olson and H. M. Berman, *J. Am. Chem. Soc.*, 1996, **118**, 519.
125. S. Blanalt-Feidt, S. O. Doronina and J. P. Behr, *Tetrahedron Lett.*, 1999, **40**, 6229.
126. R. Ishikawa, C. Kojima, A. Ono and M. Kaishono, *Magn. Reson. Chem.*, 2001, **39**, S159.
127. (a) M. Remin, E. Darżynkiewicz, I. Ekiel and D. Shugar, *Biochim. Biophys. Acta*, 1976, **435**, 405; (b) I. Ekiel, M. Remin, E. Darżynkiewicz and D. Shugar, *Biochim. Biophys. Acta*, 1979, **562**, 177.
128. (a) F. Jordan and B. Pullman, *Theor. Chim. Acta*, 1968, **9**, 242; (b) H. Berthod and B. Pullman, *Biochim. Biophys. Acta*, 1971, **232**, 595; (See also Volume 246, p. 359); (c) A. Saran, B. Pullman and A. Perahia, *Biochim. Biophys. Acta*, 1972, **287**, 211; (d) B. Pullman and H. Berthod, in: *Conformations of biological molecules and polymers, The Jerusalem Symposium on Quantum Chemistry and Biochemistry, Jerusalem, 1972*, E. D. Bergmann and B. Pullman (eds.), Volume 5, Academic Press, New York, 1973, p. 209.
129. (a) A. Saran, B. Pullman and D. Perahia, *Biochim. Biophys. Acta*, 1973, **299**, 497; (b) A. Saran, D. Perahia and B. Pullman, *Theor. Chim. Acta*, 1973, **30**, 31; (c) A. Saran, B. Pullman and D. Perahia, *Biochim. Biophys. Acta*, 1974, **349**, 189; (d) A. Saran, C. Mitra and B. Pullman, *Biochim. Biophys. Acta*, 1978, **517**, 255.
130. (a) C. Mitra and A. Saran, *Biochim. Biophys. Acta*, 1978, **518**, 19; (b) A. Saran and C. L. Chatterjee, *Biochim. Biophys. Acta*, 1980, **607**, 490; (c) A. Saran and L. N. Patnaik, *Int. J. Quant. Chem.*, 1981, **20**, 357; (d) C. L. Chatterjee and A. Saran, *Int. J. Quant. Chem. Quant. Biol. Symp.*, 1981, **8**, 129; (e) A. Saran, *Int. J. Quant. Chem.*, 1982, **20**, 439; (f) L. N. Patnaik and A. Saran, *J. Biol. Phys.*, 1984, **12**, 12; (g) A. Saran and C. L. Chatterjee, *Int. J. Quant. Chem.*, 1984, **25**, 1943.
131. S. Kang, *J. Mol. Biol.*, 1971, **58**, 297.
132. M. Remin and D. Shugar, *J. Am. Chem. Soc.*, 1973, **95**, 8146.
133. (a) A. JaworskI and I. Ekiel, *Int. J. Quant. Chem.*, 1976, **16**, 615; (b) A. Jaworski, I. Ekiel and D. Shugar, *J. Am. Chem. Soc.*, 1978, **100**, 4357.
134. D. W. Miles, P. K. Redington, D. L. Miles and H. Eyring, *Proc. Natl. Acad. Sci. USA*, 1981, **78**, 7521.
135. J. L. Alderfer, R. E. Loomis and T. J. Zielinski, *Biochemistry*, 1982, **21**, 2738.
136. (a) D. W. Boerth and F. X. Harding, *J. Am. Chem. Soc.*, 1985, **107**, 2952; (b) D. W. Boerth and P. K. Bhowmik, *J. Phys. Chem.*, 1989, **93**, 3327.
137. K. Arora and T. Schlick, *Chem. Phys. Lett.*, 2003, **378**, 1.

138. N. A. Murugan and H. W. Hugosson, *J. Phys. Chem. B*, 2009, **113**, 1012.
139. A. E. V. Haschemeyer and A. Rich, *J. Mol. Biol.*, 1967, **27**, 329.
140. H. R. Wison and A. Rahman, *J. Mol. Biol.*, 1971, **56**, 129.
141. M. Levitt and A. Warshel, *J. Am. Chem. Soc.*, 1978, **100**, 2607.
142. J. Wibrkiewicz-Kuczera and A. Rabczenko, *J. Chem. Soc. Perkin Trans.*, 1985, **2**, 789.
143. (a) L. Nilsson and M. Karplus, *J. Comput. Chem*, 1986, **7**, 591; (b) A. Lahiri, J. Sarzynska, L. Nilsson and T. Kulinski, *Theor. Chem. Acc.*, 2007, **117**, 267.
144. T. Schlick, C. Peskin, S. Broyde and M. Overton, *J. Comput. Chem.*, 1987, **8**, 1199.
145. Y. Swarna Latha and N. Yathindra, *Biopolymers*, 1992, **32**, 249.
146. J. Plavec, V. Fabre-Buet, V. Uteza, A. Grouiller and J. Chattopadhyaya, *J. Biochem. Biophys. Methods*, 1993, **26**, 317.
147. S. N. Rao, *Biophys. J.*, 1998, **74**, 3131.
148. K. Wiechelman and E. R. Taylor, *J. Biomol. Struct. Dyn.*, 1998, **15**, 1181.
149. N. Foloppe and A. D. McKerell, *Biophys. J. Phys. Chem. B*, 1998, **102**, 6669; (See also *Biophys. J.*, 1999, **76**, 3206).
150. S. K. Mishra and P. C. Mishra, *J. Comput. Chem.*, 2002, **23**, 530.
151. (a) A. S. Serianni, S. J. Wu and I. Carmichael, *J. Am. Chem. Soc.*, 1995, **117**, 8645; (b) C. A. Podlasek, W. A. Stripe, I. Carmichael, M. Shang, B. Basu and A. S. Serianni, *J. Am. Chem. Soc.*, 1996, **118**, 1413.
152. X. P. Xu, W. L. A. K. Chiu and S. C. F. Au-Yeung, *J. Am. Chem. Soc.*, 1998, **120**, 4230.
153. (a) L. Trantírek, R. Štefla, J. E. Masse, Feigon Juli and V. Sklenář, *J. Biomol. NMR*, 2002, **23**, 1; (b) M. L. Munzarova and V. Sklenář, *J. Am. Chem. Soc.*, 2002, **124**, 10666; (c) M. L. Munzarova and V. Sklenář, *J. Am. Chem. Soc.*, 2003, **125**, 3649; (d) M. Hennig, M. L. Munzarova, W. Bermel, L. G. Scott, V. Sklenář and J. R. Williamson, *J. Am. Chem. Soc.*, 2006, **128**, 5851.
154. K. Miyata, A. Kobori, R. Tamamushi, A. Ohkubo, H. Taguchi, K. Seio and M. Sekine, *Eur. J. Org. Chem.*, 3626.
155. P. Cysewski, D. Bira and K. Bialkowski, *J Mol. Struct. (Theochem)*, 2004, **678**, 77.
156. O. Plashkevych, S. Chatterjee, D. Honcharenko, W. Pathmasiri and J. Chattopadhyaya, *J. Org. Chem.*, 2007, **72**, 4716.
157. S. Barbe and M. Le Bret, *J. Phys. Chem. A*, 2008, **112**, 989.
158. J. J. Barchi, R. G. Karki, M. C. Nicklaus, M. A. Siddiqui, C. George, I. A. Mikhailopulo and V. E. Marquez, *J. Am. Chem. Soc.*, 2008, **130**, 9048.
159. (a) P. O. P. Ts'o, I. S. Melvin and A. C. Olson, *J. Am. Chem. Soc.*, 1963, **85**, 1289; (b) P. O. P. Ts'o and S. I. Chan, *J. Am. Chem. Soc.*, 1964, **86**, 4176; (c) S. I. Chan, M. P. Schweizer, P. O. P. Ts'o and G. K. Helmkamp, *J. Am. Chem. Soc.*, 1964, **86**, 4182; (d) M. P. Schweizer, S. I. Chan and P. O. P. Ts'o, *J. Am. Chem. Soc.*, 1965, **87**, 5241; (e) P. O. P. Ts'o, *Ann. NY Acad. Sci.*, 1969, **153**, 785.
160. (a) L. F. Newcomb and S. H. Gellman, *J. Am. Chem. Soc.*, 1994, **116**, 4993; (b) S. H. Gellman, T. S. Haque and L. F. Newcomb, *Biophys. J.*, 1995, **71**, 3523.
161. R. A. Friedman and B. Honig, *Biophys. J.*, 1995, **69**, 1528; (See also Volume 71, p. 3525).
162. R. Luo, H. S. R. Gilson, M. J. Potter and M. K. Gilson, *Biophys. J.*, 2001, **80**, 140.
163. P. O. P. Ts'o, in: *Basic Principles in Nucleic Acid Chemistry*, (P. O. P. Ts'o ed.), Volume 1, Academic Press, New York, 1974, p. 453.
164. V. A. Bloomfield, D. M. Crothers and I. Tinoco, *Physical Chemistry of Nucleic Acids*. Harper & Row Publishers, New York, 1974.
165. J. Šponer, J. Leszczyński and P. Hobza, *J. Phys. Chem.*, 1996, **100**, 5590.
166. J. Norberg and L. Nilsson, *Biophys. J.*, 1998, **74**, 394.
167. M. P. Schweizer, in: *Encyclopedia of Nuclear Magnetic Resonance*, D. M. Grant and R. K. Harris (eds.), Volume 5, Wiley, Chichester, 2002, p. 3332.
168. P. J. Elving and J. W. Webb, in: *The Jerusalem Symposium on Quantum Chemistry and Biochemistry, Jerusalem, 1971*, E. D. Bergmann and B. Pullman (eds.), Volume 4, The Israel Academy of Sciences and Humanities, Jerusalem, 1972, p. 371.
169. M. Guéron, C. Chachaty and S. Tran-Dinh, *Ann. NY Acad. Sci.*, 1973, **222**, 303.
170. S. Tran-Dinh and C. Chachaty, *Biochim. Biophys. Acta*, 1973, **335**, 1.
171. C. L. Fisk, E. D. Becker, H. T. Miles and T. J. Pinnavaia, *J. Am. Chem. Soc.*, 1982, **104**, 3307.

172. (a) J. Florián, J. Šponer and A. Warshel, *J. Phys. Chem. B*, 1999, **103**, 884; (b) J. Sponer, J. Leszczynski and P. Hobza, *Biopolymers*, 2002, **61**, 3.
173. P. G. Jasien and G. Fitzgerald, *J. Chem. Phys.*, 1990, **93**, 2554.
174. F. E. Hruska, C. L. Bell, T. A. Victor and S. S. Danyluk, *Biochemistry*, 1968, **7**, 3721.
175. S. M. Wang and N. C. Li, *J. Am. Chem. Soc.*, 1968, **90**, 5069.
176. P. Dea, M. P. Schweizer and G. P. Kreishman, *Biochemistry*, 1974, **13**, 1862.
177. F. Garland and S. D. Chrlstlan, *J. Phys. Chem.*, 1975, **79**, 1247.
178. E. Westhof, H. Plach, I. Cuno and H.-D. Lüdemann, *Nucleic Acids Res.*, 1977, **4**, 939.
179. P. R. Mitchell and H. SigeL, *Eur. J. Biochem.*, 1978, **88**, 149.
180. G. V. Fazakerley, D. Hermann and W. Guschlbauer, *Biopolymers*, 1980, **19**, 1299.
181. P. Martel, *J. Phys. Chem.*, 1985, **89**, 230.
182. G. Birnbaum, M. Budêšínský and J. Beránek, *Can. J. Chem.*, 1987, **65**, 271.
183. G. Birnbaum, K. L. Sadana, M. T. Razi, T. Lee, R. Sebastian, G. W. Buchko and F. E. Hruska, *Can. J. Chem.*, 1988, **66**, 1628.
184. B. P. Cho and F. E. Evans, *Biochem. Biophys. Res. Commun.*, 1991, **180**, 273.
185. N. A. Corfù and H. SigeL, *Eur. J. Biochem.*, 1991, **199**, 659.
186. J. T. Davis, S. Tirumala, J. R. Jenssen, E. Radler and D. Fabris, *J. Org. Chem.*, 1995, **60**, 4167.
187. A. M. Ababneh, C. C. Large and S. Georghiou, *Biophys. J.*, 2003, **85**, 1111.
188. M. Polak, K. L. Seley and J. Plavec, *J. Am. Chem. Soc.*, 2004, **126**, 8159.
189. M. Chahinian, H. B. Seba and B. Ancian, *Chem. Phys. Lett.*, 1998, **285**, 337.
190. B. Ancian, D. Canet and P. Mutzenhardt, *Chem. Phys. Lett.*, 2001, **336**, 410.
191. H. B. Seba, P. Thureau, B. Ancian and A. Thevand, *Magn. Reson. Chem.*, 2006, **44**, 1109.
192. S. Shimokawa, T. Yokono and J. Sohma, *Biochim. Biophys. Acta*, 1976, **425**, 349.
193. A. P. Zens, T. A. Bryson, R. B. Dunlap, R. R. Fisher and P. D. Ellis, *J. Am. Chem. Soc.*, 1976, **98**, 7559.
194. S. Bonaccio, D. Capitani, A. L. Segre, P. Walde and P. L. Luisi, *Langmuir*, 1997, **13**, 1952.
195. R. A. Newmark and C. R. Cantor, *J. Am. Chem. Soc.*, 1968, **90**, 5010.
196. V. N. Viswanadhan, M. R. Reddy, R. J. Bacquet and M. D. Erion, *J. Comput. Chem.*, 1993, **14**, 1019.
197. R. T. West, L. A. Garza, W. R. Winchester and J. A. Walmsley, *Nucleic Acids Res.*, 1994, **22**, 5128.
198. J. T. Davis, *Angew. Chem. Int. Ed.*, 2004, **43**, 668.
199. C. Aimé, S. Manet, T. Satoh, H. Ihara, K. Y. Park, F. Godde and R. Oda, *Langmuir*, 2007, **23**, 12875.
200. P. Stilbs, *Prog. NMR. Spectrosc.*, 1987, **19**, 1.
201. C. S. Johnson, Jr., *Prog. NMR. Spectrosc.*, 1999, **34**, 203.
202. W. S. Price, *Concepts Magn. Reson.*, 1997, **9**, 299; (See also Volume 10, p. 197).
203. Y. Cohen, L. Avram and L. Frish, *Angew. Chem. Int. Ed. Engl.*, 2005, **44**, 520.
204. (a) R. Rymdén and P. Stilbs, *Biophys. Chem.*, 1985, **21**, 145; (b) I. Stokkeland and P. Stilbs, *Biophys. Chem.*, 1985, **22**, 65; (See also 1986, Volume 24, p. 61).
205. M. Geringer, H. Gruber and H. Sterk, *J. Phys. Chem.*, 1991, **95**, 2525.
206. J. Lapham, J. P. Rife, P. B. Moore and D. M. Crothers, *J. Biomol. NMR*, 1997, **10**, 255.
207. S. L. Mansfield, D. A. Jayawickrama, J. S. Timmons and C. K. Larive, *Biochim. Biophys. Acta*, 1998, **1382**, 257.
208. W. S. Price, F. Tsuchiya and Y. Arata, *J. Am. Chem. Soc.*, 1999, **121**, 11503.
209. T. Evan-Salem, L. Frish, F. W. B. van Leeuwen, D. N. Reinhoudt, W. Verboom, M. S. Kaucher, J. T. Davis and Y. Cohen, *Chem. Eur. J.*, 2000, **13**, 1969.
210. M. D. Diaz and S. Berger, *Carbohydr. Res.*, 2000, **329**, 1.
211. A. Burini, J. P. Fackler, R. Galassi, A. Macchioni, M. A. Omary, M. A. Rawashdeh-Omary, B. R. Pietroni, S. Sabatini and C. Zuccaccia, *J. Am. Chem. Soc.*, 2002, **124**, 4570.
212. S. Viel, L. Mannina and A. L. Segre, *Tetrahedron Lett.*, 2002, **43**, 2515.
213. C. Anselmi, F. Bernardi, M. Centini, E. Gaggelli, N. Gaggelli, D. Valensin and G. Valensin, *Chem. Phys. Lipids*, 2005, **134**, 109.
214. M. S. Kaucher, Y. F. Lam, S. Pieraccini, G. Gottarelli and J. T. Davis, *Chem. Eur. J.*, 2005, **11**, 164.
215. C. Sanna, C. La Mesa, L. Mannina, P. Stano, S. Viel and A. L. Segre, *Langmuir*, 2006, **22**, 6031.
216. M. Cappannelli, E. Gaggelli, M. Jeżowska-Bojczuk, E. Molteni, A. Mucha, E. Porciatti, D. Valensin and G. Valensin, *J. Inorg. Biochem.*, 2007, **101**, 1005.
217. N. F. Ramsey, *Phys. Rev.*, 1950, **77**, 567; (See also 1950, Volume 78, p. 699).

218. (a) M. Karplus and J. A. Pople, *J. Chem. Phys.*, 1963, **38**, 2083; (b) J. A. Pople, *Mol. Phys.*, 1964, **7**, 301.
219. J. W. Emsley, J. Feeney and L. H. Sutcliffe, *High Resolution NMR Spectroscopy*. Pergamon Press, Oxford, 1965.
220. J. B. Stothers, *Carbon-13 NMR Spectroscopy*. Academic Press, New York, 1972.
221. (a) G. C. Levy and R. L. Lichter, *Nitrogen-15 Nuclear Magnetic Resonance Spectroscopy*. Wiley, New York, 1979; (b) G. J. Martin, M. L. Martin and J. P. Gouesnard, *in: NMR Basic Principles and Progress*, P. Diehl, E. Fluck, and R. Kosfeld (eds.) Volume 18. Springer-Verlag, Berlin, 1981.
222. (a) W. G. Klemperer, *Angew. Chem. Int. Ed. Engl.*, 1978, **17**, 246; (b) J. P. Kintzinger, *in: NMR Basic Principles and Progress*, P. Diehl, E. Fluck, and R. Kosfeld (eds.), Volume 17, Springer-Verlag, Berlin, 1981.
223. E. Fermi, *Z. Phys.*, 1930, **60**, 320.
224. (a) M. Karplus, *J. Chem. Phys.*, 1959, **30**, 11; (b) M. Karplus, *J. Am. Chem. Soc.*, 1963, **85**, 2870.
225. M. Blümel, J. M. Schmidt, F. Löhr and H. Rüterjans, *Eur. Biophys. J.*, 27.
226. (a) C. A. G. Haasnoot, F. A. A. M. dE Leeuw and C. Altona, *Tetrahedron*, 1980, **36**, 2783; (b) C. A. G. Haasnoot, F. A. A. M. de Leeuw, H. P. M. de Leeuw and C. Altona, *Org. Magn. Reson.*, 1981, **15**, 43; (c) J. van Wijk, B. D. Huckriede, J. H. IppeL and C. Altona, *Methods Enzymol.*, 1992, **211**, 286.
227. J. Kowalewski, *Prog. NMR. Spectrosc.*, 1977, **11**, 1.
228. K. A. K. Ebraheem and G. A. Webb, *Prog. NMR. Spectrosc.*, 1977, **11**, 149.
229. A. C. de Dios, *Prog. NMR. Spectrosc.*, 1996, **29**, 229.
230. (a) H. Fukui, *Prog. NMR. Spectrosc.*, 1997, **31**, 17; (b) H. Fukui, *Prog. NMR. Spectrosc.*, 1999, **35**, 267.
231. (a) T. Helgaker, M. Jaszuński and K. Ruud, *Chem. Rev.*, 1999, **99**, 293; (b) T. Helgaker, M. Jaszuśńki and M. Pecul, *Prog. NMR. Spectrosc.*, 2008, **53**, 249.
232. P. Geerlings, F. De Proft and W. Langenaeker, *Chem. Rev.*, 2003, **103**, 1793.
233. (a) C. D. Jardetzky and O. Jardetzky, *J. Am. Chem. Soc.*, 1960, **82**, 222; (See also p. 236); (b) C. D. Jardetzky, *J. Am. Chem. Soc.*, 1960, **82**, 229.
234. M. P. Schweizer, S. I. Chang, G. K. Helmkamp and P. O. P. Ts'o, *J. Am. Chem. Soc.*, 1964, **86**, 696.
235. G. S. Reddy, L. Mandell and J. H. Goldstein, *J. Chem. Soc.*, 1414.
236. P. O. P. Ts'o, N. S. Kondo, M. P. Schweizer and D. P. Hollis, *Biochemistry*, 1969, **8**, 997.
237. M. P. Schweizer, A. D. Broom, P. O. P. Ts'o and D. P. Hollis, *J. Am. Chem. Soc.*, 1968, **90**, 1042.
238. P. O. P. Ts'o, N. S. Kondo, R. K. Robins and A. D. Broom, *J. Am. Chem. Soc.*, 1969, **91**, 5625.
239. H.-D. Lüdemann, E. Westhof and I. Cuno, *Z. Naturforsch C.*, 1976, **31**, 135.
240. F. Jordan and H. Niv, *Biochim. Biophys. Acta*, 1977, **476**, 265.
241. D. Plochocka, A. Rabczenko and D. B. Davies, *Biochim. Biophys. Acta*, 1977, **476**, 1.
242. (a) M. P. Schweizer and R. K. Robins, *in: Conformation of Biological Molecules and Polymers, The Jerusalem Symposium on Quantum Chemistry and Biochemistry, Jerusalem, 1972*, E. D. Bergmann and B. Pullman (eds.), Volume 5, Academic Press, New York, 1973, p. 329; (b) M. P. Schweizer, J. T. Witkowski and R. K. Robins, *J. Am. Chem. Soc.*, 1971, **93**, 277; (c) M. P. Schweizer, E. B. Banta, J. T. Witkowski and R. K. Robins, *J. Am. Chem. Soc.*, 1973, **95**, 3770.
243. H. Follmann, B. Pfeil and H. Witzel, *Eur. J. Biochem.*, 1977, **77**, 451.
244. L. Gatlin and J. C. Davis, *J. Am. Chem. Soc.*, 1964, **84**, 4464.
245. V. Nair and D. A. Young, *J. Org. Chem.*, 1985, **50**, 406; (See also *Magn. Reson. Chem.*, 1987, **25**, 937).
246. (a) B. P. Cho, F. F. Kadlubar, S. J. Culp and F. E. Evans, *Chem. Res. Toxicol.*, 1990, **3**, 445; (b) B. P. Cho and F. E. Evans, *Nucleic Acids Res.*, 1991, **19**, 1041.
247. (a) R. E. Schirmer, J. H. Noggle, J. P. Davis and P. A. Hart, *J. Am. Chem. Soc.*, 1970, **92**, 3266; (b) R. E. Schirmer, J. P. Davis, J. H. Noggle and P. A. Hart, *J. Am. Chem. Soc.*, 1972, **94**, 2561.
248. F. E. Hruska, *J. Am. Chem. Soc.*, 1971, **93**, 1795.
249. (a) F. E. Hruska, A. A. Grey and I. C. P. Smith, *J. Am. Chem. Soc*, 1970, **92**, 214; (See also p. 4088); (b) H. Dugas, B. J. Blackburn, R. K. Robins, R. Deslauriers and I. C. P. Smith, *J. Am. Chem. Soc.*, 1971, **93**, 3468; (c) T. Schleich, B. J. Blackburn, R. D. Lapper and I. C. P. Smith, *Biochemistry*, 1972, **11**, 137; (d) R. Deslauriers and I. C. P. Smith, *Can. J. Chem.*, 1973, **51**, 833.
250. (a) T. R. Krugh, *J. Am. Chem. Soc.*, 1973, **95**, 4761; (b) B. P. Cho and M. A. McGregor, *Nucleosides Nucleotides*, 1994, **13**, 481.
251. D. W. Jones, T. T. Mokoena, D. H. Robinson and G. Shaw, *Tetrahedron*, 1981, **37**, 2995.

252. E. Westhof, O. Röder, I. Croneiss and H.-D. Lüdemann, *Z. Naturforsch C.*, 1975, **30**, 131.
253. R. A. Wevers, U. F. H. Engelke, S. H. Moolenaar, C. Brautigam, J. G. N. de Jong, R. Duran, R. A. de Abreu and A. H. van Gennip, *Clin. Chem.*, 1999, **45**, 539.
254. R. Rico-Gómez, A. Rodríguez-González, J. Ríos-Ruíz, F. Nájera and J. M. López-Romero, *Eur. J. Org. Chem.*, 2003, **20**, 4023.
255. H. Sierzputowska-Gracz, R. H. Guenther, P. F. Agris, W. Folkman and B. Golankiewicz, *Magn. Reson. Chem.*, 1991, **29**, 885.
256. R. Narukulla, D. E. Shuker, V. Ramesh and Y. Z. Xu, *Magn. Reson. Chem.*, 2007, **46**, 1.
257. (a) F. E. Hruska, K. K. Ogilvie, A. A. Smith and H. Wayborn, *Can. J. Chem.*, 1971, **49**, 2449; (b) D. J. Wood, F. E. Hruska, R. J. Mynott and R. H. Sarma, *Can. J. Chem.*, 1973, **51**, 2571.
258. (a) K. N. Slessor and A. S. Tracey, *Can. J. Chem.*, 1971, **49**, 2874; (b) K. N. Slessor and A. S. Tracey, *Carbohydr. Res.*, 1973, **27**, 407.
259. P. J. Bolon and V. Nair, *Magn. Reson. Chem.*, 1996, **34**, 243.
260. F. E. Evans and R. A. Levine, *Biopolymers*, 1987, **26**, 1035.
261. J. Gambino, T. F. Yang and G. E. Wright, *Tetrahedron*, 1994, **59**, 11363.
262. S. P. Samijlenko, I. V. Alexeeva, L. H. Palchykivs'ka, I. V. Kondratyuk, A. V. Stepanyugin, A. S. Shalamay and D. M. Hovorun, *Spectrochim. Acta A*, 1999, **55**, 1133.
263. (a) S. Tran-Dinh, J. M. Neumann, J. M. Thiéry, T. Huynh-Dinh, J. Igolen and W. Guschlbauer, *J. Am. Chem. Soc.*, 1977, **99**, 3267; (b) H. D. Lüdemann and E. Westhof, *Z. Narurforsch. C*, 1977, **32**, 528.
264. J. G. Dalton, A. L. George, F. E. Hruska, T. N. McGaig, K. K. Ogilvie, J. Peeling and D. J. Wood, *Biochim. Biophys. Acta*, 1977, **478**, 261.
265. (a) B. J. Blackburn, A. A. Grey, I. C. P. Smith and F. E. Hruska, *Can. J. Chem.*, 1970, **48**, 2866; (b) A. A. Grey, I. C. P. Smith and F. E. Hruska, *J. Am. Chem. Soc.*, 1971, **93**, 1765.
266. (a) H. M. P. Chui, J. P. Desaulniers, S. A. Scaringe and C. S. Chow, *J. Org. Chem.*, 2002, **67**, 8847; (b) J. P. Desaulniers, H. M.-P. Chui and C. S. Chow, *Bioorg. Med. Chem.*, 2005, **13**, 6777; (c) Y. C. Chang, J. Herath, T. H.-H. Wang and C. S. Chow, *Bioorg. Med. Chem.*, 2008, **16**, 2676.
267. (a) B. P. Cross and T. Schleich, *Biopolymers*, 1973, **12**, 2381; (b) W. Guschlbauer, S. Tran-Dinh, M. Blandin and J. Catlin, *Nucleic Acids Res.*, 1974, **1**, 855.
268. H.-D. Lüdemann, O. Röder, E. Westhof, E. von Goldammer and A. Müller, *Biophys. Struct. Mech.*, 1975, **1**, 121.
269. C. Rossi, M. P. Picchi, E. Tiezzi, G. Corbini and P. Corti, *Magn. Reson. Chem.*, 1990, **28**, 348.
270. T. L. Brown and C. P. Cheng, *Faraday Discuss. Chem. Soc.*, 1979, **13**, 75.
271. (a) H. D. Lüdemann, E. Westhof and O. Röder, *Eur. J. Biochem.*, 1974, **49**, 143; (b) H. Plach, E. Westhof, H. D. Lüdemann and R. Mengel, *Eur. J. Biochem.*, 1977, **80**, 295.
272. E. M. Nottoli, J. B. Lambert and R. E. Letsinger, *Org. Magn. Reson.*, 1977, **9**, 499.
273. (a) A. J. Jones, D. M. Grant, M. W. Winkley and R. K. Robins, *J. Am. Chem. Soc. Chem.*, 1970, **92**, 4079; (b) A. J. Jones, D. M. Grant, M. W. Winkley and R. K. Robins, *J. Phys Chem.*, 1970, **74**, 2684; (c) A. J. Jones, M. W. Winkley, D. M. Grant and R. K. Robins, *Proc. Natl. Acad. Sci. USA*, 1970, **65**, 27.
274. (a) D. J. Wood, F. E. Hruska and K. K. Ogilvie, *Can. J. Chem.*, 1974, **52**, 3353; (b) D. J. Wood, K. K. Ogilvie and F. E. Hruska, *Can. J. Chem.*, 1975, **53**, 2781; (c) A. L. George, F. E. Hruska, K. K. Ogilvie and A. Holy, *Can. J. Chem.*, 1978, **56**, 1170; (d) B. D. Allore, A. Queen, W. J. Blonski and F. E. Hruska, *Can. J. Chem.*, 1983, **61**, 2397.
275. (a) M. T. Chenon, R. J. Pugmire, D. M. Grant, R. P. Panzica and L. B. Townsend, *J. Heterocycl. Chem.*, 1973, **10**, 431; (b) M. T. Chenon, R. P. Panzica, J. C. Smith, R. J. Pugmire, D. M. Grant and L. B. Townsend, *J. Am. Chem. Soc.*, 1976, **98**, 4736.
276. (a) H. H. Mantsch and I. C. P. Smith, *Biophys. Res. Commun.*, 1972, **46**, 808; (b) H. H. Mantsch and I. C. P. Smith, *Can. J. Chem.*, 1973, **51**, 1384.
277. I. Luyten, K. W. Pankiewicz, K. A. Watanabe and J. Chattopadhyaya, *J. Org. Chem.*, 1998, **63**, 1033.
278. (a) B. Golankiewicz and W. Folkman, *Magn. Reson. Chem.*, 1985, **23**, 920; (b) B. Golankiewicz and W. Folkman, *Nucleic Acids Res.*, 1983, **11**, 5243; (c) B. Golankiewicz, W. Folkman, H. Rosemeyerl and F. Seela, *Nucleic Acids Res.*, 1987, **15**, 9075.
279. M. L. Hamm, S. Rajguru, A. M. Downs and R. Cholera, *J. Am. Chem. Soc.*, 2005, **127**, 12220.
280. J. H. Prestegard and S. I. Chan, *J. Am. Chem. Soc.*, 1969, **91**, 2843.
281. R. J. Cushley, B. L. Blitzer and S. R. Lipsky, *Biochem. Biophys. Res. Commun.*, 1972, **48**, 1482.

282. (a) P. A. Hart and J. P. Davis, *J. Am. Chem. Soc.*, 1969, **91**, 512; (b) J. P. Davis, *Tetrahedron*, 1972, **28**, 1155; (c) J. P. Davis and P. A. Hart, *Tetrahedron*, 1972, **28**, 2883; (d) P. A. Hart and J. P. Davis, in: *Conformation of Biological Molecules and Polymers, The Jerusalem Symposium on Quantum Chemistry and Biochemistry, Jerusalem, 1972*, E. D. Bergmann and B. Pullman (eds.), Volume 5, Academic Press, New York, 1973, p. 297.

283. (a) P. A. Hart and J. P. Davis, *Biochem. Biophys. Res. Commun.*, 1969, **34**, 733; (b) P. A. Hart and J. P. Davis, *J. Am. Chem. Soc.*, 1971, **93**, 753; (See also 1972, Volume 94, p. 2572).

284. R. K. Nanda, R. Tewari, G. Govil and I. C. P. Smith, *Can. J. Chem.*, 1974, **52**, 371.

285. J. M. Neuman, J. M. Bernassau, M. Guéron and S. Tran-Dinh, *Eur. J. Biochem.*, 1980, **108**, 457.

286. M. Legraverend, J. M. Lhoste and E. Bisagni, *Tetrahedron*, 1984, **40**, 709.

287. J. Uzawa and K. Anzai, *Can. J. Chem.*, 1984, **62**, 1555.

288. (a) H. Rosemeyer, G. Tóth and F. Seela, *Nucleosides Nucleotides*, 1989, **8**, Z587; (b) G. Tóth, H. Rosemeyer and F. Seela, *Nucleosides Nucleotides*, 1989, **8**, 1091; (c) H. Rosemeyer and F. Seela, *Nucleosides Nucleotides*, 1990, **9**, 417.

289. M. P. Groziak, A. Koohang, W. C. Stevens and P. D. Robinson, *J. Org. Chem.*, 1993, **58**, 4054.

290. T. Katsura, K. Ueno and K. Furusawa, *Magn. Reson. Chem.*, 1993, **31**, 1039.

291. B. A. Schweitzer and E. T. Kool, *J. Org. Chem.*, 1994, **59**, 7238.

292. A. Kumar, S. B. Katti, H. Rosemeyer and F. Seela, *Nucleosides Nucleotides*, 1996, **15**, 1595; (See also 1997, Volume 16, p. 507).

293. (a) A. Mele, G. Salani, F. Viani and P. Bravo, *Magn. Reson. Chem.*, 1997, **35**, 168; (b) A. Mele, B. Vergani, F. Viani, S. V. Meille, A. Farina and P. Bravo, *Eur. J. Org. Chem.*, 187.

294. S. Tran-Dinh, W. Guschlbauer and M. Guéron, *J. Am. Chem. Soc.*, 1972, **94**, 7903.

295. C. Chachaty, T. Zemb, G. Langlet, S. Tran-Dinh, H. Buc and M. Morange, *Eur. J. Biochem.*, 1976, **62**, 45.

296. P. A. Hart, *J. Am. Chem. Soc.*, 1976, **98**, 3735.

297. L. Yiu-Fai and G. Kotowicz, *Can. J. Chem.*, 1977, **55**, 3620.

298. (a) C. Chachaty and G. Langlet, *FEBS Lett.*, 1976, **68**, 181; (b) C. Chachaty, B. Perly, A. Forchioni and G. Langlet, *Biopolymers*, 1980, **19**, 1211.

299. S. I. Chan and G. P. Kreishman, *J. Am. Chem. Soc.*, 1970, **92**, 1102.

300. W. Massefski and A. G. Redfield, *J. Magn. Reson.*, 1988, **78**, 150.

301. F. J. M. Van de Ven, C. A. G. Haasnoot and C. W. Hilbers, *J. Magn. Reson.*, 1985, **61**, 181.

302. M. M. W. Moonen, C. W. Hilbers, G. A. Van der Marel, J. H. Van Boom and S. S. Wijmenga, *J. Magn. Reson.*, 1991, **94**, 101.

303. D. R. Davis, *J. Magn. Reson.*, 1991, **94**, 401.

304. G. Zhu, D. Live and Bax Ad, *J. Am. Chem. Soc.*, 1994, **116**, 8370.

305. J. H. Ippel, S. S. Wijmenga, R. de Jong, H. A. Heus, C. W. Hilbers, E. de Vroom, G. A. van der Marel and J. H. van Boom, *Magn. Reson. Chem.*, 1996, **34**, S156.

306. R. E. Hurd and B. R. Reid, *Nucleic Acids Res.*, 1977, **4**, 2747.

307. R. W. Chambers, *Prog. Nucleic Acid. Res. Mol. Biol.*, 1996, **5**, 349.

308. (a) G. Govil and A. Saran, *J. Theor. Biol.*, 1971, **30**, 621; (See also Volume 33, p. 399); (b) A. Saran and G. Govil, *J. Theor. Biol.*, 1971, **33**, 407.

309. K. Dill and A. Allerhand, *J. Am. Chem. Soc.*, 1979, **101**, 4378.

310. C. Landis and V. S. Allured, *J. Am. Chem. Soc.*, 1991, **113**, 9493.

311. T. S. Lin, J. C. Cheng, K. Ishiguro and A. C. Sartorelli, *J. Med. Chem.*, 1983, **28**, 1194.

312. A. D. Buckingham, *Can. J. Chem.*, 1960, **38**, 300.

313. J. I. Musher, *J. Chem. Phys.*, 1962, **37**, 34.

314. J. G. Batchelor, *J. Am. Chem. Soc.*, 1974, **97**, 3410.

315. (a) E. D. Becker, H. T. Miles and R. B. Bradley, *J. Am. Chem. Soc.*, 1965, **87**, 5575; (b) R. R. Shoup, H. T. Miles and E. D. Becker, *J. Am. Chem. Soc.*, 1967, **89**, 6200; (c) R. R. Shoup, E. D. Becker and H. T. Miles, *Biochem. Biophys. Res. Commun.*, 1971, **43**, 1350; (d) R. R. Shoup, H. T. Miles and E. D. Becker, *J. Phys. Chem.*, 1972, **76**, 64; (e) R. R. Shoup, E. D. Becker and M. McNeel, *J. Phys. Chem.*, 1972, **76**, 71.

316. (a) M. Raszka and N. O. Kaplan, *Proc. Natl. Acad. Sci. USA*, 1972, **69**, 2025; (b) M. Raszka, *Biochemistry*, 1974, **13**, 4616.

317. J. D. Engel and P. H. von Hippel, *Biochemistry*, 1974, **13**, 4143.

318. (a) T. P. Pitner and J. D. Glickson, *Biochemistry*, 1975, **14**, 3083; (b) T. P. Pitner, H. Sternglanz, C. E. Bugg and J. D. Glickson, *J. Am. Chem. Soc.*, 1975, **97**, 885.

319. D. M. G. Martin and C. B. Reese, *J. Chem. Soc. Chem. Commun.*, 1279.

320. F. Jordan, *J. Org. Chem.*, 1982, **47**, 2748.

321. B. Ancian, Unpublished results.

322. (a) C. B. Reese and B. Safhill, *J. Chem. Soc. Perkin*, 1972, **1**, 2937; (b) S. Needle, M. R. Sanderson, A. Subbiah, J. B. Chattopadhyaya, R. Kuroda and C. B. Reese, *Biochim. Biophys. Acta*, 1979, **565**, 379; (c) J. Hovinen, C. Glemarec, A. Sandström, C. Sund and J. Chattopadhyaya, *Tetrahedron*, 1991, **47**, 4603.

323. C. H. Schwalbe and W. Saenger, *J. Mol. Biol.*, 1973, **75**, 129.

324. P. Prusiner, T. Brennan and M. Sundaralingam, *Biochemistry*, 1973, **12**, 1196.

325. P. Singh and D. J. Hodgson, *J. Chem. Soc. Chem. Commun.*, 1973, **489**; (See also *J. Am. Chem. Soc.*, 1974, **96**, 1239, also p. 5276).

326. F. E. Hruska, *Can. J. Chem.*, 1971, **49**, 2111.

327. (a) F. E. Evans and R. H. Sarma, *J. Biol. Chem.*, 1974, **249**, 4754; (b) C. H. Lee, F. E. Evans and R. H. Sarma, *J. Biol. Chem.*, 1975, **250**, 1290.

328. W. Saenger and K. H. Scheit, *J. Mol. Biol.*, 1970, **50**, 153.

329. G. H. Y. Lin, M. Sundaralingam and S. K. Arora, *J. Am. Chem. Soc.*, 1971, **93**, 1235.

330. U. Thewalt and C. E. Bugg, *J. Am. Chem. Soc.*, 1972, **94**, 8892.

331. N. B. Hanna, K. Ramasamy, R. K. Robins and G. R. Revankar, *J. Heterocyclic. Chem.*, 1988, **25**, 1899.

332. M. Kadokura, T. Wada, K. Seio and M. Sekine, *J. Org. Chem.*, 2000, **65**, 5104.

333. J. Brasuń, A. Matera, E. Sochacka, J. Swiatek-Kozlowska, H. Kozlowski, B. P. Operschall and H. Sigel, *J. Biol. Inorg. Chem.*, 2008, **13**, 663.

334. H. O. Sintim and E. T. Kool, *J. Am. Chem. Soc.*, 2006, **128**, 396.

335. L. M. Phillips and J. K. Lee, *J. Org. Chem.*, 2005, **70**, 1211.

336. E. M. Basilio Janke, A. Dunger, H. H. Limbach and K. Weisz, *Magn. Reson. Chem.*, 2001, **39**, S177.

337. J. Pitha and K. H. Scheit, *Biochemistry*, 1975, **14**, 554.

338. E. Plesiewicz, E. Stępień, K. Bolewska and K. L. Wierzchowski, *Nucleic Acids Res.*, 1976, **3**, 1295.

339. (a) R. U. Lemieux, T. L. Nagabhushan and B. Paul, *Can. J. Chem.*, 1972, **50**, 773; (b) L. T. J. Delbaere, M. N. G. James and R. U. Lemieux, *J. Am. Chem. Soc.*, 1973, **95**, 7866.

340. G. W. Buchanan and M. J. Bell, *Can. J. Chem.*, 1983, **61**, 2445.

341. D. B. Davies, P. Rajani, M. McCoss and S. S. Danyluk, *Magn. Reson. Chem.*, 1985, **23**, 72.

342. (a) D. B. Davies, *Stud. Biophys.*, 1976, **55**, 29; (b) D. B. Davies, P. Rajani and H. Sadikot, *J. Chem. Soc. Chem. Perkin*, 1985, **2**, 279.

343. J. Uzawa and M. Uramoto, *Org. Magn. Reson.*, 1979, **12**, 612.

344. M. J. R. van Dongen, S. S. Wijmenga, R. Eritja, F. Azorin and C. W. Hilbers, *J. Biomol. NMR*, 1966, **8**, 207.

345. (a) P. C. Kline and A. S. Serianni, *J. Am. Chem. Soc.*, 1990, **112**, 7373; (b) T. Bandyopadhyay, J. Wu, W. A. Sripe, I. Carmichael and A. S. Serianni, *J. Am. Chem. Soc.*, 1997, **119**, 1737.

346. R. J. Cushley, I. Wempen and J. J. Fox, *J. Am. Chem. Soc.*, 1968, **90**, 709.

347. D. B. Davies, M. McCoss and S. S. Danyluk, *J. Chem. Soc. Chem. Commun.*, 536.

348. D. B. Davies and S. S. Danyluk, *Biochemistry*, 1974, **13**, 4417.

349. J. van Wijk, B. D. Huckriede, J. H. IppeL and C. Altona, *Methods Enzymol.*, 1992, **211**, 286.

350. F. A. A. M. de Leeuw and C. Altona, *J. Comput. Chem.*, 1983, **4**, 428.

351. D. O. Cicero, G. Barbato and R. Bazzo, *Tetrahedron*, 1995, **51**, 10303.

352. D. A. Pearlman, D. A. Case, J. W. Caldwell, W. S. Ross, T. E. Cheatham, S. DeBolt, D. Ferguson, G. Seibel and P. Kollman, *Comput. Phys. Commun.*, 1995, **91**, 1.

353. D. R. Langley, *J. Biomol. Struct. Dyn.*, 1998, **16**, 487.

354. A. D. MacKerell, D. Bashford, M. Bellott, R. L. Dunbrack, J. D. Evanseck, M. J. Field, S. Fischer, J. Gao, H. Guo, S. Ha, D. Joseph-McCarthy L. Kuchnir, *et al.*, *J. Phys. Chem. B*, 1998, **102**, 3586.

355. X. Daura, A. E. Mark and W. F. van Gunsteren, *J. Comput. Chem.*, 1998, **19**, 535.

356. W. L. Jorgensen, D. S. Maxwell and J. Tirado-Rives, *J. Am. Chem. Soc.*, 1988, **110**, 1657.

357. C. Giessner-Prettre and B. Pullman, *Biochem. Biophys. Res. Commun.*, 1976, **70**, 578.

358. D. B. Davies and S. S. Danyluk, *Can. J. Chem.*, 1970, **48**, 3112.

359. M. Remin and D. Shugar, *Biochem. Biophys. Res. Commun.*, 1972, **48**, 636.
360. B. D. Davies and A. Rabczenko, *J. Chem. Soc. Perkin*, 1975, **2**, 1703.
361. F. E. Hruska, D. J. Wood, T. N. M. cCaig, A. A. Smith and A. Holy, *Can. J. Chem.*, 1974, **52**, 497.
362. (a) R. D. Harris and W. M. McIntyre, *Biophys. J.*, 203; (b) M. Sundaralingam and L. H. Jensen, *J. Mol. Biol.*, 1965, **13**, 914; (c) M. Sundaralingajm, *J. Am. Chem. Soc.*, 1965, **87**, 599.
363. (a) G. D. Wu, A. Serianni and R. Barker, *J. Org. Chem.*, 1983, **48**, 1750; (b) A. S. Serianni and R. Barker, *J. Org. Chem.*, 1984, **49**, 3292; (c) M. J. King-Morris and A. S. Serianni, *J. Am. Chem. Soc.*, 1987, **109**, 3501.
364. P. C. Kline and A. S. Serianni, *Magn. Reson. Chem.*, 1988, **26**, 120.
365. P. C. Kline and A. S. Serianni, *Magn. Reson. Chem.*, 1990, **28**, 124.
366. P. C. Kline and A. S. Serianni, *J. Org. Chem.*, 1992, **57**, 1772.
367. J. M. Neumann, J. Borrel, J. M. Thiéry, W. Guschlbauer and S. Tran-Dinh, *Biochim. Biophys. Acta*, 1977, **479**, 427.
368. R. U. Lemieux, *Can. J. Chem.*, 1961, **39**, 115.
369. F. Seela and W. Bussmann, *Nucleosides Nucleotides*, 1985, **4**, 391.
370. M. Remin, I. Ekiel and D. Shugar, *Eur. J. Biochem.*, 1975, **53**, 197.
371. S. Neidle, M. R. Sanderson, A. Subbiah, J. B. Chattopadhyaya, R. Kuroda and C. B. Reisse, *Biochim. Biophys. Acta*, 1979, **565**, 379.
372. P. C. Manor, W. Saenger, D. B. Davies, K. Jankowski and A. Rabczenko, *Biochim. Biophys. Acta*, 1974, **340**, 472.
373. F. E. Hruska, J. G. Dalton and M. Remin, *Can. J. Chem.*, 1979, **57**, 2191.
374. G. I. Birnbaum, J. Gisiewicz, E. J. Gabe, T. S. Lin and W. P. Prusoff, *Can. J. Chem.*, 1987, **65**, 2135.
375. S. Neidle, L. Urpi, P. Serafinowski and D. Whitby, *Biochem. Biophys. Res. Commun.*, 1989, **161**, 910.
376. L. W. Tari and A. S. Secco, *Can. J. Chem.*, 1992, **70**, 894.
377. A. Camerman, D. Mastropaolo and N. Camerman, *Proc. Natl. Acad. Sci. USA*, 1987, **84**, 8239.
378. (a) S. Wolfe, *Acc. Chem. Res.*, 1972, **5**, 102; (b) G. Birnbaum, M. Cygler, K. A. Watanabe and J. J. Fox, *J. Am. Chem. Soc.*, 1982, **104**, 7626.
379. C. H. Townes and B. P. Dailey, *J. Chem. Phys.*, 1949, **17**, 782.
380. (a) D. T. Edmonds and P. A. Speight, *J. Magn. Reson.*, 1972, **6**, 265; (b) S. R. Rabbani, D. T. Edmonds and P. Gosling, *J. Magn. Reson.*, 1987, **72**, 422.
381. (a) P. Tsang, R. R. Vold and R. L. Vold, *J. Magn. Reson.*, 1987, **71**, 276; (b) R. O. Day, J. L. Ragle and Y. Yoshida, *J. Magn. Reson.*, 1987, **72**, 562.
382. J. N. Lanosinska, J. Kasprzak, E. Bojarska and Z. Kazimierczuk, *Nucleosides Nucleotides*, 1999, **18**, 1075.
383. (a) R. E. Slusher and E. L. Hahn, *Phys. Rev. Lett.*, 1964, **12**, C508; (b) R. E. Slusher and E. L. Hahn, *Phys. Rev.*, 1968, **166**, 332.
384. (a) B. Cordier, D. Grandclaude, A. Retournard, L. Merlat and D. Canet, *Mol. Phys.*, 2005, **103**, 2593; (b) D. Canet, L. Merlat, B. Cordier, D. Grandclaude, A. Retournard and M. Ferrari, *Mol. Phys.*, 2006, **104**, 1391; (c) M. Ferrari, N. Hiblot, A. Retournard and D. Canet, *Mol. Phys.*, 2007, **105**, 3005; (d) M. Ferraria, A. Retournarda and D. Canet, *J. Magn. Reson.*, 2007, **188**, 275; (e) N. Hiblot, B. Cordier, M. Ferrari, A. Retournard, D. Grandclaude, J. Bedet, S. Leclerc and D. Canet, *C. R. Chimie*, 2008, **11**, 568.
385. G. W. Buchanan and J. B. Stothers, *Can. J. Chem.*, 1982, **60**, 787.
386. G. Barbarella, M. L. Capobianco, A. Carcuro, F. P. Colonna, A. Garbese and V. Tugnoli, *Can. J. Chem.*, 1988, **66**, 2492.
387. G. W. Buchanan and M. J. Bell, *Magn. Reson. Chem.*, 1985, **24**, 493.
388. (a) V. Markowski, G. R. Sullivan and J. D. Roberts, *J. Am. Chem. Soc. Chem.*, 1977, **99**, 714; (b) N. C. Gonnella, H. Nakanishi, J. B. Holtwick, D. S. Horowitz, K. Kanamori, N. J. Leonard and J. D. Roberts, *J. Am. Chem. Soc. Chem.*, 1983, **105**, 2050.
389. J. A. Happe and M. Morales, *J. Am. Chem. Soc.*, 1966, **88**, 2078.
390. G. Barbarella, A. Bertoluzza and V. Tugnoli, *Magn. Reson. Chem.*, 1987, **25**, 864.
391. P. Büchner, W. Maurer and H. Rüterjans, *J. Magn. Reson.*, 1978, **29**, 45.
392. (a) J. Reuben, *J. Am. Chem. Soc.*, 1969, **91**, 5725; (b) D. J. Sardella and J. B. Stothers, *Can. J. Chem.*, 1969, **47**, 3089; (c) T. E. St. Amour, M. I. Burgar, B. Valentine and D. Fiat, *J. Am. Chem. Soc.*, 1981,

103, 1128; (d) R. Díez, J. San Fabiín, I. P. Gerothanassis, A. L. Esteban, J. L. M. Abboud, R. H. Contreras and D. G. de Kowalewski, *J. Magn. Reson.*, 1997, **124,** 8.

393. M. I. Burgar, D. Dhawan and D. Fiat, *Org. Magn. Reson.*, 1982, **20,** 184.
394. M. Petersheim, V. W. Miner, J. A. Gerlt and J. H. Prestegard, *J. Am. Chem. Soc.*, 1983, **105,** 6357.
395. H. M. Schwartz, M. Mc Coss and S. S. Danyluk, *Tetrahedron Lett.*, 1980, **21,** 3837.
396. H. M. Schwartz, M. Mc Coss and S. S. Danyluk, *J. Am. Chem. Soc.*, 1983, **105,** 5901.
397. H. M. Schwartz, M. Mc Coss and S. S. Danyluk, *Magn. Reson. Chem.*, 1985, **23,** 885.
398. H. Iwahashi and Y. Kyogoku, *J. Am. Chem. Soc.*, 1977, **99,** 7761; (See also 1980, Volume 102, p. 2913).
399. K. A. K. Ebraheem and G. A. Webb, *J. Magn. Reson.*, 1977, **25,** 399; (See also 1978, Volume 30, p. 211).
400. S. Scheiner, *Biopolymers*, 1983, **22,** 731.
401. R. L. Lipnick and J. D. Fissekis, *Biochim. Biophys. Acta*, 1980, **608,** 96.
402. D. M. Cheng, L. S. Kan, P. O. P. Ts'o, Y. Takatsuka and M. Ikehara, *Biopolymers*, 1983, **22,** 1427.
403. (a) H. Spiesecke and W. G. Schneider, *Tetrahedron Lett.*, 1961, **2,** 468; (b) P. C. Lauterbur, *J. Chem. Phys.*, 1965, **43,** 360.
404. R. Ghose, J. P. Marino, K. B. Wiberg and J. H. Prestegard, *J. Am. Chem. Soc.*, 1994, **116,** 8827.
405. C. Giessner-Prettre and B. Pullman, *J. Theor. Biol.*, 1977, **65,** 171; (See also p. 189).
406. C. J. Chang, D. J. Ashworth, L. J. Cheru, J. DaSilva Gomez, C. G. Lee, P. W. Mou and R. Narayan, *Org. Magn. Reson.*, 1984, **22,** 671.
407. C. Giessner-Prettre and B. Pullman, *J. Am. Chem. Soc.*, 1982, **104,** 70.
408. V. G. Malkin, O. L. Malkina, M. E. Casida and D. E. Salahub, *J. Am. Chem. Soc.*, 1994, **116,** 5898.
409. A. P. Dejaegere and D. A. Case, *J. Pys. Chem. A*, 1998, **102,** 5280.
410. (a) M. Ebrahimi, P. Rossi, C. Rogers and G. S. Harbison, *J. Magn. Reson.*, 2001, **150,** 1; (b) P. Rossi and G. S. Harbison, *J. Magn. Reson.*, 2001, **151,** 1.
411. S. B. Petersen and J. J. Led, *J. Am. Chem. Soc.*, 1981, **103,** 5308.
412. L. M. Rhodes and P. R. Schimmel, *Biochemistry*, 1971, **10,** 4426.
413. P. R. Hemmes, L. Oppenheimer and F. Jordan, *J. Phys. Chem.*, 1974, **96,** 6023.
414. F. E. Hruska, A. Mak, H. Singh and D. Shugar, *Can. J. Chem.*, 1973, **51,** 1099.
415. P. F. Agris and S. C. Brown, *Methods Enzymol.*, 1995, **261,** 270.
416. O. Röder, H.-D. Lüdemann and E. von Goldammer, *Eur. J. Biochem.*, 1975, **53,** 517.
417. E. Duchardt, L. Nilsson and J. Schleucher, *Nucleic Acids Res.*, 2005, **36,** 4211.
418. J. R. Lyerla and G. C. Levy, in: *Carbon-13 NMR Spectroscopy*, (G. C. Levy ed.), Volume 1, Wiley-Interscience, New York, 1974, p. 79.
419. Z. Neiman and F. Bergmann, *J. Chem. Soc. Chem. Commun.*, 1002.
420. R. Michalczyck and I. M. Russu, *Biophys. J.*, 1999, **76,** 2679.
421. C. S. Johnson, *Adv. Magn. Reson.*, 1966, **1,** 33.
422. K. G. Rao and C. N. Rao, *J. Chem. Soc. Perkin Trans.*, 1973, **2,** 889.
423. (a) S. Berger and J. D. Roberts, *J. Am. Chem. Soc.*, 1974, **96,** 6757; (b) G. J. Martin, J. P. Gouesnard, J. Dorie, C. Rabiler and M. L. Martin, *J. Am. Chem. Soc.*, 1977, **99,** 1381; (c) K. Kanamori and J. D. Roberts, *J. Am. Chem. Soc.*, 1983, **105,** 4698; (d) K. A. Haushalter, J. Lau and J. D. Roberts, *J. Am. Chem. Soc.*, 1996, **118,** 8891.
424. (a) L. Katz and S. Penman, *J. Mol. Biol.*, 1966, **15,** 220; (b) K. R. Shelton and J. M. Clark, *Biochemistry*, 1967, **6,** 2735; (c) M. Tomasz, J. Olson and C. M. Mercado, *Biochemistry*, 1972, **11,** 1235.
425. C. Mandal, N. R. Kallenbach and S. W. Englander, *J. Mol. Biol.*, 1979, **135,** 391.
426. H. Fritzsche, L. S. Kan and P. O. P. Ts'o, *Biochemistry*, 1981, **20,** 6118; (See also 1983, Volume 22, p. 277).
427. (a) B. McConnell, M. Raszka and M. Mandal, *Biochem. Biophys. Res. Commun.*, 1972, **47,** 692; (b) B. McConnell and P. C. Seawell, *Biochemistry*, 1972, **11,** 4382; (c) B. McConnell, *Biochemistry*, 1978, **17,** 3168.
428. P. Bendel, *J. Magn. Reson.*, 1985, **64,** 232.
429. E. Quignard, B. Buu and G. V. Fazakerley, *J. Magn. Reson.*, 1986, **67,** 342.
430. P. Mirau and F. Bovey, *J. Magn. Reson.*, 1987, **71,** 201.
431. (a) S. F. Bellon, D. Chen and E. R. Johnston, *J. Magn. Reson.*, 1987, **73,** 168; (b) R. E. Engler, E. R. Johnston and C. G. Wade, *J. Magn. Reson.*, 1988, **77,** 377.

432. D. G. Cross, A. Brown and H. F. Fisher, *Biochemistry*, 1975, **14**, 2745.
433. A. Kettani, M. Guéron and J. L. Leroy, *J. Am. Chem. Soc.*, 1997, **119**, 1108.
434. (a) S. Forsén and R. A. Hoffman, *J. Chem. Phys.*, 1963, **39**, 2892; 1964, **40**, 1189; (b) S. Forsén and R. A. Hoffman, *Acta Chem. Scand.*, 1963, **17**, 1787; (c) R. A. Hoffman and S. Forsén, *J. Chem. Phys.*, 1966, **45**, 2049.
435. B. Tiffon, B. Ancian and J. E. Dubois, *Anal. Chim. Acta*, 1981, **124**, 415.
436. (a) A. Allerhand, D. M. Doddrell and R. Komoroski, *J. Chem. Phys.*, 1971, **55**, 189; (b) A. Allerhand and R. Komoroski, *J. Am. Chem. Soc.*, 1973, **95**, 8228.
437. D. E. Woessner, *J. Chem. Phys.*, 1962, **36**, 1.
438. S. Shibata, *Magn. Reson. Chem.*, 1992, **30**, 371.
439. L. G. Werbelow and D. M. Grant, *Adv. Magn. Reson.*, 1977, **9**, 189.
440. R. Konrat, H. Sterk and J. Kalcher, *J. Chem. Soc. Faraday Trans.*, 1980, **86**, 265.
441. A. Tsutsumi, *Mol. Phys.*, 1979, **37**, 111.
442. (a) C. Dalvit and G. Bodenhausen, *Chem. Phys. Lett.*, 1989, **161**, 554; (b) N. Tjandra and A. Bax, *J. Am. Chem. Soc.*, 1987, **119**, 8076; (c) J. Dittmer, C. H. Kim and G. Bodenhausen, *J. Biomol. NMR*, 2003, **26**, 259; (d) S. Ravindranathana, C. H. Kimb and G. Bodenhausen, *J. Biomol. NMR*, 2003, **27**, 365.
443. A. Kumar, R. C. R. Grace and P. K. Madhu, *Prog. Nucl. Magn. Reson.*, 2000, **37**, 191.
444. K. Pichumani, T. Chandra, X. Zou and K. L. Brown, *J. Phys. Chem. B*, 2006, **110**, 5.
445. V. Sychrovský, N. Müller, B. Schneider, V. Smrečki, V. Špirko, J. Šponer and L. Trantírek, *J. Am. Chem. Soc.*, 2005, **127**, 14663.
446. L. Zhu, B. R. Reid, M. Kennedy and G. P. Drobny, *J. Magn. Reson. A*, 1994, **111**, 195.
447. (a) K. Akasaka, T. Imoto and H. Hatano, *Chem. Phys. Lett.*, 1973, **21**, 398; (b) K. Akasaka, T. Imoto, S. Shibata and H. Hatano, *J. Magn. Reson.*, 1975, **18**, 328.
448. T. V. Maltseva, A. Földesi and J. Chattopadhyaya, *Magn. Reson. Chem.*, 1998, **36**, 227.
449. J. Plavec, P. Roselt, A. Földesi and J. Chattopadhyaya, *Magn. Reson. Chem.*, 1998, **36**, 732.
450. A. Abragam, *Principles of Nuclear Magnetism*. Clarendon Press, Oxford, 1961.
451. B. Ancian, B. Tiffon and J. E. Dubois, *Chem. Phys. Lett.*, 1979, **65**, 281.
452. M. E. Moseley and P. Stilbs, *Can. J. Chem.*, 1978, **56**, 1302; 1979, **57**, 1074.
453. N. Bloembergen, E. M. Purcell and R. V. Pound, *Phys. Rev.*, 1948, **73**, 679.
454. M. L. Newby and N. L. Greebaum, *Proc. Natl. Acad. Sci. USA*, 2002, **99**, 12697.
455. T. L. Hwang, S. Mori, A. J. Shaka and P. C. M. van Zijl, *J. Am. Chem. Soc.*, 1997, **119**, 6203.
456. (a) M. Angulo, C. Hawat, H. H. Hofman and S. Berger, *Org. Bioanal. Chem.*, 2003, **1**, 1049; (b) M. Angulo and S. Berger, *Anal. Bioanal. Chem.*, 2004, **378**, 1555.
457. (a) K. Stott, J. Stonehouse, J. Keeler, T. L. Hwang and A. J. Shaka, *J. Am. Chem. Soc.*, 1995, **117**, 4199; (b) T. L. Hwang and A. J. Shaka, *J. Magn. Reson. A*, 1995, **112**, 275.
458. P. Acharya and J. Chattopadhyaya, *J. Org. Chem.*, 2002, **67**, 1852.
459. P. L. Rinaldi, *J. Am. Chem. Soc.*, 1983, **105**, 5167.
460. C. Yu and G. C. Levy, *J. Am. Chem. Soc.*, 1983, **105**, 6994; 1984, **106**, 6533.
461. P. Thureau, B. Ancian, S. Viel and A. Thévand, *Chem. Commun.*, 2006, **200**, 1884.
462. (a) J. Gulllerez, B. Tlffon, B. Ancian, J. Aubard and J. E. Dubols, *J. Phys. Chem.*, 1983, **87**, 3015; (b) B. Tiffon, J. Guillerez and B. Ancian, *Magn. Reson. Chem.*, 1985, **23**, 460.
463. H. B. Seba and B. Ancian, *Chem. Commun.*, 996.
464. S. Macura and R. R. Ernst, *Mol. Phys.*, 1980, **41**, 95.
465. C. L. Perrin, *J. Magn. Reson.*, 1979, **82**, 619.
466. (a) K. E. Kövér and G. Batta, *J. Magn. Reson.*, 1986, **69**, 344; (b) K. E. Kövér and G. Batta, *Prog. NMR Spectrosc.*, 1987, **19**, 223.
467. (a) P. Bigler and C. Müller, *J. Magn. Reson.*, 1988, **79**, 45; (b) K. E. Kövér and G. Batta, *J. Magn. Reson.*, 1988, **79**, 206.
468. F. Stallmach and P. Galvosas, *Annu. Rep. NMR Spectrosc.*, 2007, **61**, 52.
469. (a) P. S. Pregosin, P. G. Anil Kumar and I. Fernández, *Chem. Rev.*, 2005, **105**, 2977; (b) P. S. Pregosin, Progress. *Prog. NMR Spectrosc.*, 2006, **49**, 261.
470. P. Thureau, A. Thévand, B. Ancian, P. Escavabaja, G. S. Armstrong and V. A. Mandelshtam, *Chem. Phys. Chem.*, 2005, **6**, 1510.
471. M. Holtz, X. Mao, D. Seiferling and A. Sacco, *J. Chem. Phys.*, 1996, **104**, 669.

472. A. R. Waldeck, P. W. Kuchel, A. J. Lennon and B. E. Chapman, *Prog. Nucl. Magn. Reson. Spectrosc.*, 1997, **30,** 39.
473. P. Timmerman, J. L. Weidmann, K. A. Jolliffe, L. J. Prins, D. N. Reinhoudt, S. Shinkai, L. Frish and Y. Cohen, *J. Chem. Soc. Perkin Trans.*, 2000, **2,** 2077.
474. O. Mayzel, A. Gafni and Y. Cohen, *Chem. Commun.*, 911.
475. P. Petrova, C. Monteiro, C. Hervé du Penhoat, J. Koča and A. Imberty, *Biopolymers*, 2001, **58,** 617.
476. (a) F. Perrin, *J. Phys. Radium*, 1936, **7,** 11; (b) M. M. Tirado and J. G. Garcia de la Torre, *J. Chem. Phys.*, 1979, **71,** 2581; (c) W. Elmer and R. Pecora, *J. Chem. Phys.*, 1991, **94,** 2324.
477. P. T. Callaghan, *Principles of Nuclear Magnetic Resonance Microscopy.* Oxford University Press, Oxford, 1991.
478. C. T. W. Moonen, P. Van Gelderen, G. W. Vuister and P. C. M. Van Zijl, *J. Magn. Reson.*, 1992, **97,** 419.
479. M. Andrec and J. H. Prestegard, *J. Biomol. NMR.*, 1997, **9,** 136.
480. E. J. Cabrita and S. Berger, *Magn. Reson. Chem.*, 2002, **40,** S122.
481. D. Wu, A. Chen and C. S. Johnson, *J. Magn. Reson. A*, 1995, **115,** 260.
482. P. Thureau, Ph.D. Thesis. Université de Provence, Marseille, 2006.
483. (a) V. A. Mandelshtam, *Prog. Nucl. Magn. Reson. Spectrosc.*, 2001, **38,** 159; (b) G. S. Armstrong, N. M. Loening, J. E. Curtis, A. J. Shaka and V. A. Mandelshtam, *J. Magn. Reson.*, 2003, **163,** 139.
484. J. G. Kärger, H. Pfeifer and W. Heink, *Adv. Magn. Reson.*, 1988, **12,** 1.
485. C. S. Johnson, *J. Magn. Reson. A*, 1993, **102,** 214.
486. E. J. Cabrita, S. Berger, P. Bräuer and J. Kärger, *J. Magn. Reson.*, 2002, **157,** 124.
487. B. Ancian, B. Tiffon and J. E. Dubois, *J. Magn. Reson.*, 1979, **34,** 647.
488. M. P. Gaigeot and M. J. Sprik, *J. Phys. Chem. B*, 2002, **106,** 12376.
489. S. Woutersen, Y. Mu, G. Stock and P. Hamm, *Chem. Phys.*, 2001, **266,** 137.
490. D. Eisenberg and W. Kauzmann, *The Structure and Properties of Water.* Clarendon Press, Oxford, 1969.

CHAPTER 3

Time-Domain NMR Applied to Food Products

J. van Duynhoven,* **A. Voda,*** **M. Witek,**[†] and **H. Van As**[†]

* Unilever Research and Development, Vlaardingen, The Netherlands
† Wageningen University, Wageningen, The Netherlands

Annual Reports on NMR Spectroscopy, Volume 69
ISSN 0066-4103, DOI: 10.1016/S0066-4103(10)69003-5

145

Abstract

Time-domain NMR is being used throughout all areas of food science and technology. A wide range of one- and two-dimensional relaxometric and diffusometric applications have been implemented on cost-effective, robust and easy-to-use benchtop NMR equipment. Time-domain NMR applications do not only cover research and development but also quality and process control in the food supply chain. Here the opportunity to further downsize and tailor equipment has allowed for "mobile" sensor applications as well as online quality inspection. The structural and compositional information produced by time-domain NMR experiments requires adequate data-analysis techniques. Here one can distinguish model-driven approaches for hypothesis testing, as well as explorative multi-variate approaches for hypothesis generation. Developments in hardware and software will further enhance measurement speed and reveal more detailed structural features in complex food systems.

Key Words: Relaxometry, Diffusometry, Benchtop, Quality control, Online, Multi-variate, Two-dimensional, Moisture, Fat.

ABBREVIATIONS

1,2D	One-, two-dimensional
AG	Allerhand–Gutowski
CPMG	Carr–Purcell–Meiboom–Gill
CR	Cross-relaxation or Carver–Richards
CWFP	Continuous wave free precession
D3.3	Volume-weighted average droplet size
DSD	Droplet size distribution
DTLD	Direct tri-linear decomposition
fgCPMG	Field gradient CPMG
FFC	Fast field cycling
FID	Free induction decay
FID-CPMG	Combined FID and CPMG experiment
FID+CPMG	Reconstituted FID and CPMG experiment
FLI	Fast Laplace inversion

GRAM	Generalised rank annihilation
GPD	Gaussian phase distribution
GS	Goldman–Shen
ILT	Inverse Laplace transform
IR	Inversion recovery
JB	Jeener–Broekaert
LM	Luz Meiboom
PFGMSE	Pulse field gradient multiple spin–echo
PFGSE	Pulse field gradient spin–echo
MC	Murday–Cotts
MEM	Maximum entropy method
MMME	Multiple modulation multiple echoes
MOUSE	Mobile universal surface explorer
MLR	Multi-linear regression
MVA	Multi-variate analysis
NNLS	Non-negativity constrained least squares
OPA	Outer-product analysis
PARAFAC	Parallel factor analysis
PCA	Principal component analysis
PCR	Principal components regression
PLS	Partial least squares
PLSR	Partial least squares regression
rf	Radiofrequency
SE	Spin–echo
SFC	Solid fat content
SGP	Short gradient pulse
SldE	Solid echo
SL	Spin-lock
STE	Stimulated echo
S/V	Surface-to-volume ratio
UPen	Universal penalty
WHC	Water holding capacity

1. INTRODUCTION

In the last decades, NMR has become a widely appreciated measurement tool within food science and technology.[1] The major reason for the widespread use of NMR is its versatility in assessing a wide range of compositional and structural features in foods. Within foods, NMR is used not only for molecular structure elucidation but also as a quantitative analytical tool. Besides delivering quantitative information on which compounds are present, NMR also has a track record on resolving their nano-, meso- and microstructural arrangement.[2] For many of these questions, high-field, frequency-domain NMR instruments are deployed, but time-domain (TD) instruments operating at relatively low fields have also found widespread use. This is primarily due to their relatively low cost, ease of operation and ability to provide quantitative information on product structure and composition within short turnaround times. Thus TD-NMR instrumentation can be found throughout all areas of food science and technology, as is indicated in Figure 1. Within food research, TD-NMR instruments are used for resolving relations between properties of food and its composition and (nano-, meso-and microstructure). The relatively short experimentation times have made TD-NMR methods particularly suitable for both hypothesis generation and testing. Within food development, the insights gained in research should be translated in product prototypes, and ultimately into an industrial-scale manufacturing process. The aforementioned properties of TD-NMR are particularly appealing to food technologists who appreciate the ease of use and rapid turnaround of compositional and structural information that aids in their decision making. Finally, in the foods supply chain, the need for rapid and accurate measurement of specific compositional and structural quality parameters has made TD-NMR a widespread and appreciated tool for quality and process control.[3,4] In this review, we will cover compositional as well as structural applications of TD-NMR in all three areas of food science and technology. This is an active research area which is indicated by the large number of reviews that appeared in the past decade (Table 1). We will emphasise low-field TD-NMR applications that can be implemented on benchtop instruments or on further downsized sensors for in-line or mobile inspections of product processing or quality control. TD-NMR has found a wide application range within food products, and for the sake of conciseness we will focus on relaxometric and diffusometric ^1H NMR observations in the time domain. Hence, studies on nuclei such as ^{17}O or ^2H, which are mostly carried out in the frequency domain, will not be discussed; magnetic resonance imaging (MRI) applications will also not be covered in this review. Cross-relaxation[5,6]

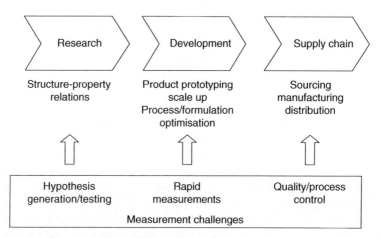

Figure 1   Schematic representation of the application areas of time-domain NMR in food science and technology.

TABLE 1 Reviews of time-domain NMR technologies and their applications in food science and technology (since 2000)

Subject	Date	References
Quality and process control by TD-NMR	2001–2009	3, 4, 7–10
Water/moisture and fat analysis by TD-NMR	2006	11
Multi-variate analysis	2002	12
Magnet designs	2009	13
Unilateral mobile NMR	2008	14
Emulsion characterisation	2007, 2009	15, 16
2D Diffusion and relaxation	2006	17
Fruit inspection by TD-NMR	2006	18
NMR relaxometry and meat science	2009	19
NMR relaxometry and dairy products	2009	20
2D TD-NMR applied to dairy products	2009	21
Novel low-field NMR techniques in food science	2006	22
Practical aspects of TD-NMR	2007	23

(also known as magnetisation transfer or z-spectroscopy) is considered as a frequency-domain experiment and is outside of the scope of this review as well.

It may seem that TD-NMR is hampered by the relatively low and inhomogeneous fields that are typically used, as well as compromises in specifications of electronics and probeheads. In this review, the ingenuity of NMR spectroscopists in exploiting the opportunities of low-field TD-NMR will be described. Many of the apparent shortcomings have been circumvented by innovations in pulse programming, data processing and data analysis. For example, distinguishing chemical species by means of their chemical shift in the frequency domain is hampered by the relative low and inhomogeneous B_0 fields of most TD instruments. Therefore, in many applications, differences in relaxation and diffusional behaviour have been exploited to distinguish, for example, oil and water. Data processing and analysis are critical for exploiting TD-NMR data to the full; hence a separate section is dedicated to these topics. Subsequently, we provide an elaborate overview of TD-NMR applications in food science and technology and conclude with a vision on future developments.

2. HISTORY

The development of the first commercial dedicated pulsed time-domain benchtop NMR instrument started in 1970 with an industrial collaboration between an instrument manufacturer (Bruker) and a foods company (Unilever). At that time, the fat and oil industry was in a strong need to replace cumbersome dilatometric methods for determination of solid content in fats. Earlier work in the 1960s on assessment of solid fat content by frequency-domain wideline

NMR[24,25] was successfully pursued in pulsed NMR mode by Bruker. The collaboration resulted in the first commercial benchtop pulsed NMR instrument for relaxometric determination of solid fat content (SFC) in fat blends.[26] The NMR SFC method rapidly conquered the market, and this development was picked up by other manufacturers like Oxford. In the decade that followed, the range of relaxometric methods widened to other food applications such as the determination of fat content in chocolate and oil content in seeds. At that time, the industry recognised that an NMR droplet sizing method developed by Packer and Rees[27] was well amenable for implementation in the benchtop mode. This also resulted in benchtop NMR instrumentation for droplet sizing in food emulsions. Although initially the benchtop NMR instruments were developed for routine quality and process control, food scientists in academia and industry rapidly recognised the potential of these relatively low-cost instruments for research purposes. On the one hand, this resulted in a wide range of applications of NMR relaxometry and diffusometry in foods. On the other hand, we also witness technological developments, such as unilateral and portable NMR for non-invasive assessment of material properties.[14] The developments at the application and technological side have resulted in a number of small to medium enterprises that offer dedicated solutions for industry and academia. Thus, time-domain NMR has become an active research area at the interface of industry and academia, and is a vital member of the "Mobile NMR" community.

3. EQUIPMENT

3.1. Magnets

In the most common TD-NMR implementation, static magnetic B_0 fields are generated by electromagnets or permanent magnets, consisting of a yoke holding two poles at a distance of a few centimetres. Thus, B_0 field strengths between 0.12 and 1.4 T can be obtained, corresponding to proton Larmor frequencies in the range of 5–60 MHz. These field strengths gain sufficient sensitivity for observing abundant species in foods such as water and oil. Such magnets are relatively cheap and can be installed on laboratory benches. Currently, most of the commercial benchtop magnets do not have a shimming unit attached and exhibit relatively poor homogeneity, typically resulting in T_2^* of a few milliseconds for most liquids. Within this review, we will occasionally also mention TD-NMR experiments that were carried out on high-field (>100 MHz) NMR spectrometers, where one can benefit from superior homogeneity, sensitivity and dynamic range. For most applications, this is not a cost-effective solution, so we will focus on routes to overcome current limitations of low-field magnets. Some examples of shimmed low-field permanent magnets have been described, achieving sufficient resolution for frequency-domain NMR,[28,29] but this has not been pursued widely. Commonly, one relies on minimising field drift by keeping magnets at a constant temperature; typically no field lock is supplied. Air gaps for most magnets are between 25 and 100 mm and allow for sample (tube) diameters between 10 and 100 mm.[13] The largest air gaps are wide enough to pass intact apples on a small conveyor belt.[30] With large air gaps, one however has to compromise on a lower Larmor frequency, where stabilisation and homogeneity are then easier to achieve. Recently, so-called Halbach magnet arrangements[31–33] have been used to design more homogeneous magnets that would be suitable for benchtop[34] implementation. Open-access Halbach magnet designs have also been proposed for online non-invasive inspection of intact food products.[35] Both ideas have not reached commercial implementation yet, however. Another approach to non-invasive inspection of intact products is the deployment of single-sided magnets with built-in measurement coils for transmitting and receiving rf signals.[14] One-side magnet geometries allow for recording of NMR signals in the sensor mode for a relatively thin slice near the magnet.[36] This precludes application to structurally heterogeneous products, however.[22,36]

3.2. Electronics

In the earliest commercial benchtop NMR instruments, rf fields were generated by relatively cheap electronics and free induction decays (FIDs) were acquired by simple diode detectors.[37] Contemporary instruments have much more advanced designs, allowing sophisticated multi-pulse experiments, and the NMR signal can be recorded with adequate digital resolution in a linear manner in phase-sensitive mode. For diffusometric experiments, accessories can be supplied that allow delivery of the strong currents needed for driving pulsed-field gradient coils. A cost-effective method for supplying strong and short linear currents was the use of capacitors, but nowadays increasingly linear gradient amplifiers are also used in commercial benchtop instruments.

Probehead requirements are different for so-called absolute and relative relaxometric measurements. For many oil or moisture content measurements, one needs to record the absolute value of the magnetisation, and this requires good rf homogeneity over a large sample volume. Typically, here one accepts longer dead times and pulse widths. For relative measurements, where one, for example, considers the decay of magnetisation, short pulse widths and dead times are required, and consequently probeheads with lower Q-factors. Here one often compromises with respect to rf homogeneity.[13] For many applications, one needs thermostatted probeheads, which is achieved by water or gas flow. In some cases, probeheads are simply maintained at magnet temperature; rapid measurement of pre-conditioned samples then prevents any significant change in temperature. For pulsed field gradient applications, both shielded as well as non-shielded gradient coils are in use. In case non-shielded gradient coils are used, one either has to use time-consuming gradient balancing procedures, or take recourse to cumbersome but effective calibration procedures.[38]

4. ONE-DIMENSIONAL RELAXOMETRY AND DIFFUSOMETRY

4.1. Relaxometry

Many of the pulse sequences in benchtop NMR relaxometry found their first application already decades ago.[39] The FID is the most basic transversal (T_2) relaxometric experiment, but at current low-field benchtop NMR instruments its applicability is compromised by the strong inhomogeneity of commercial permanent magnets now in use. Hence, the FID is mostly used to assess the rapid sub-millisecond (T_2) relaxation behaviour of solid crystalline and glassy phases where molecular mobility is low. The solid-echo (SldE) sequence is considered to describe more accurately solid phases, but it has never found wide applications in foods. For phases where molecular mobility is higher, such as in most semi-solid and liquid phase, one typically uses the Hahn spin–echo (SE) of Carr–Purcell–Meiboom–Gill (CPMG) sequences. These transversal relaxation experiments refocus dephasing due to B_0 inhomogeneity and allow assessment of relaxation times in the milliseconds to seconds range. The refocusing of magnetisation in SE and CPMG experiments is still sensitive to diffusion and exchange, and this can be exploited to obtain structural information on hydration and micro-scale morphology[40–43] (Section 6.2).

Besides transversal (T_2) relaxation experiments, one often also assesses longitudinal (T_1) relaxation behaviour by inversion recovery (IR) or saturation recovery (SR) sequences. In particular, the IR sequence is much more time consuming than T_2 measurements such as CPMG. For simple systems, rapid SR experiments and simple data-treatment procedures have been proposed,[44] but for more complex systems more sophisticated two-dimensional approaches are required (Section 5). In 1D measurements, differences in T_1 values in complex foods have effectively been exploited for selective suppression of the refocused magnetisation[45] in more advanced methods like SE T_1-Null[46] or CPMG T_1-Null.[47] Fast field cycling (FFC) offers the possibility of rapid assessment of the dependence of T_1 on static field strength. Although these NMR dispersion (NMRD) profiles contain a wealth of information on rapid dynamic processes, the number of FFC applications within foods has been limited.[48]

An attempt to improve the sensitivity when using a low-field NMR machine is steady-state free precession (SSFP).[49] An increase in the signal-to-noise ratio for a given acquisition time is gained in comparison with conventional acquisitions as FID or SE. A variation of this method is continuous wave free precession (CWFP).[50] The method employs a train of 90° pulses with a period that is much shorter than T_2^*. In this case, spin dephasing between pulses is negligible and a continuous-wave periodic signal is achieved.[50] The amplitude of the signal depends on both transverse and longitudinal relaxation times and this information is gained in a single scan experiment. The method is suitable for multi-component systems if the discrimination is possible via difference in relaxation times. Drawbacks exist due to static and rf field inhomogeneity that accumulate during repetitive excitations. In the fast motion limit, where $T_1 = T_2$, the decay from the quasi-stationary to steady-state regime occurs within a very short time.[51]

In combination with a static magnetic field gradient, SSFP allows the real-time measurement of flow.[52] Under appropriate conditions, the method is an important candidate to fast, low-field routine determinations in one-component systems but its potential has not been extensively investigated in heterogeneous food products.

To discriminate biochemical compositions in solid-like materials, the Jeener–Broekaert (JB) experiment to measure dipolar T_{1D} relaxation times has been applied to plant cell wall materials.[53–56] The process of exchanging magnetisation between rigid and mobile fractions in the plant system has been studied using the Goldman–Shen (GS) cross-relaxation experiment.[55] Compositional and molecular dynamic information in the solid phase can be also obtained from the proton spin–lattice relaxation time in the rotating frame ($T_{1\rho}$) as measured by a spin-lock (SL) sequence. These approaches are commonly used in technical polymer studies but have found only a moderate number of food applications.[57–59]

4.2. Diffusometry

The commonly used diffusometric experiments combine de- and refocusing pulsed field gradient (PFG) pulses with spin–echo (SE) or stimulated echo (STE) measurements. The motivation to apply the methods in food science arises from the fact that the observed signal attenuation is proportional to the self-diffusion constant (D) and three adjustable NMR parameters: the amplitude (G), the duration (δ) and the time interval between the de- and refocusing pulses (Δ) of the magnetic field gradient pulses. The self-diffusion constant can be derived from the exponential attenuation of signal in a field gradient as a function of $\gamma^2 G^2 \delta^2 (\Delta - \delta/3)$. Within heterogeneous systems, the spatial variance of the magnetic susceptibility induces background magnetic field gradients, which often hampers accurate determination of diffusion coefficients. This can be circumvented by approaches that employ bipolar field gradients such as the 13-interval PFGSTE sequence.[60] Alternatively, one can deploy the PFG multiple spin–echo (MSE) sequence where two PFGs and a phase cycled CPMG-like pulse train are intertwined.[61,84]

Semi-permeable barriers between (sub)cellular compartments in plants cause restriction of water self-diffusion, resulting in disturbance of the exponential PFG decay.[62] In well-defined food systems, the average displacement of molecules over a limited, specified range can be quantified in terms of barrier properties, using a combination of q-space microscopy[63] and microstructural modelling approaches (Section 6.2). To selectively probe size dimensions in heterogeneous food systems with multi-exponential relaxation decays, PFG-SE and -STE sequences have been supplied with filters to suppress unwanted contributions. Such filters relied either on differences in T_1[64] or on self-diffusion coefficients.[46,47,65] Most diffusometric NMR applications within foods rely on pulsed field gradients, but the use of a constant field gradient in combination with CPMG (fgCPMG) has also been described.[66]

5. TWO-DIMENSIONAL NMR RELAXOMETRY AND DIFFUSOMETRY

When the aforementioned 1D measurements are applied to complex heterogeneous food materials, one inevitably encounters the challenge of resolving components with similar relaxation times. Hence a number of methods have been
proposed that aim at resolving components in food products and materials
(see Table 2) in two or even three relaxation and/or diffusion time dimensions,
resulting in correlated relaxation times and/or diffusion coefficients.

5.1. $T_{1(\rho)}-T_2$ correlations

The first attempts to measure two-dimensional T_1 and T_2 correlations in the time
domain used a combination of saturation or inversion recovery (SR or IR) and FID
or CPMG. In the SR/IR-FID experiment, combined with a 2D "spin grouping"
approach,[67,68] the entire FID is monitored for every saturation or inversion recovery time.[69,70] Similar approaches have also been used to correlate spin–lattice
relaxation in the rotating frame ($T_{1\rho}$) with T_2 by combining a spin lock (SL) with

TABLE 2　Overview of 2D sequences built up from 1D relaxometric and diffusometric building
blocks and resulting correlation spectra

Sequence	Correlation	References
$T_{1(\rho)}-T_2$ correlations		
SR-FID	T_1-T_2	69,70
SR-CPMG	T_1-T_2	72
IR-CPMG	T_1-T_2	71, 74, 76
IR-PFGSE-CPMG	$T_1-(D\text{-filter})-T_2$	81
SL-FID	$T_{1\rho}-T_2$	69
SL-CPMG	$T_{1\rho}-T_2$	70
$D-T_2$ correlations using PFG		
PFGSE-CPMG	$D-T_2$	89, 21, 83
PFGMSE-CPMG	$D-T_2$	61, 84
PFGSTE-CPMG	$D-T_2$	21, 82
13-Interval PFGSTE-CPMG	$D-T_2$	85
IR-PFGSTE-CPMG	$D-(T_1\text{-filter})-T_2$	86
$D-T_2$ correlations using constant field gradients		
Multi-grade CPMG	$D-T_2$	90, 91
T_1-T_2 and $D-T_2$ correlations in inhomogeneous fields		
STE-CPMG in fringe field	$D-T_2$	
SE-CPMG in fringe field	$D-T_2$	21, 92
IR-CPMG in fringe field	T_1-T_2	
Exchange correlations		
PFGSE-store-PFGSE	$D-D$	95
CPMG-store-CPMG	T_2-T_2	93, 94, 96

an FID[69] or a CPMG[70] sequence. In SR/IR-CPMG sequences, a CPMG pulse sequence monitors the transverse magnetisation at echo times t_2 after a saturation/inversion recovery time during t_1, where spin systems evolves under the influence of spin–lattice relaxation. The SR/IR-CPMG sequence produces CMPG amplitude data sets at every t_1 time. T_1–T_2 correlations measured by IR-CPMG[70,71] were initially obtained from a 2D version of NNLS fitting, which demanded significant computer memory.[71] This firstly impeded widespread use of the IR-CPMG experiment in foods. Less extensive data manipulation was needed for IR- or SR-CPMG experiments if the respective CPMG decays for every saturation time were first subjected to 1D SPLMOD or CONTIN analysis in the T_2 direction to obtain T_1 value of the fractions with different T_2 values.[72] The SR-CPMG approach has been used to discriminate water fractions in compartmentalised plant tissues.[72,73]

Once a fast 2D Laplace inverse (FLI) algorithm[74] became available, the potential of combining two dimensions for resolving different components and water compartments in foods was early recognised.[17,75] The combined IR-CPMG experiment for measuring T_1–T_2 correlations[74,76] has already found widespread use in food science.[22] Apart from the most common application to resolve different food components by their respective relaxation times, there have been attempts to make conclusions on exchange processes. Exchange is observed in 2D T_1–T_2 correlation spectra by off-diagonal peaks when $T_1 \neq T_2$,[76–79] but so far conclusions have often been speculative. IR-CPMG protocols with different CPMG pulse spacings and spectrometer frequencies were explored for distinguishing different water compartments in plant systems.[80,81] At longer CPMG pulse spacings, the diffusion of water through local field gradients leads to faster transversal relaxation and thus structural characteristics like air spaces in the cellular tissue of plant materials may become evident. Because of the strong dependence of T_1 on the spectrometer frequency, T_1–T_2 spectra may be better resolved at a lower frequency where T_1 differences are amplified,[80] while at a higher frequency T_2 differences are magnified.[46] To better resolve minor components in complex foods, water-suppressed T_1–T_2 measurements have been carried out using the IR-PFGSE-CPMG sequence. In this method, the PFGSE step is inserted after the inversion step so that rapidly diffusing water molecules are dephased.[81]

5.2. T_2–D correlations using pulsed-field gradients

Correlation of diffusion and T_2 has been achieved by combining common diffusometric approaches, such as PFGSE or PFGSTE, with a CPMG sequence.[82,83] D–T_2 correlations have also been obtained by sequences that are not sensitive to background gradients such as PFGMSE-CPMG[61,84] or 13-interval PFGSTE-CPMG.[85] By adding a T_1-filter, specific components can be emphasised in the IR-PFGSTE-CPMG sequence.[86] In these experiments, diffusion and T_2 relaxation are sampled in gradient strength and echo time domain, respectively. Subsequently, D–T_2 correlations can be obtained using different data analysis approaches (Section 6). Discrete data analysis approaches (e.g. NNLS, SPLMOD) have been useful to distinguish the diffusion coefficient values of water fractions associated with different compartments,[84,87] as well as surface/volume ratios of cells.[86,88] Since the introduction of the 2D fast Laplace inversion (FLI) algorithm, D–T_2 correlation sequences[89] are gaining popularity for discriminating components in complex food materials on the basis of differences in diffusion behaviour.[21,46,47,81]

Another D–T_2 method to probe microstructure relies on a multi-grade CPMG[90,91] sequence, where CPMG is used to measure the water transverse relaxation time distribution and a ramped external constant field gradient (applied throughout the CPMG sequence) is used to give diffusive weightings to each relaxation time component. Although this approach allows assessment of the effective water diffusivity without effect from the background gradients, it has not found application in food science yet.

5.3. T_1–T_2 and T_2–D correlations in inhomogeneous fields

By recording STE-CPMG or SE-CPMG and IR-CPMG experiments in strongly inhomogeneous fields as, for example, present in the fringe-field of supercon-ducting magnets or in one-sided magnets, D–T_2 and T_1–T_2 maps can be obtained.[92] The sequences consist of two parts: the initial diffusion (STE/SE) or relaxation (IR) editing followed by refocusing the CPMG sequence. By vary-ing the time between the first and the second 90° pulse (STE) or the initial echo spacing (SE) and the inversion recovery time, diffusion and T_1 relaxation can be correlated. In the second part, the magnetisations are repeatedly refocused by the series of many 180° pulses with short echo spacing to obtain the relaxation information only. The analysis of the spin dynamics in the pulse sequences has been done in the formalism of possible coherence pathways, which includes the effect of diffusion, relaxation and strong field inhomogeneity. The application of appropriate phase cycling methods in the diffusion editing part of the experi-ments allowed selection of a single coherent path from many possible coherence pathways to calculate the spectrum and dependence on diffusion of the magne-tisation at the end of the editing sequence. Thus diffusional attenuation is uniform across the whole spectrum. To reduce the multiple possible coherence pathways, short echo spacing time with effective rotation axis approach in the second part of the experiments is used in that approach. Recently, STE-CPMG and IR-CPMG experiments in the presence of a strong static gradient outside of the superconducting magnet have been applied to study the heterogeneity of food products.[75]

5.4. 2D exchange experiments

Information on inter-compartmental exchange dynamics can be derived from the CPMG–store–CPMG experiment.[93] During the storage periods, magnetisation is held along the longitudinal direction, during which exchange can occur.[94] The obtained cross-peaks in the T_2–T_2 spectra have been recently used to detect inter-compartmental diffusion in cellular foods.[81] D–D correlations can be achieved by two PFGSE sequences with stimulated echoes, separated by an exchange time where the gradient strengths of the two different parts are varied independently.[89] Diffusion exchange (D–D) results have been obtained for foods systems[95] in which exchange between compartments with different diffusion behaviour was observed.

6. DATA-ANALYSIS APPROACHES

6.1. Signal processing

Since protonated materials in the probehead introduce a small but discernible rapidly decaying background signal, a correction for this is sometimes applied. The dephasing of the NMR signal due to B_0 inhomogeneity is typically cancelled by refocusing 180° pulses, as in CPMG pulse trains. This is not a useful solution when one wants to observe sub-millisecond relaxation times. Whereas for solid signals ($T_2 < 20 \ \mu s$) effects of B_0 inhomogeneity are hardly noticeable, this is not the case for semi-solid signals ($20 \ \mu s < T_2 < 200 \ \mu s$). By assuming that the effect of B_0 inhomogeneity is a straightforward deconvolution with a Gaussian function, one can simply obtain the true FID. In most spectrometers, data are recorded in quadrature mode, but many researchers apply magnitude transformation to correct for phase errors in the acquisition. This introduces a bias in the exponential fitting,[97] and algorithms have been proposed to phase-correct the quadrature data.[97,98]

6.2. Model-driven analysis

6.2.1. Free induction decays obtained from rigid phases

Several models have been applied to fit transversal relaxation data obtained by TD-NMR. For line shapes originating from crystalline lipids[99,100] or carbohydrates[101,102] with strong dipolar couplings, one commonly uses the "Abragam" sinc function[103]:

$$M(t) = A \ \exp\left(\frac{a^2 t^2}{2}\right) \ \sin(bt)/(bt) + B \ \exp\left(\frac{-t^2}{c^2}\right) + d \qquad (1)$$

where the second magnetic moment M_2 is given by

$$M_2 = a^2 + \frac{1}{3}b^2 \qquad (2)$$

For crystals with a high degree of mobility, and also for solids in a glassy state, one commonly uses a Gaussian line shape for fitting. The feasibility of quantitative assessment of different crystalline polymorphs, semi-solid and liquid phase has been demonstrated. In such approaches, one should avoid the risk of over-fitting by imposing realistic constraints in the fitting procedure.[99]

6.2.2. Diffusometric droplet sizing

The so-called short gradient pulse (SGP) and Gaussian phase distribution (GPD) approximations have been used to develop analytical formalisms for describing PFG decays of liquids that experience restricted diffusion inside spherical droplets.[63,104] The GPD approximation is most appropriate for droplet size distribution (DSD) determinations on benchtop NMR instruments.[16] The Murday and Cotts[105] (MC) equation is based on the GPD approximation:

$$R(\Delta, G, \delta, D, a) = \exp\left[-\frac{2\Delta}{T_2} - 2G^2\gamma^2 \sum_{m=1}^{\infty} \frac{1}{\alpha_m^2(\alpha_m^2 a^2 - 2)} \times f(\Delta, G, \delta, D)\right]$$

$$f(\Delta, G, \delta, D) = \left\{\frac{2\delta}{\alpha_m^2 D} - \frac{2 + \exp(-\alpha_m^2 D(\Delta - \delta)) - 2\exp(-\alpha_m^2 D\delta) - 2\exp(-\alpha_m^2 D\Delta) + \exp(-\alpha_m^2 D(\Delta + \delta))}{(\alpha_m^2 D)^2}\right\}$$

(3)

In this equation, Δ, δ, γ and T_2 have been defined before, a is the droplet radius, and α_m is the mth positive root of the Bessel function equation:

$$J_{3/2}(\alpha a)/\alpha a = J_{5/2}(\alpha a)$$

(4)

The restricted diffusion equation can be extended[27] to describe a distribution of droplet sizes according to a unimodal lognormal distribution $P(a)$:

$$P(a) = \frac{1}{2a\sigma\sqrt{2\pi}} \exp\left[-\frac{(\ln 2a - \ln D_{3.3})^2}{2\sigma^2}\right]$$

(5)

Here $D_{3.3}$ is the volume-weighted mean droplet diameter, which means that 50% of the total volume of the dispersed phase is present in droplets with a diameter smaller than $D_{3.3}$, whereas the other 50% is present in droplets with a diameter larger than $D_{3.3}$, and σ is the standard deviation of the logarithm of the droplet diameter.[106,107] As an alternative to imposing an algebraic form of the DSD, model-free approaches have also been described that require no *a priori* assumptions. Three approaches have been demonstrated that rely on a generating function method,[108] a numerical routine base to solve a Fredholm integral problem[109] and a regularisation algorithm.[110]

6.2.3. Transversal relaxation-time dispersion and exchange

By studying transversal relaxation behaviour of water as a function of CPMG inter-pulse delay, one can obtain structural information on hydration and micro-scale morphology[40–43,111] of food materials. A characteristic sigmoid-shaped curve for the water proton relaxation T_2 rate as a function of CPMG inter-pulse delay is observed if exchange occurs between sites with different Larmor frequencies. An example is shown in Figure 2, where one can note that the dispersion in T_2 as a function of CPMG inter-pulse delay is more clearly observed at higher field where differences in Larmor frequencies are larger. For describing T_2 dispersion curves, one can depart from models where spins jump between distinct sites with different Larmor frequencies ("chemical site exchange"). The model of Luz and Meiboom (LM[112]) assumes fast chemical exchange between two sites with

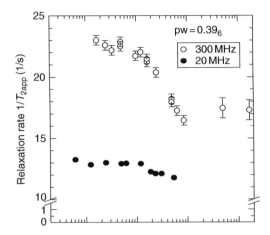

Figure 2 T_2 dispersion curves for a casein gel at 20 and 300 MHz (reprinted with permission from Reference 43. Copyright 2005, American Institute of Physics).

different Larmor frequencies and uniform T_2. The Allerhand and Gutowski (AG[113]) model also covers slow chemical exchange, and was further extended by Carver and Richards (CR[114,115]) to allow for differences in transversal relaxation rates. Alternatively, one can also imagine "diffusion/susceptibility exchange", where sites (not necessarily chemically different) are separated by local magnetic field gradients due to microstructural magnetic susceptibility mismatches. Whereas in "chemical site exchange" one discerns distinct chemical shifts, spins cover a range of Larmor frequencies in "diffusion/susceptibility exchange". Whether the full range of Larmor frequencies can be covered depends on the length scale of the local field gradients. In the "diffusive exchange model"[115,116], spins diffuse between spatially separated compartments with different Larmor frequencies and transversal relaxation times. Diffusive exchange has been elaborated in plant-tissue-specific models, which will be described in the next sections.

6.2.4. Vacuolated plant materials
In general, multi-exponential decay curves are observed for water relaxation measurements in (vacuolated) plant material. The different relaxation times can be assigned more or less uniquely to either water in the vacuole (longest T_1 and T_2), cytoplasm ($T_1 > T_2$, both shorter than vacuolar T_1 and T_2) or cell wall/extracellular space (T_2 depends strongly on the water content in this compartment, and ranges from about one millisecond and higher).[73,84,117] Diffusive exchange within compartments and exchange between compartments, passing membranes, affect the observed relaxation times.[118] The observed T_2 (and T_1) of vacuolar water has been demonstrated to depend on the bulk T_2 in the vacuole ($T_{2,bulk}$), and the surface-to-volume ratio (S/V) of the vacuole[118]:

$$1/T_{2,obs} = (H \times S/V) + 1/T_{2,bulk} \tag{6}$$

The proportionality constant H is directly related to the actual tonoplast membrane permeability for water.[117,118] The equation holds also for water in (xylem) vessels, where H now represents the loss of magnetisation at the vessel wall,[119] demonstrating that T_2 of vessel water directly relates to vessel radius. For a proper interpretation of $T_{2,obs}$ in terms of membrane permeability, we need to know S/V (cf Equation (6)). This information can be obtained by measuring the apparent diffusion coefficient D_{app} as a function of the diffusion time Δ. For diffusion in a confined compartment, free diffusion is observed at short diffusion times. At increasing Δ, the diffusion becomes restricted, but the averaging of local properties over a large enough distance does not occur yet. In that regime, D_{app} depends linearly on Δ, and the slope is determined by S/V of the compartment.

6.2.5. The numerical plant cell model

The problem that still needs to be solved is the translation of the data analysis results (e.g. T_1, T_2, D values) with respect to important intrinsic properties, like compartment sizes (a or r) and membrane permeability (P) for cells. In the last decades, much insight has been gained on how plant tissue microstructures are reflected in NMR relaxation parameters and self-diffusion behaviour of water and solutes.[120] Already two decades ago it was found that NMR relaxation of water is governed by fast chemical exchange between water and exchangeable protons on biopolymers (or solutes) and diffusion through internally generated field gradients at the air/cell interface.[40] Depending on cell morphology and membrane permeabilities, diffusive exchange between the various vacuolar, cytoplasmic and extracellular water compartments spatially averages water proton magnetisations.[73,118]

One- and two-dimensional three-compartment diffusion/relaxation models have been introduced[73,117,121,122] that discerned vacuole, cytoplasm and extracellular regions, characterised by D_i and $T_{2,i}$ constants, separated by semi-permeable tonoplast (P_1) and plasmalemma (P_2) membranes (Figure 3). The calculations simulate PFG-SE experiments and are based on solving the partial differential equations with respect to 2D spin magnetisation density in the presence of field gradients used in PFG experiments, spin relaxation and a proper set of initial and boundary conditions, which are based on the Fick's second law of diffusion.[117,121,122] The resulting spin–echo amplitude $S(G\delta, \Delta, \tau)$ depends on the three independent variables: $G\delta$, the wave vector corresponding to the pulsed gradient area; Δ, the diffusion time; and τ, the 90–180 pulse spacing. Analysis of this data set can either be done by a general approach as described above (see Section 6.2.4) or, alternatively by using the numerical plant cell model to simulate the data.[117,122]

Recently, an alternative approach has been presented to fit correlated D–T_2 experiments directly with intrinsic 1D geometric model parameters, like compartment size, membrane permeability and compartment relaxation times and diffusion coefficients.[123] Fitting of 2D or even 3D geometries is computationally more demanding and has been used to obtain intrinsic values for apple and carrot tissue as a function of processing.[41,42]

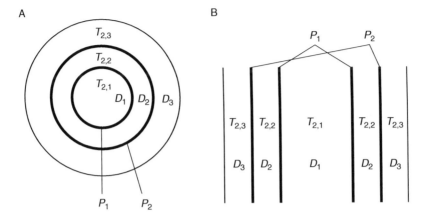

Figure 3   A schematic representation of a three-compartment plant cell model in cylindrical (A) and plane parallel (B) geometry. The labels D_i, T_{2i}, P_i, refer to the intrinsic water self-diffusion coefficients, water proton relaxation times, and membrane permeabilities for vacuole (1), cytoplasm (2) and extracellular (3) compartments (reprinted from Reference 122, Copyright 2002, with permission from Elsevier).

6.3. Discrete relaxometric and diffusometric decays

In many cases, one assumes that relaxometric and diffusometric decays consist of a sum of a limited number of exponentially decaying terms. This is, for example, the case for transversal CPMG decays and longitudinal IR decays. For relatively simple mixtures of oil and water, diffusometric decays are often bi-exponential. Several conventional approaches for such decays are available, but often ill-conditioning and local minima are encountered. For NMR decays, SPLMOD[124] has found widespread use, which has a particular advantage that it can simultaneously fit a range of decays that share common relaxation characteristics (coupled analysis). For this purpose, several other solutions have also been described such as MATRIXFIT[98] and NNLS.[125,126] Decays consisting of exponentially decaying terms can also be resolved using multi-way multi-variate methods, such as SLICING[127] and DOUBLESLICING.[128] In these approaches, a bi-dimensional array is built where the different rows represent the original decay shifted with different time lags. The resulting matrix can be analysed by so-called tri-linear data-analysis methods. This may be seem a cumbersome method, but a main advantage of (DOUBLE) SLICING is its efficiency in computation time.

In all aforementioned approaches, one is dealing with one or more decays in which only one relaxation or diffusometric mechanism is active. For complex heterogeneous systems, this frequently leads to components that cannot be resolved, either because relative populations are small or because populations have small differences in relaxation or diffusion constants. Such populations can be resolved by 2D measurements in which two relaxation principles, such as T_2–T_1 or T_2–$T_{1\rho}$ are combined. CracSpin[129] has been proposed as an approach to resolve discrete "spin groups" which share common relaxometric parameters.[68,69]

6.4. Continuously distributed relaxometric and diffusometric decays

Due to the compositional and structural heterogeneity of foods systems, nuclear spins experience various environments and exhibit a distribution of relaxation or diffusion constants. In this case, the NMR signal decay can better be treated as an integral response of a continuous distribution function rather than a limited number of discrete components. The solution of the continuum approach is non-unique, meaning that a variety of results exist that fit the data equally well.[130] Such data inversions are ill-posed problems where the solution is sensitive to small experimental errors. Several methods for data inversion based on statistical regularisation techniques have emerged. From historical reasons, it is worth mentioning that various studies were dedicated to test the performance and accuracy of different approaches to invert 1D data sets. Linear and non-linear schemes were employed to solve inversion problems. Inverse Laplace transformation (ILT) was typically performed with the NNLS fitting procedure under certain regularisation constraints.[131] Linear inversion techniques were also applied to NMR relaxation data to extract continuous as well as discrete distributions.[132,133] Constrained regularisation is a more common approach for inversion of relaxometric data, and the methods of Butler et al.[134] and Provencher (CONTIN)[135,136] have found most widespread use. Both methods employ a constrained regularisation algorithm for the inversion of noisy linear algebraic and integral equations. The solution is based on weighted least squares and a regularisation factor that imposes parsimony or statistical prior knowledge such as non-negativity. In the maximum entropy method (MEM), the inversion of continuous distributions of relaxation data is performed under maximum entropy control. Within the NMR community, the method was first applied in high-resolution frequency domain for noise reduction and resolution enhancement.[137] TD applications in food area followed, and the method proved efficient for rapid structural characterisation of various food systems.[138,139] A more recent method for data inversion is UPEN (Uniform PENalty inversion of multi-exponential decay data), which uses a negative feedback in the smoothing for the computed distributions, thus allowing stronger smoothing for a broad line than for a narrow one.[140,141] The method brings improvements with respect to constraints such as the non-negative constraint, which is rarely required or not needed at all in the case of sharp lines. Moreover, the input data spacing can be arbitrary. An extensive validation exercise has been carried out on simulated and real data.[140,142] A novel approach to estimate the solution of the inversion problem consists of a statistical sampling of solutions that are consistent with the experimental data by means of a Monte Carlo algorithm.[130] Thus the uncertainty distribution of the inverted data can be computed.

Until recently, poor performance of available algorithms impeded deployment of 2D TD-NMR methods to systems with continuous relaxation and diffusion behaviour. A breakthrough was the introduction of time-efficient and stable algorithms[74,89] for FLI of 2D data sets. The size of the 2D matrix variables associated with the experimental data set and the inversion kernel are reduced using singular-value decomposition. The 2D input and output matrices are transformed to 1D vectors by consecutive ordering of the matrix rows or columns. Thus the problem becomes 1D and the computation makes use of a generic 1D Laplace inversion.

6.5. Multi-variate data analysis approaches

A major advantage of NMR is that there is virtually no bias when detecting compounds. In the benchtop implementation, one compromises sensitivity, but bulk species such as water, lipids and carbohydrates can be detected without any difficulty. Although TD-NMR may lack chemical sensitivity, its particular strength lies in its sensitivity to detect differences in molecular mobility. The unbiased nature of NMR is only partially exploited by the strongly hypothesis/assumption-led data analysis approaches outlined in the previous section. In the last decade, the unbiased and multi-variate nature of TD-NMR has been exploited in a range of explorative chemometric studies. In one approach, first time constants and populations were extracted from a TD data set. Subsequently, these values were explored by multi-linear regression (MLR) against physical or compositional parameters.[125,144] Such an approach critically depends on a good fitting procedure, as well as adequate precautions against overfitting.[125] More commonly, the TD data themselves are subjected to multi-variate analysis (MVA), where principal components analysis (PCA) is an established workhorse for exploring patterns in large data sets[12,145] (see overview in Table 4). The PC model can in principle be used to regress against quality parameters,[146] but partial least squares (PLS) has become more established for this purpose.[144,145] PCA and PLS are applied to data sets that have a "matrix" structure. It has been argued that by expanding such bi-linear "matrix" data sets into a tri-linear "cube" structure, one should be able to better resolve patterns in large arrays of exponential decays.[12,147] The transformation of a bi-linear data set into a tri-linear one is illustrated in Figure 4. Several approaches have been proposed for obtaining a tri-linear data structure, DECRA[148] being the most basic one and SLICING and POWERSLICING providing enhanced performances.[143] For the subsequent three-way analysis of the tri-linear data sets, generalised rank annihilation (GRAM), PARAlel FACtor analysis (PARAFAC) and direct tri-linear decomposition (DTLD[98]) have been applied. Whereas GRAM can only be combined with DECRA,[148,149] DTLD can be applied to the different SLICING schemes for creating tri-linear data sets. Most applications of three-way analysis have been performed on data sets that were created from bi-linear ones. Truly tri-linear data sets can be provided by 2D T_2–D matrices,[150] which are directly suitable for three-way PARAFAC analysis.

7. BENCHTOP TIME-DOMAIN NMR APPLICATIONS

7.1. Solid fat content

The first (1975[26]) commercial TD-NMR application to foodstuff was the assessment of solid content in fats. A sample was placed in the field of a permanent (0.4 T) magnet, magnetisation was excited with a single pulse and the FID was

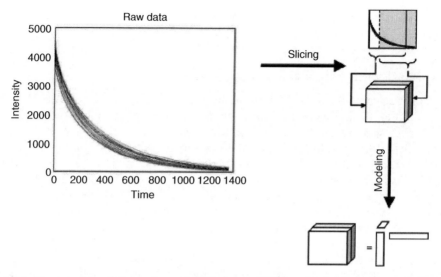

Figure 4 The principle of slicing. A set of NMR decays (left figure) is held in a matrix with each row corresponding to one measurement. This matrix is split so that the leftmost columns are in one table and the rightmost in another (upper right figure). These slices are generally overlapping, and when the lag is 1, the overlap is almost complete. The left slice has the original columns 1 to $J - 1$ and the right slice has columns 2 to J. The tables are put behind each other resulting in a tri-linear data set (lower right figure). (Reprinted from Reference 143, Copyright 2003, with permission from Elsevier).

acquired with a simple but effective diode detector.[37] The solid content was derived by a rather coarse data-processing procedure that involved recording the signal intensities at short (around 10 μs) and long (around 70 μs) relaxation time intervals.[37] Assuming that all solid signal decays before the second interval, the amount of solid phase can be calculated. Standards with known solid-to-liquid ratios are needed as calibration to correct for the non-negligible effect of the dead time. This is known to introduce a small systematic error in the SFC of fats with rigid crystals and rapidly decaying NMR line shapes. As an alternative, the so-called indirect SFC method samples the liquid part of the FID (around 70 μs) in the native and molten state, and thus calculates the amount of material. The current consensus is that the direct SFC method is precise ("small standard deviation") but slightly less accurate ("true") for specific lipid crystal forms, and the indirect method is more accurate but less precise.[151] For food technologists involved in rapid testing of product prototypes or quality/process control, precision is often of less importance than accuracy ("trueness") and thus the rapid direct SFC method is mostly preferred over the more cumbersome and less precise indirect SFC method. Both the direct and indirect SFC methods are based on rather crude treatments of the FID and do not exploit its full line shape. This has been addressed in several recent studies that demonstrate that, when lipid polymorphism is adequately taken into account,[99,152] the SFC can be determined without the use of standards (direct SFC) or melting steps (indirect SFC).

7.2. Shelf-life stability of foods

During their shelf-life, foods typically lose quality as a result of chemical, microbial and physical events. Examples of chemical deterioration are oxidation, discolouration and hydrolysis, which can lead to loss of sensorial quality and nutritional value. In time, also undesirable physical properties can develop such as caking, stickiness, chewiness and sogginess. Furthermore, products can be spoiled by microbial growth. In order to describe these events, two general concepts have been developed in food science. For predicting chemical and microbial stability,[153] one commonly deploys the concept of water activity (a_w), which is considered a good measure of the chemical potential of water[154] and can simply be derived from vapour pressure measurements. Attempts to relate water activity to molecular mobility as measured by NMR relaxometry have been successful only in some well-defined systems.[155–157] A systematic NMR and DSC investigation on freeze-dried meat at different a_w values revealed the pitfalls in interpretation of both techniques[158] in terms of molecular mobility. It has been stated that no fundamental relation exists between a_w and NMR relaxation; they just respond similarly to changing states of water.[154] For predicting physical and chemical stability, the glass/rubber transition temperature (T_g) has been proven useful. In most food systems, water is the major plasticiser of food polymers and its mobility as derived from NMR relaxation behaviour agrees well with a glass/rubber transition.[159,160] In food materials with low to intermediate moisture content, transversal relaxation becomes very efficient and the plasticising effect of water can be described within a theoretical framework.[161] Whereas T_g considers water-induced mobility on a structural and macromolecular level,[162] a_w is related to the molecular mobility of water itself.[163] Hence, for making prediction of food stability,[164] it has been recommended to consider both a_w and T_g.[165–167] The use of state diagrams based on transversal NMR relaxation has been proposed as an independent route to T_g and a_w.[168] This concept has successfully been applied to hardening and caking issues of powdered food ingredients and products.[169–171] Glass/rubber transitions also determine stability of foods in the frozen state. In most frozen food materials, non-equilibrium ice formation occurs with freeze-concentration of dissolved solutes in an amorphous liquid phase. Upon further decrease in temperature, the viscosity of this unfrozen amorphous phase increases and a solid glass is formed. This glass/rubber transition relates to many physical and chemical changes occurring in the freeze-concentrated phase. NMR relaxometry has been used to monitor the presence of "non-freezable" water in a range of model systems[172–175] and complex food products.[176–178]

7.3. Moisture and fat content

Benchtop NMR has become an accepted analytical tool for quantitative assessment of oil and moisture content in foods. The most classic application is the quantitative assessment of oil and moisture content in low-moisture ($<10\%$) food materials such as intact seeds.[179,180] By combining FID and spin–echo experiments, a set of signals is obtained from which oil and bound moisture content can be accurately and precisely derived.[181] Typically, samples need to be weighed and calibration measurements need to be performed on samples with known oil and bound moisture content.[180] Approaches that do not require sample weighing and calibration measurements have also been presented and are based on simple physical assumptions[182,183] or multi-variate models.[184]

In samples with abundant free (mobile) water such as in meat, fish and dairy products, the aforementioned relaxometric approaches fail in distinguishing the two mobile phases (oil and water).[185] In one approach, the contribution from mobile water is removed by microwave drying, and subsequently liquid oil is assessed by considering the remaining "mobile" part of the FID.[186–188] A commercial solution is available that integrates both the microwave drying step and the NMR measurement.[189] In a second approach, signals of oil and water are distinguished by combining relaxometric and diffusometric NMR measurements.[190,191] Whereas the relaxometric CPMG decay comprises contributions from both oil and water, diffusometric experiments can be tuned in such a manner that the resulting decay contains only signal from oil. The "one shot" m-PFGSE sequence[192] has been presented as an accurate and precise method to obtain a signal that corresponds to oil only. By subsequently considering the combined signal of water and oil in the CPMG decay, the moisture content also can be obtained.[193] As an alternative, a combined T_1 and T_2 measurement was carried out to produce a highly information dense data set that required multi-variate modelling to yield quantitative information.[4] CWFP has been proposed as an approach for rapid assessment of moisture in seeds and meat, but no commercial application has appeared yet. This is also the case for 2D T_1–T_2 and D–T_2 experiments, for which feasibility of separating contributions of fat and water in a range of dairy products[21,75] and meat[194] has been demonstrated.

7.4. Microstructural features in food emulsions

Figure 5 represents different cases how diffusion within the dispersed or continuous phase of food emulsions can be exploited to give microstructural information.[16] Intra-droplet restricted self-diffusion (Figure 5A) is most commonly exploited to infer DSDs. Mostly, droplet self-diffusion and inter-droplet diffusion (Figure 5B and C) are considered as nuisances in determination of DSDs. In many cases, these phenomena have been exploited for extracting microstructural features of food emulsions. Within double emulsions (W/O/W or O/W/O), several of the aforementioned diffusion mechanisms are active (Figure 5D). Within continuous phases, the droplet phase can hinder self-diffusion (Figure 5E), which can be exploited for deriving the long-range order of droplets.

The restricted self-diffusion of a liquid confined in a spherical droplet can be exploited to assess the DSD of a food emulsion. DSD determination by benchtop NMR instruments is gaining popularity (Table 3, References 16,104) because this method is precise, rapid and non-invasive, and can be run routinely by non-experts on commercial benchtop NMR instruments. Hence, DSD determination by benchtop NMR does not suffer from the drawbacks of other methods[195] that require cumbersome and invasive sample preparation procedures and

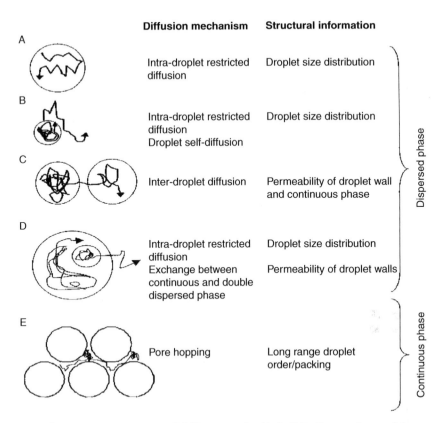

Figure 5 Schematic representation of different modes (A, B, C, D, E) to probe emulsion microstructure by PFG NMR (reprinted from Reference 16, Copyright 2009, with permission from Elsevier).

TABLE 3 Overview of applied benchtop PFG NMR droplet sizing studies

	Property	Field strength (MHz)	References
O/W	Temperature cycling stability	20	201–203
	Oxidative stability	20	204
	Oil droplets in cheese	60	205
W/O	Crystallisation inside droplets	20	206, 207
	Storage stability	20	208, 209
	Water droplets in butters	20	210
	Water droplets in margarines	20	197

trained operators. Implementation of DSD determination at the low-field strength of benchtop NMR equipment is not trivial, since water and oil signals cannot be resolved. TD filtering techniques are needed to suppress the signal of the continuous phase. In the case of W/O emulsions, the continuous water phase is successfully suppressed by means of an inverse recovery sequence that nulls the oil signal.[196-198] For O/W emulsions, such a relaxation filter cannot be used because of the dispersion of the relaxation times of the water phase. Hence, typically a diffusion filter is used that exploits the difference between the self-diffusion constants of water and oil.[65] Alternatively, multi-dimensional T_2–D measurements[89] can be used for separating oil and water signals. The PFG diffusion edited (PFG-DE)[199] sequence was indeed able to resolve dispersed/non-dispersed water and oil in brine/crude oil emulsions (at a field strength of 2 MHz), but no application to food emulsions has been demonstrated yet.

Most commercial benchtop NMR instruments are capable of providing pulsed field gradients between 2 and 4 T/m. This is sufficient for observing restricted self-diffusion of water molecules in W/O food emulsions. On some commercial systems, PFG performance is compromised by specifications of amplifiers and probeheads, and time-consuming gradient balancing procedures are required to obtain accurate PFG decays.[200] A significant gain in measurement time can be obtained by implementing an adequate calibration procedure.[38]

7.5. Functionality of food ingredients

The strengths of benchtop NMR relaxometry in assessment of phase-compositional and hydration phenomena have been exploited in a range of studies on food ingredient functionality (Table 4A). Phase behaviour of lipids and carbohydrates was one of the first applications in this area. Lipids can crystallise into three main forms: the metastable α polymorph and the more stable β and β' polymorphs. The potential of NMR to discriminate lipid crystal polymorphs was already recognised in a wideline study as early as 1957.[211] Since then, the effect of lipid polymorphism on ^{1}H NMR relaxation behaviour of the FID was recognised, but until recently no systematic investigations were carried out. Different models to fit the FIDs originating from different lipid polymorphs,[99,152] have proven to be effective for discriminating and quantifying α and β polymorphs of lipids. Combining the results from transversal (FID) relaxation data with longitudinal (T_1) relaxation data was shown to enable even better discrimination between α and β lipid crystal polymorphs. So far, discriminating β and β' polymorphs has turned out to be infeasible. Besides clearly defined crystalline and liquid states, NMR was also able to observe and quantify intermediate semi-solid phases.[212-215]

In a similar fashion as for lipids, TD-NMR relaxometry has also been deployed to observe crystallinity in carbohydrate-based food materials. T_1 relaxation has been shown to be sensitive for the presence of amorphous phases in sugars.[216] Transversal relaxation was used for the assessment of mobile/solid phases in concentrated carbohydrate/water systems, and several relaxation models were found to applicable to different mobility regimes.[101,102] Detailed information on hydrogen bonding in glassy oligosaccharide/water mixtures could be obtained from transverse relaxation NMR experiments performed at different temperatures and moisture levels.[217] Although starch can be considered as a carbohydrate, it deserves a separate discussion. Within starch, carbohydrate chains of amylopectin and amylase are organised in mesoscale crystalline and amorphous lamellae, respectively. Figure 6 shows the transversal relaxation time distribution

TABLE 4 Overview of transversal relaxometric studies of (A) food materials and (B) products which deployed analysis by means of fitting with discrete exponentials (DE), model functions (MF), continuous distributions (CD) or multi-variate analysis (MVA)

Material	Application	B_0	Acquisition	Data analysis	References
(A) Ingredients					
Encapsulates	Flavour and water content	23	FID, FIDCPMG	MVA	314, 315
Lipids	Phase-composition	20	FID-CPMG	MF	99
	Crystal polymorphism	20	FID	MF	100, 152
	Level of unsaturation	20	CPMG	DE	316
	Milk fat phase composition	20	FID, IR	DE, CD	317
		23	CPMG, CPMG T_1-Null	CD, MVA	318
	Milk fat solid content	20	FID	DE	213
Carbohydrates	Sugar polymorphism	20	IR	CD	216
	Mobility in sugar glasses	300	FID	MF	102
		10–223, 100, 300	FID, FFC	MF	319, 320
		23	FID, CPMG	DE	321
	Mobility in sugar syrup	23	FID	MF	322
	Mobility in low moisture sugar	300	FID		101
	Mobility in sugar solution	100	CPMG	DE	323
	Water dynamics in polysaccharide solutions	100	CPMG	DE	324
	Polysaccharide glass transition	20	FID, IR	DE	159, 160
	Polysaccharide-protein interactions	20	CPMG	DE	325
	Thickening/gelling in gels/solutions	20	PFG-SE	DE	326

(continued)

TABLE 4 (continued)

Material	Application	B_0	Acquisition	Data analysis	References
Proteins	Interaction with polysaccharides	20	CPMG	DE	327
	Denaturation	20	CPMG	DE	328–330
	Aggregation	20	CPMG	DE	331
	Protein content	20	CPMG	DE	332
	Gluten hydration	100	FID, IR	DE	225, 227
		300	IR, CPMG	DE	333
Starch	Bound water	30	FID, IR-FID	DE, MF	334
	Swelling, gelatinisation	20	CPMG, IR, FID	DE, MF	220
	Gelatinisation	15	FID, CPMG, IR	CD	335
	Pre-harvest condition	100	FID, CPMG	CD	336
	Granule structure	100	FID, CPMG	CD	218
		22	STE, PFGSTE	DE	218
	Retrogradation	20	FID, IR	DE	224, 337
		20	CPMG	DE	338, 339
		20	CPMG	CD, MVA	340
		23	CPMG	MVA	223
	Starch gel freezing	23	FID, CPMG	CD	221
	Starch film plasticisation	15	CPMG, FID	DE	173
	Mobility in glassy starch systems	23	FID, IR, GS	MF, DE	341
		23	FID	MF	342
	Chemical modification	20	FID, CPMG	CD	343
	Enzyme treatment and gel functionality	23	CPMG	DE, CD, MVA	344
	Interactions in gels	30	CPMG, IR	DE, MF	345

Lipid content in starch lipid composites	20	CPMG, SE	DE	346
(B) Food products				
Cereals				
Maturity of corn	200	CPMG	DE	347
Firming of rice	20	CPMG	DE	348
Water mobility in dough	20	FID, CPMG	CD	230, 233
	23	FID, CPMG	CD, DE	349
Water mobility in frozen dough	20	FID, CPMG	DE	177
Mixing and resting of dough	15	IR	DE	350
	20	CPMG	DE	235
Mixing and heating of dough	15	FID, CPMG, IR	DE	239
Frozen storage of dough	20	CPMG	DE	236–238, 351
Baking of bread	23	CPMG	MVA	240
Staling of bread	20	SR, CPMG, FID	DE	241
Water mobility in bread	20	IR,FID,CPMG	DE	352, 353
Water mobility in (model) bread	20	CPMG	CD	354, 355
Phase-composition of cake	20	FID-CPMG	DE	356
Hydration and leaching of breakfast cereals	20	CPMG	DE	357
Dairy				
WHC of dairy (models)	20	FID-CPMG, IR	DE	246
	20	CPMG	DE	244
	20	FID, CPMG, IR	DE	245
Water content of cheese	5	CPMG	DE	358, 359

(continued)

TABLE 4 (continued)

Material	Application	B_0	Acquisition	Data analysis	References
	Water distribution in cheese	20	CPMG, IR	DE	360, 361
		19	CPMG	DE, CD	362
	Cheese ageing	0.01–10	FFC	MF	363
	Phase composition of cheese	20	FID-CPMG	MF	364
	Yoghurt fermentation	21	IR, CPMG	DE	243
	Renneting of curd	20	CPMG	DE, CD	138
	Renneting of casein curd	20	CPMG, PFGSE	CD	365
	Effect of casein and fat on casein gels	20	PFGSE	DE	366
	Hydration and exchange in casein gels	300	CPMG, PFGSTE	MF	43
	Phase-behaviour of ice cream	20	FID+CPMG	DE, CD	247, 248, 367, 368
	Acidification of milk drinks	20	CPMG	DE	369–371
		23	CPMG	DE, MVA, CD	372
	Coagulation an syneresis	20	CPMG	DE, CD	138, 370
	Syneresis in dairy (model)	20	FID-CPMG	MVA	373, 374
	Water status in dairy (model)	20	FID, CPMG	DE	174, 375
	Phase composition in dairy (model)	20	FID-CPMG	MF	376, 377
Confectionary	Lipid migration in chocolate	15	CPMG	DE	378
	Hardening of bars	13	FID, CPMG	DE	168, 379
	Fat and water content in caramel	20	SR-CPMG	MVA	380
	Glass transition in caramel	20	FID	IR	381
Alcoholic beverages	Sugar and alcohol content	20	CPMG	MF	382

	Alcohol content	20	CPMG	MF	383
	Mashes and wort rheology	20	FID-CPMG	DE	384
Vinegar	Ageing and authenticity	0.01–80	FFC	DE	385
Food powders	Solubility, rehydration	15	CPMG	DE	386
		10	CPMG	DE	229
	Caking	20	FID	DE	169–171
Protein films	Plasticisation by glycerol	200	FID	MF	387
Horticultural products	Apple drying/freezing	100	CPMG	CD	41
	Apple osmotic drying	15	CPMG, IR	DE, CD	265
	Apple mealiness and ripening	23, 100, 300	CPMG, IR-CPMG, PFGSE-CPMG, IR-PFGSE-CPMG, CPMG-store-CPMG	CD	81
	Apple internal browning/watercore/bruising	5	CPMG	DE	253
		5	CPMG, PFGSE	DE	252
	Apple internal browning/online quality	5	CPMG	DE	30
	Apple mealiness	100, 200	CPMG	CD	260
	Orange storage	10	CPMG, IR	DE	254
	Water mobility and genetic/seasonal origin of tomato	20	CPMG, PFGSE-CPMG	DE, CD	83
	Tomato quality/firmness	4	CPMG	DE	255
	Banana ripening	20	CPMG, PFGSE, PFGMSE	DE	87
	Apple/strawberry Brix	23	PFGSE, CPMG T_1-Null, PFGSE-CPMG	CD, DE	47
	Apple parenchyma microstructure	30	CPMG, IR, PFGSTE-CPMG, IR-CPMG, IR-PFGSTE-CPMG	DE	86
		20	CPMG, SR, SR-CPMG	DE	73

(continued)

TABLE 4 (continued)

Material	Application	B_0	Acquisition	Data analysis	References
	Apple, mung bean seedlings—diffusion constants	20	PFGMSE-CPMG	DE	84
	Avocado maturity	23, 300	CPMG, IR-CPMG, SE T$_1$-Null, PFGSE-CPMG, PFGSE	CD	46
	Diffusive/chemical exchange in courgette/apple/onion	100	CPMG	DE	40
	Bean cell wall physical structure	90	FID, SldE, CPMG, IR, JB, GS	DE	53, 55
	Bean cell wall physical structure/changes during growing	90	FID, SldE, JB, CPMG	DE	54
	Bean cell wall physical structure/chemical modification	90	SldE, JB	DE	56
	Carrot drying/freezing	100, 300	CPMG	CD	42
	Carrot drying	20	SR	MF	44, 261–263
	Carrot heating and pressure treatment	23, 100	CPMG, IR-CPMG, CPMG-store-CPMG	CD	79
	Chinese chestnut cell wall molecular mobility	100	IR, SL, SldE	DE, MF	59
	Leaf water activity	20	SR	DE	388
	Pear browning	23, 100, 300	CPMG, IR-CPMG, PFGSE-CPMG	CD	80

	Potato cooking	23	CPMG	DE, MVA, CD	268, 269
	Potato sensorial quality	23	CPMG	DE, CD, MVA	259
		23	CPMG	DE, MVA	257, 258
	Potato dry matter	23	CPMG, IR	DE, MVA	256
	Water content in fried potatoes	10	CPMG	DE, CD	389
	Potato freeze-drying	20	SR	MF	264
	Potato freezing	100	FID, CPMG	CD	178
	Potato high-pressure processing	23, 300	IR-CPMG, CPMG	CD	267
	Potato cell-wall molecular mobility	100	IR, SR, SL, SldE	DE	57, 58
	Strawberry high-pressure processing	23, 300	CPMG	CD	266
Fish	WHC	23	IR, CPMG	DE, MVA	277
	Smoking	23	CPMG	MVA	311
	Salting	20	CPMG	MVA, CD	309–311, 390
	Storage and processing	20	CPMG	DE	307, 391
	Biological variation of lipid content	20	CPMG	MVA	185
Meat	Surimi gelation	20	CPMG	DE	312, 313
	Freeze-drying	20	CPMG	CD	392
	WHC	23	CPMG	CD	274–276
	Meat quality	20	CPMG	MVA	146
	Seasonal/geographic origin	23	CPMG	MVA	185
	Moisture content	23	FID,CPMG, IR	DE, MVA	144
	Pre-slaughter diet and meat quality	23	CPMG	CD	288

(continued)

TABLE 4 (continued)

Material	Application	B_0	Acquisition	Data analysis	References
	Pre-slaughter age and meat quality	23	CPMG	CD, MVA	393, 394
	Genetic factors and meat quality	23	CPMG	CD, MVA	286, 287
	Post-mortem conditions	20	CPMG	CD, MVA	289–292
	Meat processing	23	CPMG	CD, MVA	293, 294
		30	CPMG, IR	DE	295
	Frozen storage	23	CPMG	CD	296
		23	CPMG	DE, CD, MVA	297
	Meat ageing	23	CPMG	CD, MVA	298, 299
	Freeze-drying	20	CPMG	CD	158
	Cooking	23	CPMG	CD, MVA	297, 301, 304, 305, 395
		23	CPMG	MVA	300
		20	CPMG	DE	302, 303
	Fat content in beef	85	CWFP	MF	282
Eggs	Storage	23, 300, 10-30	CPMG, IR, FFC	DE, CD	396
	Cooking	23	IR-CPMG	CD	76
Coffee beans	Water mobility	23	CPMG, FID	DE, ME, CD	397
Seeds	Oil, water, protein content	23	FID-CPMG	DE, MVA	184
	Water, oil content	20	SE, CPMG	DE	181
	Oil content	85	CWFP	MF	398

Abbreviations of acquisition schemes are explained in Sections 4 and 5, and Table 2.

for a typical starch granule suspension. For assigning the different transversal time populations one needs to take into account the multi-scale organisation of carbohydrates,[218] and descriptions in terms of bound and free water are an oversimplification.[219] Within Figure 6, the populations with relaxation times at <10, 80 ms and seconds correspond to intragranular water, extragranular water and supernatant, respectively. The intragranular water is in exchange between the amorphous growth rings and the amorphous regions within the semi-crystalline lamellae. Whether these populations can be observed separately depends on the temperature (Figure 6) and the type of starch.[218,220] These insights have been used to study structural events during melting of starch granules and during heating (gelatinisation). The resulting amorphous starch gels have a strong tendency to recrystallise. This process is known as retrogradation and has also been extensively described in a quantitative manner by NMR relaxometry in (model) starch gels[221–224] (Table 4A).

NMR relaxometry has also been applied to study hydration and plasticisation of food proteins, in particular in relation to functionality of gluten.[225–227] The aforementioned NMR state diagrams have been used to predict the performance of proteins when used in powders and bars.[168] Rehydration behaviour of dairy proteins can also be studied under dynamic dissolution conditions by monitoring transverse relaxation behaviour under stirring.[228,229] Thus insights were gained on the effects of different drying routes on dissolution and rehydration behaviour of milk powders.[229]

7.6. Functionality of heterogeneous food products

NMR at low field strengths is often denoted as a "low-resolution" technique. This may be true when one considers resolution in the frequency domain, but TD-NMR relaxometry has successfully been applied to a wide range of compositional

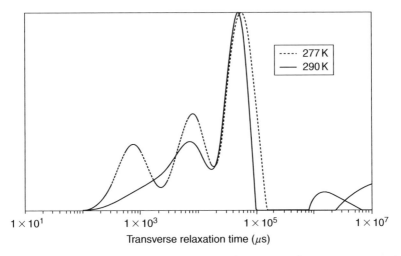

Figure 6   The distribution of water proton transverse relation times for a water-saturated packed bed of potato starch granules at 277 and 290 K (reprinted from Reference 218, Copyright 2000, with permission from Elsevier).

and structural heterogeneous food systems (Table 4B). In the next sections, the most important food product application areas will be discussed briefly. The emphasis will be on studies that aimed at resolving relations between structure and product functionality; hence, studies in which quantitative methods have routinely been applied (Sections 7.1–7.4) are out of scope.

7.6.1. Cereals

Cereals are staple foods that provide the global population with a major source of carbohydrates. Water is an important constituent of cereals, which critically determines their behaviour after harvesting, processing and during shelf-life. NMR has been applied widely to probe the dynamics of water in cereals,[231] and a major part of the TD relaxometry in this area has been focussed on relating water mobility in flour, dough and bread[230,232,233] to final product quality (Table 4). Figure 7 shows the T_2 distributions obtained for dough and its starch (carbohydrate) and gluten (protein) constituents. One can observe that gluten and starch have distinct T_2 distributions and that in dough significant averaging takes place. From such data one can conclude that starch determines the distribution of water within the dough,[230] but in most studies on the effects of dough processing and storage, changes in T_2 relaxation behaviour were related to interactions between water and gluten. The first stage in bread making involves mixing flour with water and kneading of the dough. During kneading, an elastic gluten network is formed,[234] and its strong interaction with water can be monitored by NMR relaxometry.[235] The quality of the gluten network is determined by the kneading time and the subsequent resting of the dough, and these effects also can be

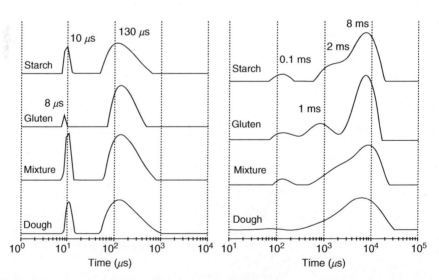

Figure 7 Continuous T_2 distributions obtained from FID (left) and CPMG (right) experiments of dough and its constituents (reprinted from Reference 230, Copyright 2007, with permission from Elsevier).

monitored by T_2 relaxation.[235] Frozen storage of dough compromises the quality of the final bread, and NMR relaxometry indicated that this can be attributed to dehydration of gluten due to ice crystal formation. This effect has been monitored in the frozen state,[177] but also after thawing, where water has been observed to become separated from the gluten network, thus increasing the transversal relaxation of water.[236–238]

During baking of the dough, the starch gelatinises and this has a profound effect on the mobility of water.[239] In "NMR-baking" experiments, CPMG decays are recorded during baking, and a MVA has provided insight into the events occurring during baking and their relation to final textural quality.[240] In the final baked bread, slow deterioration of quality occurs, where staling is a dominant factor. During staling, starch undergoes recrystallisation (retrogradation) and this also involves redistribution of water. This effect has been predominately studied in (model) starch gels,[221–224] but also in its full complexity in bread.[222,241]

7.6.2. Dairy products

Dairy products such as milk, butter, yoghurts and cheese offer a rich source of proteins, fat and minerals, and can occur in liquid, semi-solid and solid form. In this wide range of products, benchtop relaxometry has proven itself as a versatile tool to assess phase-compositional and water-redistribution phenomena.[242] In liquid milk, TD-NMR relaxometry has been used to study effects of acidification on protein hydration. When milk is further processed to diary products, NMR offers a unique and non-invasive tool to monitor and characterise the resulting dramatic structural rearrangements. A critical step in dairy production is the formation of a casein protein gel. In yoghurt, such gels are formed by lactic acid fermentation, which reduces the pH of milk, thus causing casein micelles to aggregate and form a gel network. NMR relaxometry showed a strong effect of yoghurt fermentation on water mobility, which was explained in terms of network formation and changes in casein hydration.[243] In cheese manufacturing, formation of a gel network is induced by enzymatic activity, the so-called renneting. In the resulting curd, compartmentalisation of water takes place, and the concomitant redistribution of water can be monitored in a straightforward manner by transversal NMR relaxometry[138] (Figure 8). Such studies have provided insight in the microstructural events occurring during water loss or syneresis of dairy products. Water holding capacity (WHC) remains an important quality parameter for many products, and rapid routine NMR methods have been proposed as alternatives for cumbersome and slow conventional methods.[125,228,244–246] Diffusometric water droplet sizing has become a routine quality control method for butter and margarines.[16] A similar method for assessment of oil DSDs in water-continuous dairy products is now gaining popularity.[16] Relaxometric approaches have also been used to assess the overall phase composition in dairy products. In this respect, the complex composition and frozen state of ice cream (models) presented a particular challenge. Nevertheless, benchtop NMR relaxometry has successfully been applied to study crystallisation of fat and water in ice cream (models) during freezing and frozen storage.[247–249]

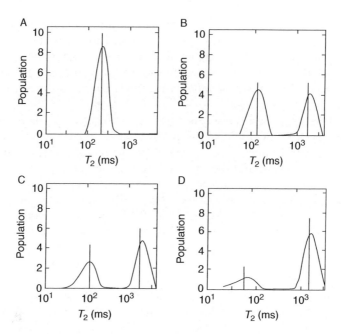

Figure 8 Continuous and discrete T_2 fits of rennetted milk before syneresis (A) and during the time course of syneresis at (B) 137 min; (C) 200 min; (D) 405 min (reprinted with permission from Reference 138. Copyright 1993, American Chemical Society).

7.6.3. Horticultural products

Fruits, vegetables, tubers and beans are challenging food stuffs with respect to maintaining quality during the chain between field (or greenhouse) and consumer or processing plant. Within this supply chain, a strong need exists for rapid, robust and cost-effective quality inspection tools.[250,251] TD-NMR holds promise in this area since it is sensitive to many of the undesirable structural events that can occur between the field and the fork. TD-NMR has already found widespread use in mechanistic studies on microstructural effects of processing routes such as air-drying, freeze-drying and osmotic drying. Both applications of NMR have benefited from early work where "numerical plant cell models" were established that relate relaxation and diffusion behaviour of water to plant tissue morphology (Section 6.2), but also descriptive multi-variate modelling has been applied.

Quality inspection. The aforementioned plant cell models opened up the possibility for non-invasive quality assessment of horticultural products. Low-field conventional, "stationary" NMR relaxometry and diffusometry have been applied to assess internal browning and watercore in apples.[252,253] Changes in T_2 components associated with different water compartments in affected apples could be associated with a movement of water from vacuoles into cytoplasm and extracellular spaces.[253] Effects on T_2 and T_1 values of juice sacs of oranges held promise for detection of freeze damage.[254] The effect of defects on the firmness of tomato has been correlated with changes in the T_2 relaxation times.[255] Dry matter content is an important quality parameter for many fruits and vegetables and could be determined in a non-invasive manner by multi-variate modelling of CPMG decays of potatoes.[256] Multi-variate models based on CPMG decays obtained from raw potatoes successfully predicted the sensory texture[257,258] and other quality parameters[259] of cooked potatoes.

Besides standard T_2 CPMG and T_1 methods, more sophisticated relaxation and diffusion NMR techniques were also tested to assess feasibility of fast online assessment of quality factors such as maturity, oil content and presence of hard lumps in avocado.[46] Two-dimensional T_2–T_1 correlation measurements proved to be adequate for determination of oil content in avocado tissue, but the long acquisition time excluded it from online implementation. A single-shot method that employed T_1-nulling of the water magnetisation offered advantages with respect to measurement time. Another single-shot technique suppressed the water signal by means of a diffusion filter and gave a satisfactory linear correlation with oil content. Water suppression by the diffusive attenuation was also applied to measure the percentage of soluble solids (Brix), which is a measure of ripeness of apple and strawberry.[47] The observation that sugar diffusivity is significantly lower than that of the vacuolar or cytoplasmic water fraction was used in a PFG method to selectively suppress water in the cellular tissue. PFGSE-CPMG has been used to probe the effect of the variety and harvest period effects of tomato.[83]

Several 1D and 2D relaxation and diffusion protocols were also tested to find the most sensitive technique for the detection of mealiness and ripening in apples.[81,260] The loss of membrane integrity upon internal browning in pear tissues was manifested by changes in relaxation time distributions in T_2–T_1 and T_2–D correlation measurements.[80] These correlation measurements demonstrated that for damaged pear tissue, proton pools were grouped into a lower amount of populations; shorter T_2 values and increased diffusion coefficients were observed.

Effects of processing on morphology. Already two decades ago, TD-NMR relaxometry was used to study air-drying[44,261–263] and freeze-drying[42] of carrots. Relaxometry could detect sublimation of the frozen core and removal of non-frozen water during freeze-drying. Also anomalous freeze-drying of potato could be detected.[264] A similar approach was used to study the osmotic dehydration of apple.[265] These studies were followed by more sophisticated approaches where the concepts of non-freezing water and three-compartment relaxation/diffusion models were deployed. CPMG relaxation measurements[178] localised non-freezing water in cellular tissue of potato in cell walls and starch granules. T_2 relaxation measurements were used to study changes in sub-cellular water compartmentation and cell membrane integrity after applying air-drying, freeze-drying, freeze–thaw processing and rehydration in apple tissue.[41] Deployment of the three-compartment plant cell model[121] showed that water loss from the vacuole was accompanied by an overall volume shrinkage of the whole cell. Coalescence of T_2 relaxation time distributions and reduction in T_2 relaxation time values were evidence of loss of membrane integrity in rehydrated apple tissue after freeze-drying or freeze–thaw processing. This approach was also applied to carrot parenchyma tissue[42] which, unlike potato[178] and apple,[41] does not contain large numbers of starch granules or intercellular air gaps. Carrot tissue distinguishes itself due to the high levels of dissolved sugar in the vacuole and this strongly affects relaxation behaviour during drying and freezing. The observed decrease in relaxation times during drying was a consequence of vacuolar shrinkage and the progressive concentration of vacuolar sugars. The latter enhances proton exchange between water and dissolved sugars. The relaxation behaviour of carrot tissue during drying could be modelled by incorporating two-site exchange into the numerical plant cell model.

Microstructural effects of high-pressure treatment were investigated by means of 1D (CPMG[266]) and 2D (T_1–T_2[267] and T_2–T_2[79]) relaxometry. The use of 2D methods also requires assignment of relaxation components to plant structural features, which is a non-trivial task.[78] Effects of potato cooking were studied using transversal (CPMG) relaxometry combined with multi-variate modelling.[268,269] Effects of banana ripening during storage were studied by relaxometry (CPMG), diffusometry (PFGSE) and a combination thereof (PFGMSE-CPMG) and results were translated into microstructural parameters using the numerical plant cell model.[87]

Effect of cell wall plasticity on physical properties. Cell walls play an important structural role in plant cells and they are also the source of dietary fibre in foods. Molecular mobility in cell wall components such as cellulose, hemicellulosic polysaccharides, pectic polysaccharides, protein and water has been characterised by NMR relaxation behaviour as measured by FID, SldE, IR and JB methods.[53,55] These NMR studies also assessed the impact of chemical fractionation on the physical properties and the extent of damage of the etiolated hypocotyl of bean cell wall fragments.[56] Similar NMR relaxation approaches indicated an increase of more rigid components of the cell wall matrix, an increase in cellulose crystallinity and more rigid association between cellulose and hemicellulose during growth of the etiolated hypocotyls of beans.[54] The parameters affecting molecular mobility of the functional groups in cell wall biopolymers have been studied in pectin materials and cell wall materials from potatoes[57,58] and Chinese water chestnut.[59] The observation of characteristic maxima in the temperature dependency of T_2, T_1 and $T_{1\rho}$ relaxation rates yields information on molecular motion of different chemical groups. Thus the impact of hydration on the cell wall components was shown, as well as the partial dissolution and/or chemical degradation disrupting the structure of the cell wall, with the cellulose component unaffected. It was found that non-freezing water was mainly associated with the pectic materials, and to a lesser extent with cellulose which is mostly in an anhydrous crystalline form.[57] A similar hydration effect was found in T_2, T_1 and $T_{1\rho}$ relaxation studies of the Chinese water chestnut.[59]

7.6.4. Fish and meat

Fish and meat have supply chains where many factors can determine quality of the final product on the plate of the consumer. WHC or drip loss is an important quality parameter which is strongly determined by the distribution of water over different structural components in meat and fish.[270,271] Juiciness and tenderness are other quality parameters that strongly depend on water distribution.[271] TD-NMR relaxometry typically reveals three distinct water populations in meat and fish: water associated with proteins ($1 < T_{2b} < 10$ ms), water trapped within the myofibrils ($40 < T_{21} < 60$ ms) and water outside the myofibrillar lattice ($140 < T_{21} < 400$ ms).[272,273] Water distributions in meat and fish are typically analysed by straightforward CPMG; in one case also fgCPMG has been applied.[66] Longitudinal relaxation has found only limited applications in meat science and is not likely to gain more attraction.[270] Most data-analysis approaches rely on fitting of discrete exponentials of continuous distributions to obtain the aforementioned water populations. Thus quantitative correlations between NMR relaxation behaviour and WHC of meat[274–276] have been established. Multi-variate approaches have also been successfully deployed to establish the relations between relaxometric decays and quality parameters of fish and meat.[146,277–280]

Quality inspection of meat and fish. Moisture and fat levels are important quality parameters for meat and fish, and NMR was early recognised as a potential measurement tool.[281] Nowadays, routine benchtop NMR method for rapid and non-invasive quantitative assessment of moisture and fat are becoming accepted in the meat and fish industry (see Section 7.3). CWFP has been presented as a more accurate alternative for conventional relaxometric assessment of fat content in meat, but routine application has not been implemented yet.[282] The industry has identified a need for assessment of meat and fish quality in the truly non-invasive "sensor" mode.[283] The feasibility of measuring moisture and fat content in meat[284] and (live) fish[285] by a means of non-invasive single-side NMR sensors has already been demonstrated.

Effect of pre-slaughter and post-mortem conditions on meat. It is well known that diet, genotype and age of animals can have a strong effect on meat quality. Genotype determines water redistribution during post-mortem chilling[286] and cooking.[287] Dietary creatine has a genotype-dependent effect on post-mortem water distribution and the final meat quality.[288] Developmental stage and muscle type have a strong effect on water distribution within muscles. Indeed, age at slaughtering determines water distribution in meat and final quality attributes. After slaughtering, the muscle goes into rigor and this involves major rearrangement in the muscle structure. The impact on water distribution can be sensed in a quantitative and detailed manner by NMR relaxometry,[273,289–291] which is illustrated in Figure 9. This has been exploited in TD-NMR studies where effects of post-mortem conditions such as pH, chilling[289] and carcass handling[292] on water distribution were studied.

Effect of processing of meat. Freeze-drying of meat is a route towards extending shelf-life, and NMR has been used to assess the fate of water.[158] The effect of industrial meat processing such as curing,[293,294] ultrasound treament[295] and tumbling on meat microstructure and final quality has also been studied by NMR. Frozen storage[296,297] and ageing[298,299] can also profoundly change the water distribution in meat. The ultimate meat processing step is cooking, which has been investigated by "NMR-cooking" studies. These explorative studies deployed dynamic in situ measurements of CPMG relaxation decays in

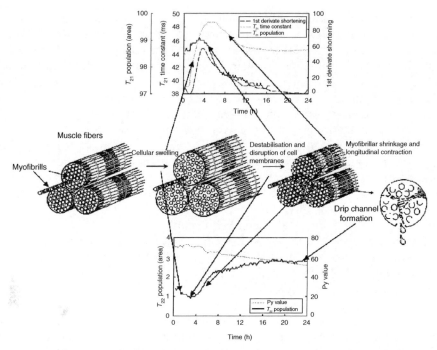

Figure 9   Illustration of proposed mechanisms in the post-mortem reorganisation of muscle fluids and its relation to observed changes on T_{21} and T_{22} populations (reprinted from Reference 273, Copyright 2004, with permission from Elsevier).

combination with MVA. Thus a new water population that develops in meat during cooking was revealed.[300] This phenomenon was assigned to protein denaturation and heat-induced changes in water mobility in studies where NMR was combined with DSC[301–303] and FT-IR.[304,305]

Effect of processing and storage of fish. Many of the insights into the microstructural basis of the NMR relaxation behaviour of meat can be translated to fish.[306] Also, in fish one mostly observes three components in the transversal relaxation curves which can be assigned in a similar manner as for meat. Not surprisingly, WHC of fish can also be related to transverse relaxation behaviour in a quantitative manner.[277] The effect of frozen storage had a clear effect on the water distribution as measured by NMR, and clear relations to textural quality were found.[307] The effect of fish preservation steps such as salting, smoking and freeze-drying have also been extensively studied by NMR relaxometry. As shown by NMR, salting can induce significant effects on the water mobility in fish,[308] and an interaction takes place with microstructural damage due to freezing.[309,310] Smoking of fish involves both salting and heat treatment, and their effects on water distribution were also studied by NMR.[311] Effects of heat- and pressure treatment during preparation of surimi were studied by NMR, which revealed effects of denaturation and aggregation of fish proteins.[312,313]

8. MOBILE TIME-DOMAIN NMR APPLICATIONS

8.1. Unilateral NMR sensors

Unilateral magnet devices present an attractive option for non-invasive assessment of compositional and microstructure of food materials. These devices allow easy sample access and are portable, which makes them attractive for quality and process control in industrial environments. A main handicap of these devices is magnet field inhomogeneity, which significantly compromises sensitivity and robustness under practical circumstances. FIDs obtained by such devices are also compromised by the very short T_2^* values and thus typically echo techniques are required. In most (commercial) applications, the unilateral NMR sensors suffer from relatively long dead times, which impede direct observation of (semi-solid) components.[399] Recent developments in unilateral magnets design[400] offer, however, larger sensitive volumes, higher sensitivity and shorter dead times.

The first applications of the unilateral NMR devices in foods were aimed at obtaining compositional information in a through-package manner, so the product could be analysed in sealed conditions. Quantitative information was successfully obtained on the fat/water content of food emulsions,[401–404] as well as oxygen content of bottled beverages.[405] A unilateral NMR sensor was also able to assess in vivo the fat content in both live and slaughtered fish, demonstrating the feasibility for online quality control.[285] Relaxometry and diffusometry using a single-side magnet sensor were employed to investigate the ripening of grapes, water content in chicken and diffusion anisotropy in asparagus and in bovine tendon.[284] The NMR MOUSE was also able to asses the microstructural quality of a food emulsion in a through-package mode.[406]

8.2. Online quality inspection

Online NMR sensors are of particular interest for quality control of food products on a large industrial scale. Most of the studies were carried out on fruits moving continuously on a conveyor belt with the aim of sorting them for internal damage or defects.[30,407] Issues with respect to industrial implementation arise because of conveyor speed, magnetic field homogeneity, rf coil design and sample polarisation time. The concept of motional relativity has been demonstrated to overcome some of the aforementioned problems. The method involves a constant rf field and the translation of the sample is exploited to obtain the NMR signal.[408,409] This brings significant advantages for online applications such as the possibility to work with conveyor speeds of meters per second; also some of the rf hardware limitations are eliminated because of continuous irradiation, that is, no eddy currents and no coil ring-down effect. Within this online approach, the FIRE sequence[409] is able to measure both T_1 and T_2 in a single shot by using an inversion recovery and signal read-out with a truncated CPMG train. Yet, no commercial food application has been described so far. Another recent online approach is based on a CWFP technique, which was applied to measure the oil content of intact seeds.[398] The method was deployed on a high-field spectrometer and more than 20,000 samples/h can be potentially analysed. So far, only proof of principle has been demonstrated, but the feasibility of deployment of CWFP in industrial environments is not yet proven.[49-51]

9. PERSPECTIVES AND CONCLUSIONS

9.1. Hardware

The current developments in magnet design will continue to deliver tailored solutions for specific applications. One can envisage more specific solutions[410,411] for single-sided unilateral magnet designs where the proof of principle of achieving frequency-domain resolution has been demonstrated. The homogeneous field provided by Halbach designs may be used for more generic applications.[412] Affordable solutions for locking and shimming[413] of such magnets will ultimately enhance signal quality and provide frequency resolution.[414] These routes overcome some intrinsic limitations of current TD-NMR methods in resolving different chemical components. Further steps in down-sizing current benchtop NMR equipment can be expected when developments in rf microcoils[415-417] are exploited.

9.2. Fast measurements

Several rapid, basic, single-shot approaches have already found practical TD-NMR applications.[46,192] Further improvement of current instrumental benchtop NMR equipment will enable application of more rapid and advanced single-shot measurements. Such techniques will be of particular interest for online and/or real-time applications. An example is the Difftrain sequence,[418] which allows rapid acquisition of diffusometric PFG decays. This method employs small flip angle excitation for multiple acquisition of longitudinally stored magnetisation. The large number of gradient decoding steps requires better gradient amplifier and coil specifications than currently available on most commercial benchtop instruments, however. Multiple modulation-multiple echo (MMME) is an elegant ultrafast single-shot[419-421] technique to measure relaxation times and diffusivities without the need for phase cycling. MMME is rather an NMR acquisition scheme portfolio than a single-pulse sequence, as it is able to provide rapid measurements of diffusion, flow and imaging. The sequence consists of a set of rf pulses with unequal time spacings that generate a maximal number of spin–echoes.[422] Due to the fact that information about relaxation and diffusion is available from a one-shot measurement, it is expected that this approach will soon enter the field of food science and technology. The CWFP technique has been demonstrated in feasibility studies on seeds[398,423] and meat,[282] and here other foods applications also can be envisaged.

9.3. In situ measurements

Currently, the hardware geometry of current benchtop NMR equipment compromises real-time measurements under dynamic conditions such as shear[424] or variable temperature.[425] The aforementioned advances in magnet and rf coil design, as well as rapid acquisition approaches, will allow further exploitation of the non-invasive nature of NMR under dynamic circumstances. Currently, most dynamic in situ magnetic resonance measurements are in spatially resolved diffusometric of relaxometric mode at high-field, that is, rheo-MRI,[426,427] but there is no compelling reason to preclude implementation at the benchtop NMR mode whether in bulk or spatially resolved mode.

9.4. Enhancing information yield from the time-domain

The current arsenal of 2D sequences will expand with experiments that can reveal specific structural features. An example is the incorporation of field cycling into 2D correlation experiments.[428] Major developments can be expected in strategies for assigning signals in the resulting 2D correlation plots.[77] Processing the acquired TD data sets into meaningful diffusometric/relaxometric correlation plots will remain a challenge. Further breakthroughs can be expected from multi-variate approaches that exploit the tri-linear structure in such data sets.[150]

ACKNOWLEDGEMENT

Ewoud van Velzen (Unilever R&D, Vlaardingen) is acknowledged for critically reading the manuscript and providing useful comments.

REFERENCES

1. J. P. M. Van Duynhoven, *Encyclopedia of Spectroscopy and Spectrometry*. Elsevier, Oxford, UK, 2010.
2. A. M. Gil, in: *Encyclopedia of Food Sciences and Nutrition*, (C. Benjamin ed.), 2003, p. 5447. Academic Press, Oxford.
3. H. Todt, W. Burk, G. Guthausen, A. Guthausen, A. Kamlowski and D. Schmalbein, *Eur. J. Lipid Sci. Technol.*, 2001, **103**, 835.
4. G. Guthausen, H. Todt, W. Burk, D. Schmalbein and A. Kamlowski, in: *Modern Magnetic Resonance*, (G. Webb ed.), 2008, Springer, The Netherlands.
5. Y. Vodovotz, L. C. Dickinson and P. Chinachoti, *J. Agri. Food Chem.*, 2000, **48**, 4948.
6. Y. Vodovotz, E. Vittadini and J. R. Sachleben, *Carbohydr. Res.*, 2002, **337**, 147.
7. P. S. Belton, *Pure Appl. Chem.*, 1997, **69**, 47.
8. A. Nordon, C. A. McGill and D. Littlejohn, *Analyst*, 2001, **126**, 260.
9. A. Nordon, C. A. McGill and D. Littlejohn, *Appl. Spectrosc.*, 2002, **56**, 75.
10. M. J. McCarthy, P. N. Gambhir and A. G. Goloshevsky, in: *NMR Imaging in Chemical Engineering*, S. Stapf and S.-I. Han (eds.), 2009, p. 471. Wiley, Weinheim, Germany.
11. H. Todt, G. Guthausen, W. Burk, D. Schmalbein and A. Kamlowski, *Food Chem.*, 2006, **96**, 436.
12. R. Bro, F. van den Berg, A. Thybo, C. M. Andersen, B. M. Jorgensen and H. Andersen, *Trends Food Sci. Technol.*, 2002, **13**, 235.
13. E. Fukushima, in: *Magnetic Resonance Microscopy: Spatially Resolved NMR Techniques and Applications*, S. Codd and J. Seymour (eds.), 2009, p. 1. Wiley, Weinheim, FRG.
14. B. Blumich, J. Perlo and F. Casanova, *Prog. Nucl. Magn. Reson. Spectrosc.*, 2008, **52**, 197.

15. M. L. Johns and K. G. Hollingsworth, *Prog. Nucl. Magn. Reson. Spectrosc.*, 2007, **50**, 51.
16. A. Voda and J. P. M. van Duynhoven, *Trends Food Sci. Technol.*, 2009, **20**, 533.
17. Y. Q. Song, *in: NMR Imaging in Chemical Engineering*, S. Stapf and S. I. Han (eds.), 2006, p. 163. Wiley, Weinheim, Germany.
18. R. R. Milczarek and M. J. McCarthy, *in: Magnetic Resonance Microscopy, Spatially Resolved NMR techniques and Applications*, S. Codd and J. Seymour (eds.), 2009, p. 289. Wiley, Weinheim, Germany.
19. H. C. Bertram, R. L. Meyer and H. J. Andersen, *Magnetic Resonance in Food Science: Challenges in a Changing World*. Royal Society of Chemistry, Cambridge, 2009, p. 241.
20. F. Mariette, *in: Modern Magnetic Resonance*, (G. A. Webb ed.), 2006, Springer, The Netherlands.
21. Y. Q. Song, *Prog. Nucl. Magn. Reson. Spectrosc.*, 2009, **55**, 324.
22. B. P. Hills, *Ann. Rep. NMR Spectrosc.*, 2006, **58**, 177.
23. L. Andrade, W. MacNaughtan and I. A. Farhat, *Magnetic Resonance in Food Science: From Molecules to Man*. Royal Society of Chemistry, Cambridge, 2007, p. 114.
24. D. Chapman, R. E. Richards and R. W. Yorke, *J. Am. Oil Chem. Soc.*, 1960, **37**, 243.
25. W. A. Bosin and R. A. Marmor, *J. Am. Oil Chem. Soc.*, 1968, **45**, 335.
26. K. van Putte and J. van den Enden, *J. Am. Oil Chem. Soc.*, 1973, **51**, 318.
27. K. J. Packer and C. J. Rees, *J. Colloid Interface Sci.*, 1972, **40**, 216.
28. J. P. Renou, A. Briguet, P. Gatellier and J. Kopp, *Int. J. Food Sci. Technol.*, 1987, **22**, 169.
29. P. Fairbrother and D. N. Rutledge, *Analusis*, 1993, **21**, 113.
30. W. Chayaprasert and R. Stroshine, *Postharvest Biol. Technol.*, 2005, **36**, 291.
31. K. Halbach, *Nucl. Instrum. Methods*, 1980, **169**, 1.
32. K. Halbach, *Nucl. Instrum. Methods Phys. Res.*, 1981, **187**, 109.
33. H. Raich and P. Blumler, *Concepts Magn. Reson. Part B Magn. Reson. Eng.*, 2004, **23B**, 16.
34. G. Moresi and R. Magin, *Concepts Magn. Reson. Part B Magn. Reson. Eng.*, 2003, **19B**, 35.
35. B. P. Hills, K. M. Wright and D. G. Gillies, *J. Magn. Reson.*, 2005, **175**, 336.
36. J. Mitchell, P. Blumler and P. J. McDonald, *Prog. Nucl. Magn. Reson. Spectrosc.*, 2006, **48**, 161.
37. K. P. A. M. van Putte and J. van den Enden, *J. Phys. E*, 1973, **6**, 910.
38. J. P. M. van Duynhoven, B. Maillet, J. Schell, M. Tronquet, G. J. W. Goudappel, E. Trezza, A. Bulbarello and D. van Dusschoten, *Eur. J. Lipid Sci. Technol.*, 2007, **109**, 1095.
39. D. N. Rutledge, *J. Chimie Phys. Phys. Chim. Biol.*, 1992, **89**, 273.
40. B. P. Hills and S. L. Duce, *Magn. Reson. Imaging*, 1990, **8**, 321.
41. B. P. Hills and B. Remigereau, *Int. J. Food Sci. Technol.*, 1997, **32**, 51.
42. B. P. Hills and K. P. Nott, *Appl. Magn. Reson.*, 1999, **17**, 521.
43. A. Gottwald, L. K. Creamer, P. L. Hubbard and P. T. Callaghan, *J. Chem. Phys.*, 2005, **122**, 34506.
44. J. P. M. Marques, D. N. Rutledge and C. J. Ducauze, *Int. J. Food Sci. Technol.*, 1991, **26**, 173.
45. K. R. Metz, P. J. Stankiewicz, J. W. Sassani and R. W. Briggs, *Magn. Reson. Med.*, 1986, **3**, 575.
46. N. Marigheto, S. Duarte and B. P. Hills, *Appl. Magn. Reson.*, 2005, **29**, 687.
47. N. Marigheto, K. Wright and B. P. Hills, *Appl. Magn. Reson.*, 2006, **30**, 13.
48. S. Baroni, *Magnetic Resonance in Food Science: Challenges in a Changing World*. Royal Society of Chemistry, Cambridge, 2009, p. 65.
49. R. B. D. Azeredo, L. A. Colnago and M. Engelsberg, *Anal. Chem.*, 2000, **72**, 2401.
50. R. B. V. Azeredo, L. A. Colnago, A. A. Souza and M. Engelsberg, *Anal. Chim. Acta*, 2003, **478**, 313.
51. T. Venancio, M. Engelsberg, R. B. V. Azeredo, N. E. R. Alem and L. A. Colnago, *J. Magn. Reson.*, 2005, **173**, 34.
52. H. Van As and T. J. Schaafsma, *J. Magn. Reson.*, 1987, **74**, 526.
53. A. L. Mackay, J. Wallace, A. Tepper and I. Taylor, *Biopolymers*, 1982, **21**, 1521.
54. J. C. Wallace, A. L. Mackay, K. Sasaki and I. E. P. Taylor, *Planta*, 1993, **190**, 227.
55. A. L. Mackay, J. C. Wallace, K. Sasaki and I. E. P. Taylor, *Biochemistry*, 1988, **27**, 1467.
56. I. E. P. Taylor, J. C. Wallace, A. L. Mackay and F. Volke, *Plant Physiol.*, 1990, **94**, 174.
57. H. R. Tang, P. S. Belton, A. Ng, K. W. Waldron and P. Ryden, *Spectrochim. Acta A Mol. Biomol. Spectrosc.*, 1999, **55**, 883.

58. H. R. Tang and P. S. Belton, *Advances in Magnetic Resonance in Food Science*. Royal Society of Chemistry, Cambridge, 1999, p. 166.
59. H. R. Tang, B. L. Zhao, P. S. Belton, L. H. Sutcliffe and A. Ng, *Magn. Reson. Chem.*, 2000, **38**, 765.
60. R. M. Cotts, M. J. R. Hoch, T. Sun and J. T. Markert, *J. Magn. Reson.*, 1989, **83**, 252.
61. D. van Dusschoten, C. Moonen, P. De Jager and H. Van As, *Magn. Reson. Med.*, 1996, **36**, 907.
62. P. N. Sen, *Concepts Magn. Reson. Part A*, 2004, **23A**, 1.
63. P. T. Callaghan, *Principles of Nuclear Magnetic Resonance Microscopy*. Clarendon Press, Oxford, UK, 1993.
64. J. C. Van den Enden, D. Waddington, H. van Aalst, C. G. Van Kralingen and K. J. Packer, *J. Colloid Interface Sci.*, 1990, **140**, 105.
65. G. J. W. Goudappel, J. P. M. van Duynhoven and M. M. W. Mooren, *J. Colloid Interface Sci.*, 2001, **239**, 535.
66. H. C. Bertram, *Magn. Reson. Imaging*, 2004, **22**, 557.
67. H. Peemoeller, *Bull. Magn. Reson.*, 1989, **11**, 19.
68. H. Peemoeller, *Bunseki*, 1997, **11**, 19.
69. H. Peemoeller and M. M. Pintar, *J. Magn. Reson.*, 1980, **41**, 358.
70. H. Peemoeller, *J. Magn. Reson.*, 1981, **45**, 193.
71. A. E. English, K. P. Whittall, M. L. G. Joy and R. M. Henkelman, *Magn. Reson. Med.*, 1991, **22**, 425.
72. J. E. Snaar and H. Van As, *J. Magn. Reson.*, 1992, **99**, 139.
73. J. E. Snaar and H. Van As, *Biophys. J.*, 1992, **63**, 1654.
74. Y. Q. Song, L. Venkataramanan, M. D. Hurlimann, M. Flaum, P. Frulla and C. Straley, *J. Magn. Reson.*, 2002, **154**, 261.
75. M. D. Hurlimann, L. Burcaw and Y. Q. Song, *J. Colloid Interface Sci.*, 2006, **297**, 303.
76. B. Hills, S. Benamira, N. Marigheto and K. Wright, *Appl. Magn. Reson.*, 2004, **26**, 543.
77. N. Marigheto, L. Venturi, D. Hibberd, K. M. Wright, G. Ferrante and B. P. Hills, *J. Magn. Reson.*, 2007, **187**, 327.
78. M. Furfaro, N. Marigheto, G. Moates, K. Cross, M. Parker, K. Waldron and B. Hills, *Appl. Magn. Reson.*, 2009, **35**, 521.
79. M. Furfaro, N. Marigheto, G. Moates, K. Cross, M. Parker, K. Waldron and B. Hills, *Appl. Magn. Reson.*, 2009, **35**, 537.
80. N. Hernandez-Sanchez, B. P. Hills, P. Barreiro and N. Marigheto, *Postharvest Biol. Technol.*, 2007, **44**, 260.
81. N. Marigheto, L. Venturi and B. Hills, *Postharvest Biol. Technol.*, 2008, **48**, 331.
82. S. Peled, D. G. Cory, S. A. Raymond, D. A. Kirschner and F. A. Jolesz, *Magn. Reson. Med.*, 1999, **42**, 911.
83. F. P. Duval, M. Cambert and F. Mariette, *Appl. Magn. Reson.*, 2005, **28**, 29.
84. D. van Dusschoten, A. de Jager and H. Van As, *J. Magn. Reson.*, 1995, **116**, 22.
85. G. Laicher, D. C. Ailion and A. G. Cutillo, *J. Magn. Reson. Ser. B*, 1996, **111**, 243.
86. T. A. Sibgatullin, A. V. Anisimov, P. A. de Jager, F. J. Vergeldt, E. Gerkema and H. Van As, *Biofysics*, 2007, **52**, 196.
87. A. Raffo, R. Gianferri, R. Barbieri and E. Brosio, *Food Chem.*, 2005, **89**, 149.
88. T. Sibgatullin, F. Vergeldt, A. Anisimov and H. Van As, *Dokl Biol. Sci.*, 2006, **411**, 488.
89. P. T. Callaghan, S. Godefroy and B. N. Ryland, *J. Magn. Reson.*, 2003, **162**, 320.
90. B. P. Hills, K. M. Wright and J. E. M. Snaar, *Magn. Reson. Imaging*, 1996, **14**, 305.
91. B. P. Hills, K. M. Wright and J. E. M. Snaar, *Magn. Reson. Imaging*, 1996, **14**, 715.
92. M. D. Hurlimann and L. Venkataramanan, *J. Magn. Reson.*, 2002, **157**, 31.
93. J. H. Lee, C. Labadie, C. S. Springer and G. S. Harbison, *J. Am. Chem. Soc.*, 1993, **115**, 7761.
94. P. McDonald and J. Korb, *Phys. Rev. E*, 2005, **72**, 011409.
95. S. Godefroy and P. T. Callaghan, *Magn. Reson. Imaging*, 2003, **21**, 381.
96. L. Monteilhet and J. Korb, *Phys. Rev. E*, 2006, **74**, 061404.
97. L. Van der Weerd, F. J. Vergeldt, D. J. Adrie and H. Van As, *Magn. Reson. Imaging*, 2000, **18**, 1151.
98. H. T. Pedersen, R. Bro and S. B. Engelsen, *J. Magn. Reson.*, 2002, **157**, 141.
99. E. Trezza, A. M. Haiduc, G. J. W. Goudappel and J. P. M. van Duynhoven, *Magn. Reson. Chem.*, 2006, **44**, 1023.

100. M. Adam-Berret, C. Rondeau-Mouro, A. Riaublanc and F. Mariette, *Magn. Reson. Chem.*, 2008, **46**, 550.
101. W. Derbyshire, M. van den Bosch, D. van Dusschoten, W. MacNaughtan, I. A. Farhat, M. A. Hemminga and J. R. Mitchell, *J. Magn. Reson.*, 2004, **168**, 278.
102. I. J. van den Dries, D. van Dusschoten and M. A. Hemminga, *J. Phys. Chem. B*, 1998, **102**, 10483.
103. A. Abragam, *The Principles of Nuclear Magnetism.* London, Oxford, 1961.
104. M. L. Johns, *Curr. Opin. Colloid Interface Sci.*, 2009, **14**, 178.
105. J. S. Murday and R. M. Cotts, *J. Chem. Phys.*, 1968, **48**, 4938.
106. M. Alderliesten, *Part. Part. Syst. Charact.*, 1990, **7**, 233.
107. M. Alderliesten, *Part. Part. Syst. Charact.*, 1991, **8**, 237.
108. L. Ambrosone, A. Ceglie, G. Colafemmina and G. Palazzo, *J. Chem. Phys.*, 1997, **107**, 10756.
109. L. Ambrosone, A. Ceglie, G. Colafemmina and G. Palazzo, *J. Chem. Phys.*, 1999, **110**, 797.
110. K. G. Hollingsworth and M. L. Johns, *J. Colloid Interface Sci.*, 2003, **258**, 383.
111. B. P. Hills, S. F. Takacs and P. S. Belton, *Food Chem.*, 1990, **37**, 95.
112. Z. Luz and S. Meiboom, *J. Chem. Phys.*, 1963, **39**, 366.
113. A. Allerhand and H. S. Gutowski, *J. Chem. Phys.*, 1965, **42**, 1587.
114. J. P. Carver and J. E. Richards, *J. Magn. Reson.*, 1972, **6**, 89.
115. B. P. Hills, K. M. Wright and P. S. Belton, *Mol. Phys.*, 1989, **67**, 1309.
116. P. S. Belton and B. P. Hills, *Mol. Phys.*, 1987, **61**, 999.
117. L. van der Weerd, M. M. A. E. Claessens, T. Ruttink, F. J. Vergeldt, T. J. Schaafsma and H. Van As, *J. Exp. Bot.*, 2001, **52**, 2333.
118. H. Van As, *J. Exp. Bot.*, 2007, **58**, 743.
119. N. M. Homan, C. W. Windt, F. J. Vergeldt, E. Gerkema and H. Van As, *Appl. Magn. Reson.*, 2007, **32**, 157.
120. B. P. Hills and C. J. Clark, *Ann. Rep. NMR Spectrosc.*, 2003, **50**, 75.
121. B. P. Hills and J. E. M. Snaar, *Mol. Phys.*, 1992, **76**, 979.
122. L. van der Weerd, S. M. Melnikov, F. J. Vergeldt, E. G. Novikov and H. Van As, *J. Magn. Reson.*, 2002, **156**, 213.
123. D. van Dusschoten, F. Vergeldt and H. Van As, *Book of Abstracts.* 2007, ICMRM9, Aachen.
124. R. W. W. Van Resandt, R. H. Vogel and S. W. Provencher, *Rev. Sci. Instrum.*, 1982, **53**, 1392.
125. A. M. Haiduc, J. P. M. van Duynhoven, P. Heussen, A. A. Reszka and C. Reiffers-Magnani, *Food Res. Int.*, 2007, **40**, 425.
126. R. Bro and S. DeJong, *J. Chemom.*, 1997, **11**, 393.
127. C. Manetti, C. Castro and J. P. Zbilut, *J. Magn. Reson.*, 2004, **168**, 273.
128. L. Andrade, E. Micklander, I. Farhat, R. Bro and S. B. Engelsen, *J. Magn. Reson.*, 2007, **189**, 286.
129. W. P. Weglarz and H. Haranczyk, *J. Phys. D Appl. Phys.*, 2000, **33**, 1909.
130. M. Prange and Y. Q. Song, *J. Magn. Reson.*, 2009, **196**, 54.
131. C. L. Lawson and R. J. Hanson, *Solving Least Squares Problems.* SIAM, Philadelphia, 1995.
132. R. S. Menon, M. S. Rusinko and P. S. Allen, *Magn. Reson. Med.*, 1991, **20**, 196.
133. K. P. Whittall and A. L. Mackay, *J. Magn. Reson.*, 1989, **84**, 134.
134. J. P. Butler, J. A. Reeds and S. V. Dawson, *J. Numer. Anal.*, 1981, **18**, 381.
135. S. W. Provencher and V. G. Dovi, *J. Biochem. Biophys. Methods*, 1979, **1**, 313.
136. S. W. Provencher, *Comput. Phys. Commun.*, 1982, **27**, 229.
137. E. D. Laue, J. Skilling, J. Staunton, S. Sibisi and R. G. Brereton, *J. Magn. Reson.*, 1985, **62**, 437.
138. C. Tellier, F. Mariette, J. P. Guillement and P. Marchal, *J. Agri. Food Chem.*, 1993, **41**, 2259.
139. F. Mariette, J. P. Guillement, C. Tellier and P. Marchal, *in: Signal Treatment and Signal Analysis in NMR*, (D. N. Rutledge ed.), 1996, p. 218. Elsevier, Paris.
140. G. C. Borgia, R. J. S. Brown and P. Fantazzini, *J. Magn. Reson.*, 2000, **147**, 273.
141. G. C. Borgia, R. J. Brown and P. Fantazzini, *J. Magn Reson*, 1998, **132**, 65.
142. G. C. Borgia, R. J. Brown and P. Fantazzini, *J. Magn. Reson.*, 1998, **132**, 65.
143. S. B. Engelsen and R. Bro, *J. Magn. Reson.*, 2003, **163**, 192.
144. A. Gerbanowski, D. N. Rutledge, M. H. Feinberg and C. J. Ducauze, *Sci. Aliments*, 1997, **17**, 309.

145. E. Micklander, L. G. Thygesen, H. T. Pedersen, F. van den Berg, R. Bro, D. N. Rutledge and S. B. Engelsen, *in: Magnetic Resonance in Food Science*, (P. S. Belton, A. M. Gil, G. A. Webb, D. N. Rutledge eds.), 2003, p. 239. Royal Society of Chemistry, Cambridge, UK.
146. R. J. S. Brown, F. Capozzi, C. Cavani, M. A. Cremonini, M. Petracci and G. Placucci, *J. Magn. Reson.*, 2000, **147**, 89.
147. R. Bro, *Crit. Rev. Anal. Chem.*, 2006, **36**, 279.
148. W. Windig and B. Antalek, *Chemom. Intell. Lab. Syst.*, 1997, **37**, 241.
149. A. Nordon, P. J. Gemperline, C. A. McGill and D. Littlejohn, *Anal. Chem.*, 2001, **73**, 4286.
150. E. Tonning, D. Polders, P. T. Callaghan and S. B. Engelsen, *J. Magn. Reson.*, 2007, **188**, 10.
151. J. P. M. van Duynhoven, G. J. W. Goudappel, M. Gribnau and V. K. S. Shukla, *AOCS Inform*, 1999, **10**, 479.
152. J. van Duynhoven, I. Dubourg, G. J. Goudappel and E. Roijers, *J. Am. Oil Chem. Soc.*, 2002, **79**, 383.
153. Y. H. Roos, R. B. Leslie and P. J. Lillford, *Water Management in the Design and Distribution of Quality Food*. Technomic Publishing, Lancaster, 2007.
154. S. J. Schmidt, *Adv. Food Nutr. Res.*, 2004, **48**, 1.
155. B. P. Hills, C. E. Manning and Y. Ridge, *J. Chem. Soc. Faraday Trans.*, 1996, **92**, 979.
156. B. P. Hills, C. E. Manning, Y. Ridge and T. Brocklehurst, *J. Sci. Food Agri.*, 1996, **71**, 185.
157. B. P. Hills, C. E. Manning and J. Godward, *in: Advances in Magnetic Resonance in Food Science*, P. S. Belton, B. P. Hills, and G. A. Webb (eds.), 1999, p. 45. Royal Society of Chemistry, Cambridge, UK.
158. L. Venturi, P. Rocculi, C. Cavani, G. Placucci, M. D. Rosa and M. A. Cremonini, *J. Agri. Food Chem.*, 2007, **55**, 10572.
159. R. R. Ruan, Z. Z. Long, A. J. Song and P. L. Chen, *Food Sci. Technol. Lebensmittel Wiss. Technol.*, 1998, **31**, 516.
160. R. Ruan, Z. Long, P. Chen, V. Huang, S. Almaer and I. Taub, *J. Food Sci.*, 1999, **64**, 6.
161. W. P. Weglarz, C. Inoue, M. Witek, H. Van As, and J. P. M. van Duynhoven, *in: Water Properties in Food, Health, Pharmaceutical and Biological Systems: ISOPOW 10*, (D. S. Reid, T. Sajjaanantakul, eds.), 2010, Wiley, New York.
162. M. S. Rahman, *Trends Food Sci. Technol.*, 2006, **17**, 129.
163. E. Vittadini and P. Chinachoti, *Int. J. Food Sci. Technol.*, 2003, **38**, 841.
164. J. Chirife and M. P. Buera, *J. Food Eng.*, 1995, **25**, 531.
165. Y. H. Roos, *J. Food Eng.*, 1995, **24**, 339.
166. Y. H. Roos, *Food Technol.*, 1995, **49**, 97.
167. C. P. Sherwin and T. P. Labuza, *in: Water Properties of Food, Pharmaceutical and Biological Materials*, (P. Buera ed.), 2006, p. 343. Chips Books, Weimar, TX.
168. X. Y. Lin, R. Ruan, P. Chen, M. S. Chung, X. F. Ye, T. Yang, C. Doona and T. Wagner, *J. Food Sci.*, 2006, **71**, R136.
169. M. S. Chung, R. Ruan, P. Chen, J. H. Kim, T. H. Ahn and C. K. Baik, *Lebensm. Wiss. Technol. Food Sci. Technol.*, 2003, **36**, 751.
170. M. S. Chung, R. Ruan, P. Chen, Y. G. Lee, T. H. Ahn and C. K. Baik, *J. Food Sci.*, 2001, **66**, 1147.
171. M. S. Chung, R. R. Ruan, P. Chen, S. H. Chung, T. H. Ahn and K. H. Lee, *J. Food Sci.*, 2000, **65**, 134.
172. S. Li, L. C. Dickinson and P. Chinachoti, *J. Agri. Food Chem.*, 1998, **46**, 62.
173. Y. R. Kim, B. S. Yoo, P. Cornillon and S. T. Lim, *Carbohydr. Polym.*, 2004, **55**, 27.
174. A. Le Dean, F. Mariette, T. Lucas and M. Marin, *Lebensm. Wiss. Technol. Food Sci. Technol.*, 2001, **34**, 299.
175. S. Ablett, C. J. Clarke, M. J. Izzard and D. R. Martin, *J. Sci. Food Agri.*, 2002, **82**, 1855.
176. S. Lee, S. Moon, J. Y. Shim and Y. R. Kim, *Food Sci. Biotechnol.*, 2008, **17**, 102.
177. J. Rasanen, J. M. V. Blanshard, J. R. Mitchell, W. Derbyshire and K. Autio, *J. Cereal Sci.*, 1998, **28**, 1.
178. B. P. Hills and G. Lefloch, *Food Chem.*, 1994, **51**, 331.
179. P. N. Gambhir, *Trends Food Sci. Technol.*, 1992, **3**, 191.
180. P. H. Krygsman and A. E. Barrett, *Oil Extraction and Analysis: Critical Issues and Comparative Studies*. AOCS Press, Champaign, 2004, p. 152.
181. G. Rubel, *J. Am. Oil Chem. Soc.*, 1994, **71**, 1057.

182. P. N. Tiwari and P. N. Gambhir, *J. Am. Oil Chem. Soc.*, 1995, **72**, 1017.
183. P. N. Tiwari and W. Burk, *J. Am. Oil Chem. Soc.*, 1980, **57**, 119.
184. H. T. Pedersen, L. Munck and S. B. Engelsen, *J. Am. Oil Chem. Soc.*, 2000, **77**, 1069.
185. D. Nielsen, G. Hyldig, J. Nielsen and H. H. Nielsen, *Lwt Food Sci. Technol.*, 2005, **38**, 537.
186. T. P. Leffler, C. R. Moser, B. J. McManus, J. J. Urh, J. T. Keeton and A. Claflin, *J. AOAC Int.*, 2008, **91**, 802.
187. J. I. T. Keeton, S. M. Eddy, C. R. Moser, B. J. McManus and T. P. Leffler, *J. AOAC Int.*, 2003, **86**, 1193.
188. E. Nagy and L. Kormendy, *Acta Alimentaria*, 2003, **32**, 289.
189. B. McManus and M. Horn, *Oil Extraction and Analysis: Critical Issues and Comparative Studies.* AOCS Press, Champaign, 2004, p. 152.
190. J. G. Seland, G. H. Sorland, H. W. Anthonsen and J. Krane, *Appl. Magn. Reson.*, 2003, **24**, 41.
191. G. Sorland, *Appl. Magn. Reson.*, 2004, **26**, 417.
192. G. H. Sorland, P. M. Larsen, F. Lundby, A. P. Rudi and T. Guiheneuf, *Meat Sci.*, 2004, **66**, 543.
193. G. H. Sorland, P. M. Larsen, F. Lundby, H. W. Anthonsen and B. J. Foss, *Magnetic Resonance in Food Science: The Multivariate Challenge.* Royal Society of Chemistry, Cambridge, UK, 2005, p. 20.
194. G. H. Sorland, F. Lundby and A. Ukkelberg, *Magnetic Resonance in Food Science: From Molecules to Man.* Royal Society of Chemistry, London, 2007, Vol. **189**.
195. J. P. M. van Duynhoven, G. J. W. Goudappel, G. van Dalen, P. C. van Bruggen, J. C. G. Blonk and A. P. A. M. Eijkelenboom, *Magn. Reson. Chem.*, 2002, **40**, S51.
196. J. C. Vandenenden, D. Waddington, H. Vanaalst, C. G. Vankralingen and K. J. Packer, *J. Colloid Interface Sci.*, 1990, **140**, 105.
197. B. Balinov, O. Soderman and T. Warnheim, *J. Am. Oil Chem. Soc.*, 1994, **71**, 513.
198. I. Fourel, J. P. Guillement and D. Lebotlan, *J. Colloid Interface Sci.*, 1994, **164**, 48.
199. C. P. Aichele, M. Flaum, T. M. Jiang, G. J. Hirasaki and W. G. Chapman, *J. Colloid Interface Sci.*, 2007, **315**, 607.
200. W. Price, *Concepts Magn. Reson.*, 1998, **10**, 197.
201. S. Kiokias and A. Bot, *Food Hydrocolloids*, 2005, **19**, 493.
202. S. Kiokias and A. Bot, *Food Hydrocolloids*, 2006, **20**, 245.
203. A. Bot, F. P. Duval, C. P. Duif and W. G. Bouwman, *Int. J. Food Sci. Technol.*, 2007, **42**, 746.
204. S. Kiokias, C. Dimakou and V. Oreopoulou, *Food Chem.*, 2007, **105**, 94.
205. P. T. Callaghan, K. W. Jolley and R. S. Humphrey, *J. Colloid Interface Sci.*, 1983, **93**, 521.
206. D. Rousseau and S. M. Hodge, *Colloids Surf. A Physicochem. Eng. Asp.*, 2005, **260**, 229.
207. S. M. Hodge and D. Rousseau, *J. Am. Oil Chem. Soc.*, 2005, **82**, 159.
208. D. Rousseau, L. Zilnik, R. Khan and S. Hodge, *J. Am. Oil Chem. Soc.*, 2003, **80**, 957.
209. S. M. Hodge and D. Rousseau, *Food Res. Int.*, 2003, **36**, 695.
210. I. Fourel, J. P. Guillement and D. Lebotlan, *J. Colloid Interface Sci.*, 1995, **169**, 119.
211. D. Chapman, R. E. Richards and R. W. Yorke, *J. Phys. Chem.*, 1957, 436.
212. D. Le Botlan, L. Ouguerram, L. Smart and L. Pugh, *J. Am. Oil Chem. Soc.*, 1999, **76**, 255.
213. D. Le Botlan and I. Heliefourel, *Anal. Chim. Acta*, 1995, **311**, 217.
214. D. J. Le Botlan and I. Helie, *Analysis*, 1994, **22**, 108.
215. D. J. LeBotlan and L. Ouguerram, *Anal. Chim. Acta*, 1997, **349**, 339.
216. D. Le Botlan, E. Casseron and E. Lantier, *Analusis*, 1998, **26**, 198.
217. K. Aeberhardt, Q. D. Bui and V. Normand, *Biomacromolecules*, 2007, **8**, 1038.
218. H. R. Tang, J. Godward and B. Hills, *Carbohydr. Polym.*, 2000, **43**, 375.
219. D. Le Botlan, Y. Rugraff, C. Martin and P. Colonna, *Carbohydr. Res.*, 1998, **308**, 29.
220. M. Ritota, R. Gianferri, R. Bucci and E. Brosio, *Food Chem.*, 2008, **110**, 14.
221. F. Lionetto, A. Maffezzoli, M. A. Ottenhof, I. A. Farhat and J. R. Mitchell, *Starch-Starke*, 2005, **57**, 16.
222. X. Wang, S. G. Choi and W. L. Kerr, *J. Sci. Food Agri.*, 2004, **84**, 371.
223. L. G. Thygesen, A. Blennow and S. B. Engelsen, *Starch-Starke*, 2003, **55**, 241.
224. I. A. Farhat, J. M. V. Blanshard and J. R. Mitchell, *Biopolymers*, 2000, **53**, 411.
225. P. Belton, I. J. Colquhoun, A. Grant, N. Wellner, J. M. Field, P. R. Shewry and A. S. Tatham, *Int. J. Biol. Macromol.*, 1995, **17**, 74.
226. P. Belton, *J. Cereal Sci.*, 1994, **19**, 115.

227. P. Belton, A. M. Gil, A. Grant, E. Alberti and A. S. Tatham, *Spectrochim. Acta A Mol. Biomol. Spectrosc.*, 1998, **54**, 955.
228. A. Davenel, P. Schuck and P. Marchal, *Milchwissenschaft Milk Sci. Int.*, 1997, **52**, 35.
229. A. Davenel, P. Schuck, F. Mariette and G. Brule, *Lait*, 2002, **82**, 465.
230. C. J. Doona and M. Y. Baik, *J. Cereal Sci.*, 2007, **45**, 257.
231. B. Hills, A. Grant and P. Belton, in: *Characterisation of Cereals and Flours*, G. Kaletunc and K. J. Breslauer (eds.), 2003, p. 409. Marcel Dekker, New York.
232. R. Ruan and P. L. Chen, *Water in Foods and Biological Materials. A Nuclear Magnetic Resonance Approach*. Technomic Publishing, Lancaster, PA, 1998.
233. R. R. Ruan, X. A. Wang, P. L. Chen, R. G. Fulcher, P. Pesheck and S. Chakrabarti, *Cereal Chem.*, 1999, **76**, 231.
234. P. S. Belton, *J. Cereal Sci.*, 2005, **41**, 203.
235. E. Esselink, H. van Aalst, M. Maliepaard, T. M. H. Henderson, N. L. L. Hoekstra and J. van Duynhoven, *Cereal Chem.*, 2003, **80**, 419.
236. J. Yi, W. L. Kerr and J. W. Johnson, *J. Food Sci.*, 2009, **74**, E278.
237. J. Yi and W. L. Kerr, *J. Food Eng.*, 2009, **93**, 495.
238. J. H. Yi and W. L. Kerr, *Lwt Food Sci. Technol.*, 2009, **42**, 1474.
239. Y. R. Kim and P. Cornillon, *Lebensm. Wiss. Technol. Food Sci. Technol.*, 2001, **34**, 417.
240. S. B. Engelsen, M. K. Jensen, H. T. Pedersen, L. Norgaard and L. Munck, *J. Cereal Sci.*, 2001, **33**, 59.
241. P. L. Chen, Z. Long, R. Ruan and T. P. Labuza, *Food Sci. Technol. Lebensmittel Wiss. Technol.*, 1997, **30**, 178.
242. R. Karoui and J. De Baerdemaeker, *Food Chem.*, 2007, **102**, 621.
243. C. Mok, *Food Sci. Biotechnol.*, 2008, **17**, 895.
244. R. Hinrichs, J. Gotz and H. Weisser, *Food Chem.*, 2003, **82**, 155.
245. R. Hinrichs, J. Gotz, M. Noll, A. Wolfschoon, H. Eibel and H. Weisser, *Int. Dairy J.*, 2004, **14**, 817.
246. R. Hinrichs, J. Gotz, M. Noll, A. Wolfschoon, H. Eibel and H. Weisser, *Food Res. Int.*, 2004, **37**, 667.
247. T. Lucas, F. Mariette, S. Dominiawsyk and D. Le Ray, *Food Chem.*, 2004, **84**, 77.
248. T. Lucas, M. Wagener, P. Barey and F. Mariette, *Int. Dairy J.*, 2005, **15**, 1064.
249. T. Lucas, D. Le Ray, P. Barey and F. Mariette, *Int. Dairy J.*, 2005, **15**, 1225.
250. J. A. Abbott, *Postharvest Biol. Technol.*, 1999, **15**, 207.
251. P. Butz, C. Hofmann and B. Tauscher, *J. Food Sci.*, 2005, **70**, R131.
252. K. M. Keener, R. L. Stroshine and J. A. Nyenhuis, *J. Am. Soc. Hortic. Sci.*, 1999, **124**, 289.
253. B. K. Cho, W. Chayaprasert and R. L. Stroshine, *Postharvest Biol. Technol.*, 2008, **47**, 81.
254. P. N. Gambhir, Y. J. Choi, D. C. Slaughter, J. F. Thompson and M. J. McCarthy, *J. Sci. Food Agri.*, 2005, **85**, 2482.
255. S. Y. S. Tu, Y. J. Choi, M. J. McCarthy and K. L. McCarthy, *Postharvest Biol. Technol.*, 2007, **44**, 157.
256. A. K. Thybo, H. J. Andersen, A. H. Karlsson, S. Donstrup and H. Stodkilde-Jorgensen, *Lebensm. Wiss. Technol. Food Sci. Technol.*, 2003, **36**, 315.
257. A. K. Thybo, I. E. Bechmann, M. Martens and S. B. Engelsen, *Food Sci. Technol. Lebensmittel Wiss. Technol.*, 2000, **33**, 103.
258. L. G. Thygesen, A. K. Thybo and S. B. Engelsen, *Lebensm. Wiss. Technol. Food Sci. Technol.*, 2001, **34**, 469.
259. V. T. Povlsen, A. Rinnan, F. van den Berg, H. J. Andersen and A. K. Thybo, *Lebensm. Wiss. Technol. Food Sci. Technol.*, 2003, **36**, 423.
260. P. Barreiro, A. Moya, E. Correa, M. Ruiz-Altisent, M. Fernandez-Valle, A. Peirs, K. M. Wright and B. P. Hills, *Appl. Magn. Reson.*, 2002, **22**, 387.
261. J. P. M. Marques, D. N. Rutledge and C. J. Ducauze, *Sci. Aliments*, 1991, **11**, 513.
262. J. P. M. Marques, D. N. Rutledge and C. J. Ducauze, *Sci. Aliments*, 1992, **12**, 613.
263. J. M. Marques, D. N. Rutledge and C. J. Ducauze, *Food Sci. Technol. Lebensmittel Wiss. Technol.*, 1991, **24**, 93.
264. J. P. M. Marques, C. Leloch, E. Wolff and D. N. Rutledge, *J. Food Sci.*, 1991, **56**, 1707.
265. P. Cornillon, *Lebensm. Wiss. Technol. Food Sci. Technol.*, 2000, **33**, 261.
266. N. Marigheto, A. Vial, K. Wright and B. Hills, *Appl. Magn. Reson.*, 2004, **26**, 521.
267. B. Hills, A. Costa, N. Marigheto and K. Wright, *Appl. Magn. Reson.*, 2005, **28**, 13.

268. M. Mortensen, A. K. Thybo, H. C. Bertram, H. J. Andersen and S. B. Engelsen, *J. Agri. Food Chem.*, 2005, **53**, 5976.
269. E. Micklander, A. K. Thybo and F. van den Berg, *Lwt Food Sci. Technol.*, 2008, **41**, 1710.
270. H. C. Bertram and H. J. Andersen, *Ann. Rep. NMR Spectrosc.*, 2004, **53**, 157.
271. H. C. Bertram and H. J. Andersen, *J. Anim. Breed. Genet.*, 2007, **124**, 35.
272. H. C. Bertram, P. P. Purslow and H. J. Andersen, *J. Agri. Food Chem.*, 2002, **50**, 824.
273. H. C. Bertram, A. Schafer, K. Rosenvold and H. J. Andersen, *Meat Sci.*, 2004, **66**, 915.
274. H. C. Bertram, H. J. Andersen and A. H. Karlsson, *Meat Sci.*, 2001, **57**, 125.
275. H. C. Bertram, S. Donstrup, A. H. Karlsson and H. J. Andersen, *Meat Sci.*, 2002, **60**, 279.
276. I. K. Straadt, M. Rasmussen, J. F. Young and H. C. Bertram, *Meat Sci.*, 2008, **80**, 722.
277. S. M. Jepsen, H. T. Pedersen and S. B. Engelsen, *J. Sci. Food Agri.*, 1999, **79**, 1793.
278. K. N. Jensen, B. M. Jorgensen, H. H. Nielsen and J. Nielsen, *J. Sci. Food Agri.*, 2005, **85**, 1259.
279. H. C. Bertram, H. J. Andersen, A. H. Karlsson, P. Horn, J. Hedegaard, L. Norgaard and S. B. Engelsen, *Meat Sci.*, 2003, **65**, 707.
280. J. Brondum, L. Munck, P. Henckel, A. Karlsson, E. Tornberg and S. B. Engelsen, *Meat Sci.*, 2000, **55**, 177.
281. J. P. Renou, J. Kopp and C. Valin, *J. Food Technol.*, 1985, **20**, 23.
282. C. C. Correa, L. A. Forato and L. A. Colnago, *Anal. Bioanal. Chem.*, 2009, **393**, 1357.
283. I. Martinez, M. Aursand, U. Erikson, T. E. Singstad, E. Veliyulin and C. van der Zwaag, *Trends Food Sci. Technol.*, 2003, **14**, 489.
284. S. Rahmatallah, Y. Li, H. C. Seton, J. S. Gregory and R. M. Aspden, *Eur. Food Res. Technol.*, 2006, **222**, 298.
285. E. Veliyulin, C. van der Zwaag, W. Burk and U. Erikson, *J. Sci. Food Agri.*, 2005, **85**, 1299.
286. H. C. Bertram, A. H. Karlsson and H. J. Andersen, *Meat Sci.*, 2003, **65**, 1281.
287. H. C. Bertram, S. B. Engelsen, H. Busk, A. H. Karlsson and H. J. Andersen, *Meat Sci.*, 2004, **66**, 437.
288. J. F. Young, H. C. Bertram, K. Rosenvold, G. Lindahl and N. Oksbjerg, *Meat Sci.*, 2005, **70**, 717.
289. E. Tornberg, M. Wahlgren, J. Brondum and S. B. Engelsen, *Food Chem.*, 2000, **69**, 407.
290. H. C. Bertram, A. K. Whittaker, H. J. Andersen and A. H. Karlsson, *J. Agri. Food Chem.*, 2003, **51**, 4072.
291. H. C. Bertram, M. Kristensen and H. J. Andersen, *Meat Sci.*, 2004, **68**, 249.
292. H. C. Bertram and M. D. Aaslyng, *Meat Sci.*, 2007, **76**, 524.
293. H. C. Bertram, R. L. Meyer, Z. Y. Wu, X. F. Zhou and H. J. Andersen, *J. Agri. Food Chem.*, 2008, **56**, 7201.
294. A. Hullberg and H. C. Bertram, *Meat Sci.*, 2005, **69**, 709.
295. J. Stadnik, Z. J. Dolatowski and H. M. Baranowska, *Lwt Food Sci. Technol.*, 2008, **41**, 2151.
296. H. C. Bertram, R. H. Andersen and H. J. Andersen, *Meat Sci.*, 2007, **75**, 128.
297. M. Mortensen, H. J. Andersen, S. B. Engelsen and H. C. Bertram, *Meat Sci.*, 2006, **72**, 34.
298. I. K. Straadt, M. Rasmussen, H. J. Andersen and H. C. Bertram, *Meat Sci.*, 2007, **75**, 687.
299. Z. Y. Wu, H. C. Bertram, A. Kohler, U. Bocker, R. Ofstad and H. J. Andersen, *J. Agri. Food Chem.*, 2006, **54**, 8589.
300. E. Micklander, B. Peshlov, P. P. Purslow and S. B. Engelsen, *Trends Food Sci. Technol.*, 2002, **13**, 341.
301. H. C. Bertram, Z. Y. Wu, F. van den Berg and H. J. Andersen, *Meat Sci.*, 2006, **74**, 684.
302. A. Rochdi, L. Foucat and J. P. Renou, *Food Chem.*, 2000, **69**, 295.
303. A. Rochdi, *Biopolymers*, 1999, **50**, 690.
304. Z. Y. Wu, H. C. Bertram, U. Bocker, R. Ofstad and A. Kohler, *J. Agri. Food Chem.*, 2007, **55**, 3990.
305. H. C. Bertram, A. Kohler, U. Bocker, R. Ofstad and H. J. Andersen, *J. Agri. Food Chem.*, 2006, **54**, 1740.
306. C. M. Andersen and A. Rinnan, *Lebensm. Wiss. Technol. Food Sci. Technol.*, 2002, **35**, 687.
307. C. Steen and P. Lambelet, *J. Sci. Food Agri.*, 1997, **75**, 268.
308. I. G. Aursand, L. Gallart-Jornet, U. Erikson, D. E. Axelson and T. Rustad, *J. Agri. Food Chem.*, 2008, **56**, 6252.
309. I. G. Aursand, E. Veliyulin, U. Bocker, R. Ofstad, T. Rustad and U. Erikson, *J. Agri. Food Chem.*, 2009, **57**, 46.
310. U. Erikson, E. Veliyulin, T. E. Singstad and M. Aursand, *J. Food Sci.*, 2004, **69**, E107.

311. H. Loje, D. Green-Petersen, J. Nielsen, B. M. Jorgensen and K. N. Jensen, *J. Sci. Food Agri.*, 2007, **87**, 212.
312. M. U. Ahmad, Y. Tashiro, S. Matsukawa and H. Ogawa, *J. Food Sci.*, 2004, **69**, E497.
313. M. U. Ahmad, Y. Tashiro, S. Matsukawa and M. Ogawa, *Fish. Sci.*, 2005, **71**, 655.
314. L. Andrade, I. A. Farhat, K. Aeberhardt, V. Normand and S. B. Engelsen, *Food Biophys.*, 2008, **3**, 33.
315. L. Andrade, *Appl. Spectrosc.*, 2009, **63**, 141.
316. D. J. LeBotlan and I. Helie, *Analusis*, 1994, **22**, 108.
317. D. Le Botlan, L. Ouguerram, L. Smart and L. Pugh, *J. Am. Oil Chem. Soc.*, 1999, **76**, 255.
318. H. C. Bertram, L. Wiking, J. H. Nielsen and H. J. Andersen, *Int. Dairy J.*, 2005, **15**, 1056.
319. B. P. Hills, Y. L. Wang and H. R. Tang, *Mol. Phys.*, 2001, **99**, 1679.
320. I. J. van den Dries, D. van Dusschoten, M. A. Hemminga and E. van der Linden, *J. Phys. Chem. B*, 2000, **104**, 10126.
321. S. Ablett, C. J. Clarke, M. J. Izzard and D. R. Martin, *J. Sci. Food Agri.*, 2002, **82**, 1855.
322. H. Kumagai, W. MacNaughtan, I. A. Farhat and J. R. Mitchell, *Carbohydr. Polym.*, 2002, **48**, 341.
323. B. P. Hills, *Mol. Phys.*, 1991, **72**, 1099.
324. B. P. Hills, C. Cano and P. S. Belton, *Macromolecules*, 1991, **24**, 2944.
325. V. Ducel, D. Pouliquen, J. Richard and F. Boury, *Int. J. Biol. Macromol.*, 2008, **43**, 359.
326. E. Brosio, A. Dubaldo and B. Verzegnassi, *Cell. Mol. Biol.*, 1994, **40**, 569.
327. J. Goetz and P. Koehler, *Lwt Food Sci. Technol.*, 2005, **38**, 501.
328. P. Lambelet, F. Ducret, J. L. Leuba and M. Geoffroy, *J. Agri. Food Chem.*, 1991, **39**, 287.
329. P. Lambelet, R. Berrocal and F. Ducret, *J. Dairy Res.*, 1989, **56**, 211.
330. P. Lambelet, R. Berrocal and F. Renevey, *J. Dairy Res.*, 1992, **59**, 517.
331. L. Indrawati, R. L. Stroshine and G. Narsimhan, *J. Sci. Food Agri.*, 2007, **87**, 2207.
332. N. D. Shiralkar, H. P. Harz and H. Weisser, *Lebensmittel Wiss. Technol.*, 1983, **16**, 18.
333. P. S. Belton, S. L. Duce and A. S. Tatham, *J. Cereal Sci.*, 1988, **7**, 113.
334. M. Witek, H. Peemoeller, J. Szymonska and B. Blicharska, *Acta Phys. Pol. A*, 2006, **109**, 359.
335. A. Gonera and P. Cornillon, *Starch Starke*, 2002, **54**, 508.
336. P. Chatakanonda, P. Chinachoti, K. Sriroth, K. Piyachomkwan, S. Chotineeranat, H. R. Tang and B. Hills, *Carbohydr. Polym.*, 2003, **53**, 233.
337. D. Lebotlan and P. Desbois, *Cereal Chem.*, 1995, **72**, 191.
338. M. A. Ottenhof and I. A. Farhat, *J. Cereal Sci.*, 2004, **40**, 269.
339. I. A. Farhat, J. M. V. Blanshard, M. Descamps and J. R. Mitchell, *Cereal Chem.*, 2000, **77**, 202.
340. B. Jaillais, M. A. Ottenhof, I. A. Farhat and D. N. Rutledge, *Vib. Spectrosc.*, 2006, **40**, 10.
341. R. Partanen, V. Marie, W. MacNaughtan, P. Forssell and I. Farhat, *Carbohydr. Polym.*, 2004, **56**, 147.
342. G. Roudaut, *Carbohydr. Polym.*, 2009, **77**, 489.
343. S. G. Choi and W. L. Kerr, *Lebensm. Wiss. Technol. Food Sci. Technol.*, 2003, **36**, 105.
344. M. R. Hansen, A. Blennow, I. Farhat, L. Nørgaard, S. Pedersen and S. B. Engelsen, *Food Hydrocolloids*, 2009, **2038**, 23.
345. H. M. Baranowska, M. Sikora, S. Kowalski and P. Tornasik, *Food Hydrocolloids*, 2008, **22**, 336.
346. J. A. Kenar, *Indust. Crops Prod.*, 2007, **26**, 77.
347. R. R. Ruan, P. L. Chen and S. Almaer, *Hortscience*, 1999, **34**, 319.
348. R. R. Ruan, C. Zou, C. Wadhawan, B. Martinez, P. L. Chen and P. Addis, *J. Food Proc. Preserv.*, 1997, **21**, 91.
349. A. Assifaoui, D. Champion, E. Chiotelli and A. Verel, *Carbohydr. Polym.*, 2006, **64**, 197.
350. Y. R. Kim, P. Cornillon, O. H. Campanella, R. L. Stroshine, S. Lee and J. Y. Shim, *J. Food Sci.*, 2008, **73**, E1.
351. E. F. J. Esselink, H. van Aalst, M. Maliepaard and J. P. M. van Duynhoven, *Cereal Chem.*, 2003, **80**, 396.
352. G. Roudaut, D. van Dusschoten, H. Van As, M. A. Hemminga and M. Le Meste, *J. Cereal Sci.*, 1998, **28**, 147.
353. G. L. Roudaut, M. Maglione, D. van Dusschoten and M. Le Meste, *Cereal Chem.*, 1999, **76**, 70.
354. X. Wang, S. G. Choi and W. L. Kerr, *Lebensm. Wiss. Technol. Food Sci. Technol.*, 2004, **37**, 377.
355. X. Wang, S. G. Choi and W. L. Kerr, *J. Sci. Food Agri.*, 2004, **84**, 371.
356. F. Le Grand, M. Cambert and F. Mariette, *J. Agri. Food Chem.*, 2007, **55**, 10947.
357. T. Lucas, D. Le Ray and F. Mariette, *J. Food Eng.*, 2007, **80**, 377.

358. M. Budiman, R. L. Stroshine and P. Cornillon, *J. Dairy Res.*, 2002, **69**, 619.
359. M. Budiman, R. L. Stroshine and O. H. Campanella, *J. Texture Stud.*, 2000, **31**, 477.
360. R. Gianferri, M. Maioli, M. Delfini and E. Brosio, *Int. Dairy J.*, 2007, **17**, 167.
361. R. Gianferri, V. D'Aiuto, R. Curini, M. Delfini and E. Brosio, *Food Chem.*, 2007, **105**, 720.
362. N. Noronha, E. Duggan, G. R. Ziegler, E. D. O'Riordan and M. O'Sullivan, *Int. Dairy J.*, 2008, **18**, 641.
363. S. Godefroy, J. P. Korb, L. K. Creamer, P. J. Watkinson and P. T. Callaghan, *J. Colloid Interface Sci.*, 2003, **267**, 337.
364. B. Chaland, F. Mariette, P. Marchal and J. De Certaines, *J. Dairy Res.*, 2000, **67**, 609.
365. A. Metais, M. Cambert, A. Riaublanc and F. Mariette, *Int. Dairy J.*, 2006, **16**, 344.
366. A. Metais, M. Cambert, A. Riaublanc and F. Mariette, *J. Agri. Food Chem.*, 2004, **52**, 3988.
367. D. Le Botlan, J. Wennington and J. C. Cheftel, *J. Colloid Interface Sci.*, 2000, **226**, 16.
368. T. Lucas, D. Le Ray, P. Barey and F. Mariette, *Int. Dairy J.*, 2005, **15**, 1225.
369. M. H. Famelart, F. Gaucheron, F. Mariette, Y. Le Graet, K. Raulot and E. Boyaval, *Int. Dairy J.*, 1997, **7**, 325.
370. F. Mariette, C. Tellier, G. Brule and P. Marchal, *J. Dairy Res.*, 1993, **60**, 175.
371. F. Mariette, P. Maignan and P. Marchal, *Analusis*, 1997, **25**, M24.
372. T. Salomonsen, M. T. Sejersen, N. Viereck, R. Ipsen and S. B. Engelsen, *Int. Dairy J.*, 2007, **17**, 294.
373. A. M. Haiduc, J. P. M. Van Duynhoven, P. Heussen, A. A. Reszka and C. Reiffers-Magnani, *Food Res. Int.*, 2007, **40**, 425.
374. A. M. Haiduc and J. van Duynhoven, *Magn. Reson. Imaging*, 2005, **23**, 343.
375. A. Le Dean, F. Mariette and M. Marin, *J. Agri. Food Chem.*, 2004, **52**, 5449.
376. F. P. Duval, J. P. M. van Duynhoven and A. Bot, *J. Am. Oil Chem. Soc.*, 2006, **83**, 905.
377. L. Day, M. Xu, P. Hoobin, I. Burgar and M. A. Augustin, *Food Chem.*, 2007, **105**, 469.
378. P. Walter, *Food Res. Int.*, 2002, **35**, 761.
379. Y. Li, K. Szlachetka, P. Chen, X. Y. Lin and R. Ruan, *Cereal Chem.*, 2008, **85**, 780.
380. T. Rudi, G. Guthausen, W. Burk, C. T. Reh and H. D. Isengard, *Food Chem.*, 2008, **106**, 1375.
381. M. S. Chung, R. R. Ruan, P. L. Chen and X. Wang, *Food Sci. Technol. Lebensmittel Wiss. Technol.*, 1999, **32**, 162.
382. C. Tellier, M. Guilloucharpin, P. Grenier and D. Lebotlan, *J. Agri. Food Chem.*, 1989, **37**, 988.
383. M. Guillou and C. Tellier, *Anal. Chem.*, 1988, **60**, 2182.
384. J. Gotz, J. Schneider, P. Forst and H. Weisser, *J. Am. Soc. Brewing Chem.*, 2003, **61**, 37.
385. S. Baroni, *J. Agri. Food Chem.*, 2009, **57**, 3028.
386. D. P. Granizo, B. L. Reuhs, R. Stroshine and L. J. Mauer, *Lwt Food Sci. Technol.*, 2007, **40**, 36.
387. C. L. Gao, M. Stading, N. Wellner, M. L. Parker, T. R. Noel, E. N. C. Mills and P. S. Belton, *J. Agri. Food Chem.*, 2006, **54**, 4611.
388. P. Gambhir, R. K. Pramila, S. Nagarajan, D. K. Joshi and P. N. Tiwari, *Cell. Mol. Biol.*, 1997, **43**, 1191.
389. H. Hickey, B. MacMillan, B. Newling, M. Ramesh, P. Van Eijck and B. Balcom, *Food Res. Int.*, 2006, **39**, 612.
390. I. G. Aursand, L. Gallart-Jornet, U. Erikson, D. E. Axelson and T. Rustad, *J. Agri. Food Chem.*, 2008, **56**, 6252.
391. P. Lambelet, F. Renevey, C. Kaabi and A. Raemy, *J. Agri. Food Chem.*, 1995, **43**, 1462.
392. R. N. M. Pitombo and G. A. M. R. Lima, *J. Food Eng.*, 2003, **58**, 59.
393. H. C. Bertram, M. Rasmussen, H. Busk, N. Oksbjerg, A. H. Karlsson and H. J. Andersen, *J. Magn. Reson.*, 2002, **157**, 267.
394. H. C. Bertram, I. K. Straadt, J. A. Jensen and M. D. Aaslyng, *Meat Sci.*, 2007, **77**, 190.
395. H. C. Bertram, M. D. Aaslyng and H. J. Andersen, *Meat Sci.*, 2005, **70**, 75.
396. L. Laghi, M. A. Cremonini, G. Placucci, S. Sykora, K. Wright and B. Hills, *Magn. Reson. Imaging*, 2005, **23**, 501.
397. M. L. Mateus, D. Champion, R. Liardon and A. Voilley, *J. Food Eng.*, 2007, **81**, 572.
398. L. A. Colnago, M. Engelsberg, A. A. Souza and L. L. Barbosa, *Anal. Chem.*, 2007, **79**, 1271.
399. S. Martini, M. L. Herrera and A. Marangoni, *J. Am. Oil Chem. Soc.*, 2005, **82**, 313.
400. A. E. Marble, I. V. Mastikhin, B. G. Colpitts and B. J. Balcom, *J. Magn. Reson.*, 2007, **186**, 100.
401. H. T. Pedersen, S. Ablett, D. R. Martin, M. J. D. Mallett and S. B. Engelsen, *J. Magn. Reson.*, 2003, **165**, 49.

402. A. Guthausen, G. Guthausen, A. Kamlowski, H. Todt, W. Burk and D. Schmalbein, *J. Am. Oil Chem. Soc.*, 2004, **81**, 727.
403. E. Veliyulin, I. V. Mastikhin, A. E. Marble and B. J. Balcom, *J. Sci. Food Agri.*, 2008, **88**, 2563.
404. O. V. Petrov, J. Hay, I. V. Mastikhin and B. J. Balcom, *Food Res. Int.*, 2008, **41**, 758.
405. H. Stork, A. Gadke and N. Nestle, *J. Agri. Food Chem.*, 2006, **54**, 5247.
406. A. M. Haiduc, E. E. Trezza, D. van Dusschoten, A. A. Reszka and J. P. M. van Duynhoven, *Lwt Food Sci. Technol.*, 2007, **40**, 737.
407. P. Chen, M. J. McCarthy, S. M. Kim and B. Zion, *Trans. ASAE*, 1996, **39**, 2205.
408. B. P. Hills and K. M. Wright, *Magnetic Resonance in Food Science—The Multivariate Challenge*. Royal Society of Chemistry, Cambridge, 2005, p. 175.
409. B. P. Hills and K. M. Wright, *J. Magn. Reson.*, 2006, **178**, 193.
410. J. Perlo, V. Demas, F. Casanova, C. A. Meriles, J. Reimer, A. Pines and B. Blumich, *Science*, 2005, **308**, 1279.
411. J. Perlo, F. Casanova and B. Blumich, *J. Magn. Reson.*, 2006, **180**, 274.
412. B. Blumich, *Chem. Phys. Lett.*, 2009, **477**, 231.
413. E. Danieli, J. Perlo, F. Casanova and B. Blumich, *Magnetic Resonance Microscopy: Spatially Resolved NMR Techniques and Applications*. Wiley, Weinheim, FRG, 2009, p. 487.
414. E. Danieli, *J. Magn. Reson.*, 2009, **198**, 80.
415. M. J. McCarthy, Y. J. Choi, A. G. Goloshevsky, J. S. De Ropp, S. D. Collins and J. H. Walton, *J. Texture Stud.*, 2006, **37**, 607.
416. A. McDowell, *Appl. Magn. Reson.*, 2008, **35**, 185.
417. M. J. McCarthy, J. H. Walton, J. S. De Ropp, S. D. Collins, M. V. Shutov and A. G. Goloshevsky, *Food Sci. Biotechnol.*, 2004, **13**, 848.
418. J. P. Stamps, B. Ottink, J. M. Visser, J. P. M. Van Duynhoven and R. Hulst, *J. Magn. Reson.*, 2001, **151**, 28.
419. E. E. Sigmund, H. Cho and Y. Q. Song, *Concepts Magn. Reson. Part A*, 2007, **30A**, 358.
420. Y. Q. Song, *Magn. Reson. Imaging*, 2005, **23**, 301.
421. Y. Q. Song and X. P. Tang, *J. Magn. Reson.*, 2004, **170**, 136.
422. Y.-Q. Song, E. E. Sigmund and H. J. Cho, in: *Magnetic Resonance Microscopy. Spatially Resolved NMR Techniques and Applications*, S. Codd and J. Seymour (eds.), 2009, p. 31. Wiley, Weinheim.
423. R. A. Prestes, L. A. Colnago, L. A. Forato, L. Vizzotto, E. H. Novotny and E. Carrilho, *Anal. Chim. Acta*, 2007, **596**, 325.
424. G. Mazzanti, E. M. Mudge and E. Y. Anom, *J. Am. Oil Chem. Soc.*, 2008, **85**, 405.
425. S. S. Narine and K. L. Humphrey, *J. Am. Oil Chem. Soc.*, 2004, **81**, 101.
426. D. Gabriele, M. Migliori, R. Di Sanzo, C. O. Rossi, S. A. Ruffolo and B. de Cindio, *Food Hydrocolloids*, 2009, **23**, 619.
427. P. Callaghan, *Curr. Opin. Colloid Interface Sci.*, 2006, **11**, 13.
428. L. Venturi, N. Woodward, D. Hibberd, N. Marigheto, A. Gravelle, G. Ferrante and B. P. Hills, *Appl. Magn. Reson.*, 2008, **33**, 213.

CHAPTER **4**

From Helical Jump to Chain Diffusion: Solid-State NMR Study of Chain Dynamics in Semi-Crystalline Polymers

Yefeng Yao and **Qun Chen**

Abstract   Recent progresses in solid-state NMR study of the helical jump and chain diffusion in semi-crystalline polymers are briefly surveyed. Special emphasis is placed on the nature of helical jump, chain diffusion and their correlation. Combination of the new solid-state NMR techniques allows dynamic elucidation with different length scales from atomistic to nanoscopic and the correlation between them. The morphological factors influencing the dynamics such as the thickness of crystal lamellae, the chain entanglements in the amorphous phase, the interphase between crystalline and amorphous regions of polymer sample are discussed. The correlation between the helical jumps

Department of Physics, Shanghai Key Laboratory of Magnetic Resonance, East China, Normal University, Shanghai, China

Annual Reports on NMR Spectroscopy, Volume 69
ISSN 0066-4103, DOI: 10.1016/S0066-4103(10)69004-7
© 2010 Elsevier Ltd.
All rights reserved.

and the chain diffusion is realised by their different temperature dependence and is further discussed in terms of the defect transfer mechanism.

Key Words: Helical jump, Chain diffusion, Solid-state NMR, Chemical shift anisotropy, Dipolar coupling, Quadrupolar interaction, Exchange, Semi-crystalline polymer, Polyethylene, Interphase, Lamellar thickness, Entanglements, Drawing behaviour.

1. INTRODUCTION

Helical jump and chain diffusion are two related chain dynamics present in the crystalline regions of semi-crystalline polymers.[1,2] It is believed that these dynamics are important for the crystal thickening,[3] ultra-high drawability[4] and α-relaxation.[5] The presence of these chain dynamics was first observed in the 1970s by dielectric and dynamic–mechanical relaxation measurements.[6,7] Afterwards, extensive work has been done to understand the nature of this dynamics and its implications on the properties of semi-crystalline polymers.[8–13] The results of these classic works have been discussed in several extensive reviews and books.[1,14–16] Now, state-of-the-art solid-state NMR provides more detailed knowledge about the nature of the helical jump and the specific chain diffusion present in polymer crystals and improves our understanding of the mechanical behaviour of the polymer at the microscopic level. In this chapter, we briefly survey the recent progress in the study of helical jump and chain diffusion. The discussion is focussed on the morphological factors influencing the dynamics such as the thickness of crystal lamellae, the chain entanglements in the amorphous phase, the interphase between crystalline and amorphous regions of the sample. In the final part of this chapter, the correlation between the helical jump and the chain diffusion—the efficiency problem and the nature of the defect transfer mechanism—is discussed.

2. NMR STRATEGIES FOR PROBING CHAIN DYNAMICS IN SOLID POLYMERIC MATERIALS

NMR provides powerful methods that allow detailed studies of molecular dynamics in the frequency range from Hz to MHz. Roughly speaking, fast dynamics (MHz) can be characterised by NMR relaxation time measurements, intermediate dynamics (kHz) by monitoring the motionally modulated NMR interactions (e.g. quadrupolar interaction (^{2}H), chemical shift anisotropy (CSA) and dipolar coupling) and slow dynamics (Hz) by exchange experiments. The following discussion includes spin–lattice relaxation measurement, probing motionally modulated/averaged NMR interaction and longitudinal spin exchange experiment, which are directly or indirectly related to the topics of this chapter.

2.1. Spin–lattice relaxation measurement

If spins are placed in a magnetic field without disturbance for a long time, they reach the thermal equilibrium state. The populations in different energy levels obey the Boltzmann distribution corresponding to the temperature of the environment. RF pulses can disturb the equilibrium state of the spins. For example, a π-pulse will invert the population distribution, whereas a $\pi/2$-pulse will equalise the populations and create coherence. The spin–lattice relaxation (or termed T_1 relaxation) is the process by which the excited spins may return to the equilibrium state. Based on the model of fluctuating local fields, the relationship between the relaxation rate $1/T_1$ and the correlation time of molecular motion τ_c can be expressed generally as follows:

$$T_1^{-1} \propto \frac{\tau_c}{1 + (2\pi\omega_L\tau_c)^2} \tag{1}$$

where ω_L is the Larmor frequency. Equation (1) shows that T_1 has a minimum which occurs when the correlation time of molecular motion τ_c is equal to $(2\pi\omega_L)^{-1}$. Or in other words, the T_1 process is sensitive to the molecular motion with the characteristic frequency of motion $(\tau_c^{-1}/2\pi)$ close to the Larmor frequency ω_L. Therefore, T_1 usually can be a measure of the high-frequency motions (MHz) in a system. For polymers, such high-frequency motions are often found in the amorphous regions.

2.2. Probing motionally modulated/averaged NMR interaction

NMR interactions generally are anisotropic. The angular dependence of the NMR signal in high magnetic fields is similar for all NMR interactions and reads as[17]

$$\omega = \omega_L + \frac{1}{2}\delta(3\cos^2\theta - 1 - \eta\sin^2\theta\cos 2\phi) \tag{2}$$

where ω_L is the Larmor frequency, δ describes the strength of the anisotropic interaction: that is, CSA, dipolar coupling and quadrupole coupling for ^2H and η is the asymmetry parameter describing the deviation of the anisotropic interaction from axial symmetry. The angles θ and ϕ are the polar angles of the magnetic field B_0 in the principal axes system of the interaction tensor, which are directly related to the geometry of tensor in the molecular frame. When molecular dynamics is considered, the polar angles become time dependent. The motional modulation effects will clearly show up when the characteristic frequency of motion $(\tau_c^{-1}/2\pi)$ matches the strength of the NMR interaction. For the intermediate molecular dynamics, the frequency of motion (kHz to hundreds of kHz) is in the frequency range of the strength of anisotropic chemical shift, dipolar coupling and quadrupolar coupling of ^2H. The information on molecular dynamics can thus be extracted by monitoring and analysing the characteristic NMR signals from the motionally modulated/averaged NMR interactions. In the following, two examples will be given to demonstrate the motionally averaged effect on CSA and ^1H–^{13}C dipolar coupling.

• *Line shape analysis on a static powder sample*

   The line shape of the static NMR powder pattern contains preliminary information about the molecular dynamics. Examples of ^{13}C CSA line shapes for the all-*trans* conformer of CH_2 group are shown in Figure 1. The three cases considered include an isotropic powder pattern, the motionally averaged ones from uniaxial rotation along the chain backbone and a fast isotropic motion.

   The left pattern in Figure 1 is often observed in crystalline regions of polyethylene, where the CH_2 units either stay statically or involve in the 180° jump motion. In both cases, the symmetry of the ^{13}C CSA tensor remains unchanged, resulting in the same line shape. The middle pattern in Figure 1 has been found for the CH_2 groups in the non-crystalline regions of solution-crystallised ultra-high molecular weight polyethylene (UHMW-PE). Such a special line shape originates from an axially symmetric tensor and has been considered as the indication of the axial rotation of a CH_2 unit along the local chain backbone present in the non-crystalline regions of sample.[18] The third pattern shown in Figure 1 exhibits an isotropic line shape. Such a line shape has been shown to be quite common in non-crystalline regions of solid polymers, resulting from the fast isotropic Brownian motion of chain segments.

• *Monitoring averaged dipolar coupling*

   Molecular motions can be monitored by detecting the motionally averaged dipolar coupling constant. The idea of this method is based on the fact that fast intermediate molecular motions ($\tau_c^{-1}/2\pi > 10^4\,\text{Hz}$) usually lead to a characteristic reduction of the observed dipolar coupling constant.[19] The extent of the coupling constant reduction yields information on fast intermediate molecular motions. Figure 2 shows examples in which the motions induce a reduction of the

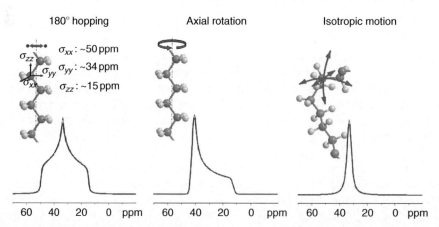

Figure 1   Simulated ^{13}C CSA patterns of CH_2 group in PE. The tensor orientation of CH_2 group in PE is indicated in the left chain model picture: σ_{xx} is parallel to the proton–proton internuclear vector, σ_{yy} is parallel to the H–C–H angle bisector and σ_{zz} is parallel to the polymer chain axis. The simulated powder patterns were obtained from WEBLAB (http://weblab.mpip-mainz.mpg.de/weblab/weblab.html).

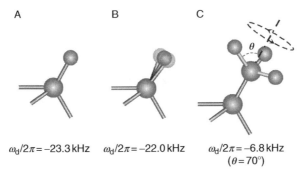

A   B   C

$\omega_d/2\pi = -23.3$ kHz $\omega_d/2\pi = -22.0$ kHz $\omega_d/2\pi = -6.8$ kHz
                                                      $(\theta = 70°)$

Figure 2   The motional-averaging effects on the dipolar coupling of a typical ^{13}C–^{1}H group: (A) a static ^{13}C–^{1}H group; (B) a ^{13}C–^{1}H group with libration; (C) a ^{13}C–^{1}H group in a methyl group ($\theta = 70°$).

dipolar coupling constant in the case of a ^{1}H–^{13}C spin pair. In the static case, the distance between a ^{13}C atom and the direct-bonded ^{1}H atom is ~0.109 nm, corresponding to a direct dipolar coupling of $|\omega_d/2\pi| = 23.3$ kHz. In most cases at normal temperatures, ^{1}H in a ^{1}H–^{13}C pair has a rapid librational motion around its original site, and the experimentally observed dipolar coupling is thus always less than 23.3 kHz. For example, the ^{1}H–^{13}C dipolar coupling of the ^{1}H–^{13}C group in L-alanine is $|\omega_d/2\pi| = 22.0$ kHz. Moreover, if more motions are involved for the ^{1}H–^{13}C pair, the dipolar coupling could be further reduced. A typical case can be found in methyl groups, where the fast rotation about its three-fold symmetry axis leads to a marked reduction of the ^{1}H–^{13}C dipolar coupling value. Assuming that the angle between the rotation axis and one of the three C–H bonds is 70,° as indicated in Figure 2C, computer simulation shows that the ^{1}H–^{13}C dipolar coupling in this case would be reduced to $|\omega_d/2\pi| = 6.8$ kHz, quite close to the experimental value.

2.3. Longitudinal spin exchange experiment

For the studies of slow motion ($\tau_c^{-1}/2\pi <$ kHz), the longitudinal exchange experiment has been proven to be a very useful method. The basic principle of the exchange experiment (shown in Figure 3) is to measure the NMR frequency of spins at different times and then to detect the slow dynamics by monitoring the change in NMR frequency. Based on this principle, various longitudinal magnetisation exchange experiments have been developed. These various exchange experiments may look very different, but they do share some common features. In the literature, the longitudinal magnetisation exchange experiments have been applied to monitor the conformation transition in amorphous regions of polymers, the jump motions in polymer crystals and the chain diffusion between amorphous and crystalline regions of polymers. Figure 3 shows a general scheme of a 2D exchange experiment that is widely used for detecting the slow dynamic process in polymer systems.

As multi-dimensional NMR experiments are always very time consuming, the longitudinal spin exchange experiment is often performed with its 1D version.

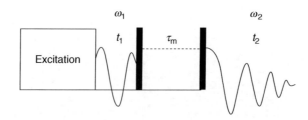

Figure 3   The general pulse scheme of a 2D exchange experiment.

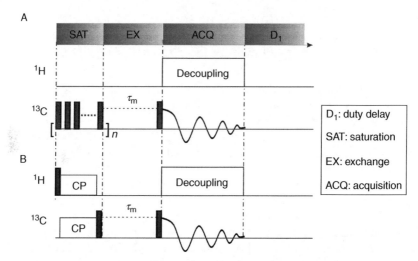

Figure 4   The pulse schemes of 1D ^{13}C exchange experiments: (A) single-pulse-based exchange experiment; (B) Cross Polarization (CP)-based exchange experiment.

In the following section, two 1D longitudinal spin exchange experiment have been used to monitor the chain diffusion process in semi-crystalline PEs. The pulse sequences of these experiments are illustrated in Figure 4. The principle of these experiments is the same as the above 2D longitudinal spin exchange experiment and thus will not be discussed here. For more details of these experiments and how these pulse sequences are able to monitor the chain diffusion in semi-crystalline polymers, the reader is referred to the literature.[20]

3. HELICAL JUMP AND CHAIN DIFFUSION IN SEMI-CRYSTALLINE POLYMERS

The helical jump process in polymer crystallites is often described as a combination of rotation and translation such that the polymer chain before and after the jump is in energetically equivalent, crystallographically allowed orientations and positions. For example, in the 3_1 helix of Polyoxymethylene (POM), a rotation by 120° combined with a translation by one –OCH$_2$– unit fulfils such a requirement.[2]

Assuming that the defect created by the jump process cannot be stably kept in the crystallites, the jump thus will lead to a decrease of chain length in the amorphous regions by one repeat unit on one side of the crystallite and an increase by the same length on the other side, which means that the chain stem achieves one diffusion step. In this sense, therefore, the local jump process and the medium-range chain diffusion are strongly coupled. However, it remained unclear for a long time whether every single jump does result in a diffusion step of the chain stem. Moreover, as the long polymer chains always traverse between crystalline and amorphous regions, the chain dynamics in the amorphous regions but neighbouring the crystallites might also couple to the dynamics in the crystalline region because of the chain connectivity. This, in fact, relates to the problem of the packing arrangement of polymer chain between crystalline and non-crystalline regions of sample, especially the chain arrangement in the interphase between crystalline and non-crystalline regions (or termed as the interfacial components). Different packing arrangements give rise to different spatial restrictions, which in turn manifest themselves in the dynamic behaviour of the polymer chains. The influencing factors on the chain dynamics in polymer crystals and the correlation between the local helical jump and the medium-range chain diffusion are the two foci of the discussion in the following sections.

3.1. Helical jump in polymer crystals

The presence of helical jump in polymer crystals was first realised from the so-called α-relaxation process in dynamic–mechanical relaxation measurements.[5,7] Later, Kentgens et al. provided a detailed description of such chain dynamics in semi-crystalline POM by 2D exchange ^{13}C NMR under magic angle spinning (MAS).[21] By using 2D–nD exchange NMR without sample rotation, Schmidt-Rohr and Spiess have shown a very detailed picture of helical jump in many polymer crystals, including POM, poly(ethylene xoide) (PEO), isotactic Polypropylene (iPP).[1] The representative work of helical jump from Spiess's group is the 2D–3D exchange spectra in a highly oriented POM sample, (–O–CH$_2$–)$_n$.[2] In such a sample, the highly oriented chains within ±4° are parallel to each other, forming the crystallites (a 9$_5$-helical structure). The only degree of freedom for the polymer chain, therefore, is the rotation of chain segments around the chain axis, which is parallel to the macroscopic drawn axes. Therefore, the patterns given by the 2D exchange NMR have a unique relation with the rotation angle around the chain axis.

Figure 5A shows the pattern of highly oriented POM without sample rotation at $T = 360$ K. In this pattern, the characteristic ridges indicate the presence of three kinds of helical jumps present in the crystallites. The explanation for such a pattern is that the helical jump process in the crystallites involves not just a flip-flop between two adjacent orientations but can lead to long-range diffusion. The simulated patterns in Figure 5B confirm this point of view. However, comparison of the ridge intensity generated by the different jump motion shows that the jump motion of ±200° jump angles seems to have a much higher possibility than those of ±400° and ±600°. This phenomenon, on one hand, is indicative of the low

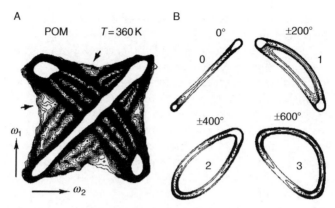

Figure 5 Elucidation of chain dynamics in the crystallites of highly oriented POM at 360 K by 2D exchange ^{13}C NMR with a mixing time of 1 s: (A) experimental spectrum; (B) simulated spectra for different rotation angles around the chain axis. The experimental spectrum shows the superposition of the patterns of one, two and three steps of 200° rotation. The three-step rotation (pattern marked by arrows) is coupled to a displacement by ±0.5 nm along the chain axis. From Ref. 4 (Figure 2).

possibility of the return jumps in the crystallites, while, on the other hand, also indicates that one local jump process does not necessarily result in one diffusion step for the involved chain stem because of the presence of the return jump.

For the simplest polymer—polyethylene—although the 180° helical jump in the crystals has been proposed for a long time, the direct experimental observation was not achieved until recently, because the 180° helical jump just inverts the chemical shift, germinal ^{1}H–^{1}H dipolar, ^{13}C–^{1}H dipolar and ^{2}H quadrupolar tensors without changing the corresponding NMR frequencies. The simple exchange NMR is thus not capable of detecting the helical jump process in polyethylene crystals. Recently, by 2D exchange spectrum of the ^{13}C–^{13}C dipolar coupling in ^{13}C–^{13}C pair labelled PE, the helical jump in PE crystals was directly observed for the first time by Schmidt-Rohr's group.[22] In that work, the 180° helical jump was monitored in terms of the reorientation of ^{13}C–^{13}C internuclear vectors in the crystallites of 5% ^{13}C–^{13}C pair labelled PE. The cartoon picture in Figure 6A illustrates the ^{13}C–^{13}C bonds reorientation process from θ_1 to θ_2. During this reorientation process, the dipolar coupling of the ^{13}C–^{13}C pair will change the orientation angle and thus the dipolar frequency. The correlation function of the chain reorientation thus can be measured directly in terms of the mixing time dependence of stimulated echoes of the dipolar coupling by the pulse sequence in Figure 6B.

Figure 7A shows the experimental dipolar exchange spectrum using the mixing time $t_m = 100$ ms. It exhibits intensity far from the diagonal, which indicates large changes in frequency and orientation due to the helical jumps during t_m. Figure 7B is the simulation spectrum of the above experiment with a 180° jump rate of 15 s^{-1}. As one can see, this simulation spectrum is very similar to the experimental one. The slight difference in the asymmetry of the spectra was attributed to the double quantum filtering.

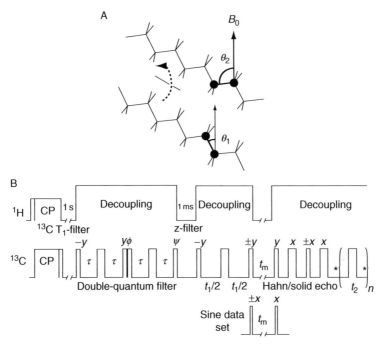

Figure 6 (A) General case of a 180° flip in an unoriented sample, producing a great change in the $^{13}C-^{13}C$ bond orientation, from θ_1 to θ_2; (B) 2D pulse sequence to detect reorientations of $^{13}C-^{13}C$ dipolar tensors. From Ref. 22 (Figures 1 and 2).

Using the same labelled PE sample, Schnell et al. demonstrated another NMR technique for directly monitoring the 180° jump in PE crystals.[23] The pulse sequence used is shown in Figure 8, which actually is a reduced 3D experiment consisting of two dipolar double-quantum (DQ) MAS experiment blocks. The dipolar coupling between the $^{13}C-^{13}C$ spin pairs is used to generate the DQ coherences in the first and second dimensions. The jump motion occurring during the mixing time can thus be monitored by probing the $^{13}C-^{13}C$ dipolar coupling tensors because of their orientation dependence. Figure 9A shows the $^{13}C-^{13}C$ DQ MAS sideband pattern from the rotor encoded DQ sideband experiment. Simulation yields a $^{13}C-^{13}C$ dipolar coupling strength of $D_{ij} = 2\pi(1.89 \pm 0.01)$ kHz. Figure 9C shows the sideband patterns obtained using the pulse sequence in Figure 8. In these sideband patterns, the centreband reflects moieties that have remained in, or returned to, their initial position during the mixing time, whereas the sidebands predominantly contain the contributions from the moieties that have undergone a reorientation during the mixing time. Simulation shows that the $^{13}C-^{13}C$ reorientation angle in the PE crystallites is $70 \pm 5°$, similar to that observed by Schmidt-Rohr.

The above works exhibit uncontroverted proofs for the presence of large-amplitude helical jumps in polymer crystals. However, as local dynamics, the helical jump of a chain segment in principle does not inevitably lead to the

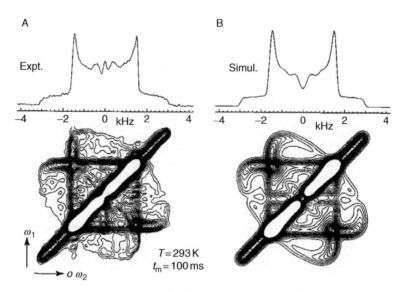

Figure 7   2D exchange spectrum of the $^{13}C-^{13}C$ dipolar coupling in $^{13}C-^{13}C$ pair labelled HDPE: (A) experimental spectrum, $T = 293$ K, $t_m = 100$ ms; (B) corresponding simulation for 180° flips. Thirty contour lines are plotted between 0.8% and 20% of the maximum intensity. The slight asymmetry of the spectrum is due to the double-quantum filter, which does not excite equally the signals for all $^{13}C-^{13}C$ orientations. Integral projections onto the ω_2 axis are shown at the top. From Ref. 22 (Figure 3).

medium-range chain diffusion. In this sense, the efficiency of the helical jump thus becomes quite intriguing. The detailed discussion about the correlation between the helical jump and chain diffusion—whether every helical jump could always efficiently result in a diffusion step of the chain stem—is given in the following sections.

3.2. Chain diffusion in semi-crystalline polymers and its influencing factors

Concerning the chain connectivity, the helical jump in polymer crystals in fact indicates the presence of chain diffusion between amorphous and crystalline regions. The first direct experimental evidence for the chain diffusion between amorphous and crystalline regions of semi-crystalline polymers was demonstrated in high molecular weight PE by Schmidt-Rohr and Spiess.[24] Using 2D MAS exchange experiments, they showed a clear exchange process occurring between the local conformations in the amorphous and crystalline regions of PE at elevated temperatures. Figure 10C is the 2D MAS exchange spectrum of UHMW-PE given in their work (acquired at 363 K with an exchange time of 1 s). The appearance of an off-diagonal peak in the spectrum indicates that the large-scale chain diffusion between crystalline and amorphous regions occurs on the time scale of several seconds.

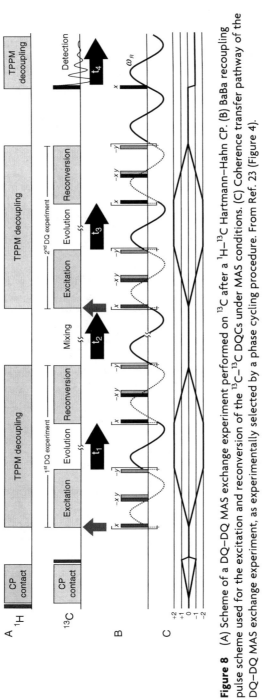

Figure 8 (A) Scheme of a DQ–DQ MAS exchange experiment performed on ^{13}C after a ^{1}H–^{13}C Hartmann–Hahn CP. (B) BaBa recoupling pulse scheme used for the excitation and reconversion of the ^{13}C–^{13}C DQCs under MAS conditions. (C) Coherence transfer pathway of the DQ–DQ MAS exchange experiment, as experimentally selected by a phase cycling procedure. From Ref. 23 (Figure 4).

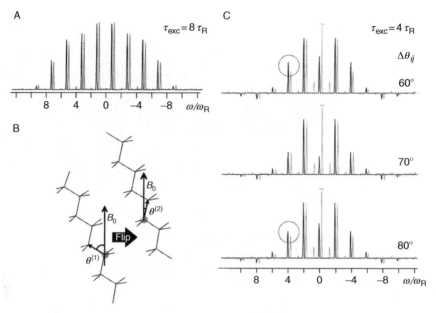

Figure 9 (A) Calculated (black) and experimental (gray) $^{13}C-^{13}C$ DQ MAS sideband pattern for the crystalline phase of poly(ethylene), yielding a $^{13}C-^{13}C$ dipolar coupling strength of $D_{ij} = 2\pi$ (1.89 ± 0.01) kHz. (B) 180° chain flip in the crystallites, giving rise to a reorientation of 112° (≡68°) of the $^{13}C-^{13}C$ dipolar tensor. (C) Calculated (black) and experimental (gray) MAS sideband patterns of 'referenced' DQ–DQ exchange experiments. The experimental pattern is recorded for a mixing time of $t_m = 600$ ms at $T = 300$ K and yields a reorientation angle of 70 ± 5° for the $^{13}C-^{13}C$ dipolar tensor. The circles mark the deviations from the patterns calculated for 60° and 80°, respectively, in the sensitive spectral region. From Ref. 23 (Figure 5).

On traversing between the crystalline and amorphous regions of semi-crystalline polymers, the topological constraints on polymer chains may show their influences on the chain diffusion. Considering the packing arrangement of the chain, the constraints include the restriction from the crystal lattice, chain entanglements in the amorphous regions and the interphase between crystalline and non-crystalline regions. Recently, these influences on the chain diffusion in semi-crystalline PEs were studied extensively by employing 1D ^{13}C exchange NMR.[20] The pulse sequence used in the work is the same as the 'Torchia' pulse sequence[25] (see Figure 11A). Provided that chain diffusion is the only possible way to depolarise ^{13}C in the crystalline regions of PE for short times t, the decay of the crystalline signal monitors directly the chain diffusion. The relation between the change of ^{13}C polarisation and chain diffusion in the experiment is illustrated in Figure 11B and C. Equation (3) describes the relationship between the normalised signal intensity and the diffusion time:

$$\frac{I_t}{I_0} = 1 - \frac{\sqrt{2nD}}{L_0} t^{1/2} \qquad (3)$$

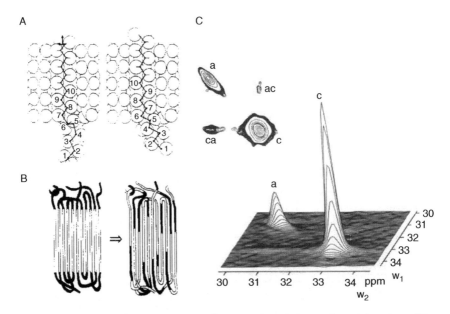

Figure 10   CP-MAS 2D exchange spectrum of UHMW-PE at 363 K, with a mixing time of 1 s. The intensity of the cross-peak (ca) is 2.5% of that of the diagonal peak (c), while the (ac) peak intensity makes up 6% of that in the diagonal peak (a). The former ratio is more decreased by T_1 relaxation, so the effect due to exchange amounts to roughly 5%. From Ref. 24 (Figure 2).

where I_0 and I_t represent the intensity of crystalline signal at the exchange time equal to 0 and t, respectively, L_0 is the lamellar thickness of crystallite, n is the dimensionality factor of the diffusion and D is the diffusion coefficient. Compared to the previous method, the big advantage of this method is its ease in quantifying the signal intensity and therefore the chain diffusion coefficient. For more details, the reader is referred to the original paper.[20]

3.2.1. Influence of lamellar thickness

Three samples have been chosen to demonstrate the influence of lamellar thickness on chain diffusion: solution-crystallised UHMW-PE (SC-PE), with a lamellar thickness $L_0 \sim 12$ nm, the annealed sample of $L_0 \sim 24$ nm and a solution-spun Dyneema® fibre with extended crystalline regions. For the annealed sample in the solid state, it is stated that topological constraints due to entanglements in the non-crystalline regions do not change, provided there is no change in the draw ratio. The decay of the crystalline signal in these samples is displayed in Figure 12 and compared to a hypothetical exponential decay of the ^{13}C magnetisation due to T_1 relaxation with $T_1 = 1500$ s. As noted before, the data do not follow a relaxation curve; instead, the decay at the initial stages follows the $t^{1/2}$ behaviour expected for chain diffusion.

Fitting the early parts of the signal decay curves yields slopes of -0.12 s$^{-1/2}$ for the original SC-PE and -0.058 s$^{-1/2}$ for the lamella-doubled SC-PE, very closely related by a factor of 2. Provided the diffusion coefficient D is the same for the two

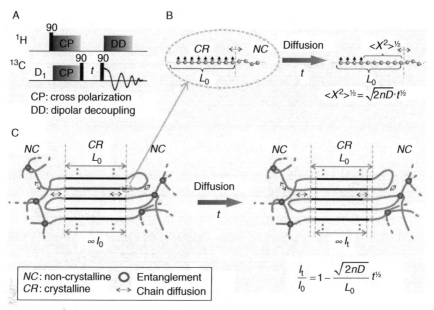

Figure 11 (A) Demonstration of the 1D ^{13}C exchange pulse sequence. (B) The cartoon pictures mimic the change of polarisation during the experiment, where the small arrows are the representation of polarisation and the spheres are the coarse grain representation of the chain segments. The polarisation in the crystalline regions is exaggerated for convenience of understanding. (C) Illustration of the polarisation decrease resulting from the chain diffusion in solid PE. At one time point, the chain segments in the crystalline regions are fully polarised (marked as black). Then, through the chain diffusion the chain segments gradually lose the polarisation. From Ref. 20 (Figure 1).

samples, it is then apparent from Equation (3) that the slopes reflect the different lamellar thicknesses prior to and after annealing. Thus, one could conclude that this is indeed the case and the observed decay in the initial stages is due to chain diffusion between the crystalline and non-crystalline regions.

The signal decay of the UHMW-PE fibre is also shown in Figure 12. Again, the straight line in this sample indicates chain diffusion, yielding a slope of -0.016, approximately eight times smaller than that of the original SC-PE. If we again assume that the chain diffusion coefficient in this fibre is similar to that in the original SC-PE, the thickness of crystallites in this fibre is then estimated to be around 100 nm, which is well within the range of crystal thicknesses reported for the commercially available solution-spun fibres. Because of the fibrillar morphology of the sample, however, L_0 in the fibre should better be interpreted as the averaged length of extended all-*trans* conformations, rather than crystal thickness.

Probably the most intriguing feature here is the fact that the data suggest a constant chain diffusion coefficient $D \approx 1.4 \times 10^{-18}$ m^2 s^{-1} at 320 K in the different samples having different lamellar thicknesses. At first sight, this seems to be at variance with the studies of the α-relaxation of PE by mechanical, dielectric and NMR spectroscopic studies, where the relaxation rate largely reduces with

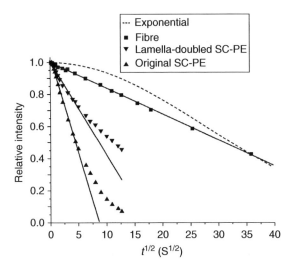

Figure 12   The decay of crystalline signal intensity in the original SC-PE (▲), the lamellar-doubled SC-PE (▼), the Dyneema® fibre (■) plotted against the square root of the exchange time. The pure exponential decay curve with a relaxation time $T_1 = 1500$ s (dash line) is shown in the plots for a comparison. The experimental temperatures were 320 K. From Ref. 20 (Figure 2).

increasing lamellar thickness. All those experiments, however, cannot differentiate between local processes such as defect mobility and medium-range translational motion. The above data probe chain diffusion directly in well-defined samples and suggest that defect-driven mechanisms are responsible for local relaxation but of minor importance only for chain diffusion.

3.2.2. Influence of chain entanglements

Two samples (SC-PE and MC-PE) have been chosen to demonstrate the influence of non-crystalline structure on chain diffusion. The MC-PE sample was prepared on cooling the melt of the SC-PE sample at a rate of 10 °C/min. Different crystallisation conditions result in different topological constraints in the non-crystalline regions. For MC-PE, disordered random coils are predominant, whereas for SC-PE *trans* conformers are more probable. But, though the two samples have different crystallinity, the crystal structures and lamellar thicknesses are the same.

Figure 13 compares the decays of the crystalline signal of SC-PE and MC-PE at 320 K. The signal in SC-PE decays much faster than in MC-PE. Since both samples have similar lamellar thicknesses, this indicates different chain diffusion coefficients D for the two samples. Using Equation (1) with $n = 1$, we calculate $D = 1.4 \times 10^{-18}$ m^2 s^{-1} for SC-PE and $D = 4.7 \times 10^{-20}$ m^2 s^{-1} for MC-PE. The apparent difference in the diffusion coefficients can be attributed to the different transition entropy that the chain has to overcome when it diffuses between crystalline and non-crystalline regions. In accordance with this, the rapid decay of the signal in MC-PE ($\approx 8\%$) at very early times is ascribed to the chain segments in the interphase between crystalline and non-crystalline regions. There the local chain orientation is preferentially along the normal to the lamella.

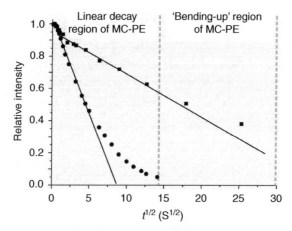

Figure 13   The decay of crystalline signal intensity in SC-PE (●) and MC-PE (■) plotted against the square root of the exchange time t for two different temperatures. The experimental temperatures were 320 K. From Ref. 20 (Figure 3).

When the polymer segments have diffused long distances, or equivalently for long times, the decay curves in Figure 13 are bend upwards. This curvature cannot be attributed to ^{13}C T_1 relaxation in the PE crystallites, since the relaxation process would speed up the decay, leading to a bending down of the curve. Instead, the bending up in Figure 13 indicates slowing down of chain diffusion, and is thus rather indicative of additional constraints imposed on this process. The origin of this feature may be related to chain entanglements, which can retard the chain diffusion. Moreover, the chain diffusion itself could lead to bending up of the curve at long times, if considered as curvilinear diffusion.[26]

3.2.3. Influence of hot drawing

Hot drawing can induce substantial changes in the morphology of PE samples, which then manifest themselves in the chain diffusion. Four samples have been examined to study this aspect, including two hot-drawn samples (prepared from SC-PE with draw ratio $\lambda=5$, 30); a Dyneema® fibre with $\lambda \gg 30$; and the original SC-PE with $\lambda=1$. The signal decays of these samples are plotted in Figure 14. With increasing draw ratio, the decay is slowed down. The original undrawn SC-PE and the sample with $\lambda=5$ show the fastest decays. When λ is increased to 30, however, the signal decay is significantly slowed down and a pronounced bending up is observed. For the fibre sample, the signal decay is strictly linear with the slowest decay.

First, one may consider the linear decay region, that is, the initial 20% decay for $\lambda=30$ and the initial 50% decay for all the other samples. Fitting the curves in these linear regions yields four slopes, that is, $-0.1192 \text{ s}^{-1/2}$ for the undrawn SC-PE, $-0.1071 \text{ s}^{-1/2}$ for $\lambda=5$, $-0.063 \text{ s}^{-1/2}$ for $\lambda=30$ and $-0.016 \text{ s}^{-1/2}$ for the fibre. Attributing the variation of these slopes of the drawn solution-crystallised samples to the change of lamellar thickness as above, the lamellar thicknesses in these

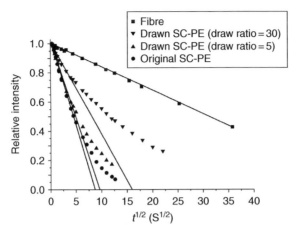

Figure 14   The decay of crystalline signal intensity in the original SC-PE (●), the drawn SC-PE sample with $\lambda=5$ (▲), the drawn SC-PE sample with $\lambda=30$ (▼), the Dyneema® fibre (■) plotted against the square root of the exchange time t. These data were acquired from the exchange experiments performed at 320 K. From Ref. 20 (Figure 4).

samples are estimated to be 13.3 nm for the slightly drawn sample ($\lambda=5$) and 22.7 nm for the drawn sample ($\lambda=30$), in remarkable agreement with results from the small angle X-ray scattering (SAXS) experiments.[27] This further confirms that the chain diffusion constant is not strongly affected by the drawing process, and that the increase in crystal thickness slows down the chain diffusion between crystalline and non-crystalline regions. It is to be noted that easy drawability in the SC-PE arises as a result of the absence of the entanglements and the presence of extended chain conformations in the non-crystalline regions, which tend to be maintained during deformation.

Following Peterlin,[28] complicated changes in the morphology of the sample during the drawing process are anticipated. Studies performed on the structural deformation suggest that, on drawing, initially the non-crystalline (amorphous) component is deformed, which is followed by the orientation of the crystallites in the drawing direction. Ultra-drawing, with draw ratios $\lambda>15$ results in conversion of the folded chain crystals to the extended chain crystals by chain sliding along the c-axis accompanied by increasing crystallinity. The ^{13}C NMR signal decay reflects the increasing length of all-*trans* conformers. From the above it is then expected that in our series of samples the deviation from simple diffusion will be particularly strong for $\lambda=30$, where the reorganisation of the crystallites is most pronounced. The bending-up curvature then most probably results from a distribution of lamellar thicknesses as also indicated by an increase in the width of the SAXS peaks, as well as from a non-uniform distribution of entanglements. Contrary to that, the fibre sample shows a linear decay against $t^{1/2}$ within the whole experimentally accessible range until the signal decays to about 55% of the initial value. This could be indicative of the near absence of chain entanglements in this fibre sample, as a result of being solution spun.

3.2.4. Influence of interfacial components between the crystalline and non-crystalline area

The interfacial components in semi-crystalline polymers, sometimes also termed as 'interphase', 'intermediate' and 'semi-ordered' in the literature, are the components bridging the non-crystalline and crystalline regions of sample.[29] The common picture of interfacial components in the literature is described as follows: Assuming the crystallites are of an infinite extent in the basal plane, then at small distances away from each crystallite surface most of the chain segments will just originate from the crystallite. The average chain orientation thus will not be as random as in the bulk amorphous matrix but will be preferentially distributed around the normal to the crystallite surface. This spatial constraint in the interfacial components reduces the motional degree of freedom of chain motion present in the non-crystalline regions, but obviously is not as strong as that from the crystal lattice. As a result, the dynamics of the interfacial components most likely will take a transitional mode between the fast isotropic Brownian motion in the amorphous regions and the well-defined helical jump in crystals. In this context, therefore, if the chain dynamics in the amorphous and crystalline regions are coupled to form the medium range diffusion, one might easily anticipate that the motion mode of the interfacial components will be important for the chain diffusion.

However, because of the semi-ordered nature of the interfacial components, monitoring such components gives a big challenge for the analysis techniques. NMR approaches for monitoring such components were mainly focussed on the relaxation measurements, because the semi-ordered structures of interfacial components often make them indistinguishable from the crystalline structure. Chen and co-workers have demonstrated a strategy to monitor the interfacial components easily.[30,31] The strategy can be simply described as follows: Through manipulation on ^{1}H magnetisation, an initial state in which ^{1}H magnetisation of the crystalline region is relatively enriched in the crystal surface area was created. Such an initial state was then mapped to ^{13}C magnetisation through cross polarization (CP). ^{13}C T_1 measurement was carried out under such an initial state subsequently. They found that relaxation curve of the crystalline region measured under such an initial state is markedly different from that obtained by the conventional method. The contribution from the component of fast relaxation was greatly enhanced, indicating that such crystalline component corresponds to the crystal surface. By combining the above strategy with the ^{1}H spin diffusion, Zhang et al. studied the proton second moment in the surface of the orthorhombic PE crystals.[32] They found that the proton second moment increases with the spin-diffusion time, reflecting a gradual decrease of the chain mobility from the surface to the inner part of the crystals. Such a gradual decrease of the chain mobility, on one hand, might indicate the gradual change of the structures and the resultant spatial constraints on the chain segments. On the other hand, this phenomenon in fact might also indicate the correlation between the chain dynamics of the crystalline and the non-crystalline chain segments in the samples.

Recently, Yao et al. studied the correlation between the local dynamics in the different regions of a sample and the influence of the local dynamics in the interfacial area on the medium-range chain diffusion.[18,33] The samples used in that work included two UHMW-PE samples crystallised from dilute solution and from melt. It was found that the chain dynamics present in the non-crystalline regions of the solution-crystallised UHME-PE is close to a local axial rotation around the chain backbone, while the chain dynamics in the non-crystalline regions of melt-crystallised sample is more like a fast isotropic motion. But interestingly, the fast isotropic dynamics in the non-crystalline regions of the melt-crystallised sample does not facilitate chain diffusion. In contrast, fast chain diffusion was observed in the solution-crystallised sample. The co-existence of the local restricted chain mobility with the enhanced chain diffusion in the solution-crystallised sample, then, is explained by the facilitating effect of the local axial rotation around the chain backbone on the chain diffusion. The chain folding structure present in the non-crystalline regions has been considered to give rise to the spatial restriction, which results in the local axial rotation.

The influence of interfacial components on the chain diffusion can be realised from the cartoon picture of the chain dynamics shown in Figure 15. In this picture, a chain stem is moving between the non-crystalline region (right side) and the crystalline region (left side). The tube represents spatial and dynamic restrictions of the PE chain from its environment and the diameter of tube indicates the degree of restriction. In this model, the chain diffusion is achieved by the sliding of the chain stem in the tube. Due to the chain connectivity, the motion of the non-crystalline chain portion is connected with that of the chain portion in the crystallite. Taking into account that the chain diffusion in the crystal is always along the chain backbone, it is easy to imagine that motions in the non-crystalline regions involving extended *trans* segments rotating around the local backbone are very effective for the chain diffusion, whereas random motions such as almost isotropic fluctuations of chain segments are not compatible with and may even be an obstacle for chain diffusion. Following this idea, reducing the motional degree

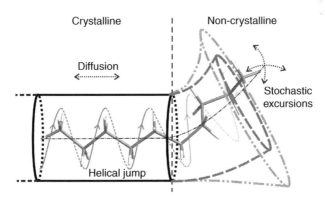

Figure 15   Illustration of chain motions in PE. The individual helical jump generates the translation of chain stem by one CH_2 unit. These initial translation steps evolve in the course of time into a diffusive motion between crystalline and non-crystalline regions, which is observed in the NMR experiments. The tubes represent spatial and dynamic restrictions of the PE chain due to its local environment. From Ref. 18 (Figure 7).

of freedom of the non-crystalline chain motion, that is, from random motion to the motion along the extended chain conformations, can favour chain diffusion. In thermodynamics terms, the reduction of the motional degrees of freedom of the chain motion in fact is lowering the entropy change involved in the translation motion from the crystalline to the non-crystalline environment. This explains why local restricted chain mobility can co-exist with the fast chain diffusion in the solution-crystallised sample. The influence of interfacial components on the chain diffusion thus can be understood along this line.

3.3. The correlation between helical jump and chain diffusion

From the above discussion, the presence of the coupling between the helical jump and the chain diffusion obviously is incontrovertible. But one question remains: that is, whether one helical jump inevitably results in one diffusion step of the chain stem. In fact, it is not logical to expect that every helical jump inevitably leads to a diffusion stem, as the jumps might be reflected by some obstacle in the crystals or annihilated by the jumps from the opposite direction, although the helical jump of the stem coupled to a translation in accordance with the crystal structure seems to be a very effective mechanism for the formation of chain diffusion. But the above experiments monitoring the helical jump, the chain diffusion, or the dielectric and mechanical relaxation separately cannot tell the relationship between movements of whole stems and local jumps of segments, which might or might not be coupled to translation of the whole stem.

The temperature dependence of the local jump rates sheds light on this issue. Figure 16 shows the local jump rates in PE crystals measured by $^{13}C-^{13}C$ dipole–dipole couplings (inverted triangle) and double quantum MAS exchange NMR (upright triangle). Using the chain diffusion coefficient D obtained from the ^{13}C

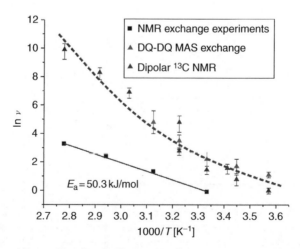

Figure 16 Arrhenius plots of local and effective jump rates in melt-crystallised PE samples. From Ref. 33 (Figure 4).

exchange experiment, one is also able to calculate the effective jump rate, which is directly associated with the chain diffusion. These effective jump rates are plotted as the squares in Figure 16.

At a given temperature, the local jump rates taken from the literature are always higher than the effective jump rates derived from chain diffusion. This indicates that a local motion of a CH_2 group in PE crystallites does not always lead to a translation of the whole polymer chain. Since the PE samples studied in these works may have slightly different properties, such as different molecular weight, degree of chain branching, or crystallinity, which might influence the chain dynamics, the direct comparison of the absolute values for the jump rates should not be over-emphasised. However, the different temperature dependence for the local and effective jump rates, where the former exhibit an increasing apparent activation enthalpy with increasing temperature whereas the latter exhibit a simple Arrhenius behaviour, cannot be attributed to possible minor differences in the chemical properties of the samples. In fact, the increasing difference between the effective jump rate determined here and the local jump rates indicates that with increasing temperature the local jump motions observed through aniso-tropic NMR interactions and by mechanical relaxation become more and more ineffective for the translational motion of the chain.

This phenomenon can be explained by taking a closer look at the mechanism of the chain diffusion process. In the literature, defect-driven mechanisms have been proposed to facilitate the translation of an extended all-*trans* chain through a PE crystallite, although the nature of defect may still be under debate in the different models. In the defect-driven scenario, a defect is created, for example, at one side of the crystal, and the translation motion is accomplished when the defect has moved to the other side and finally left the crystallite. If, however, the defect only travels inside the crystallite, is reflected back, or annihilated, it will not lead to a translational motion of the stem but will still cause local reorientation of the CH_2 units. Thus, local jump rates may differ from the effective jump rate responsible for chain diffusion. Moreover, it should be noted that the defects have to leave the crystallites and thus must pass through the interphase between the crystallite and the non-crystalline region in order to generate an overall translation of the extended all-*trans* chain segments. Therefore, the conformations in the interphase and their dynamic modes become essential for the chain translation motion. The influence of the interphase on the chain diffusion is discussed in the next section.

Based on the defect model, the non-linear behaviour of local jump rates in the Arrhenius plot of Figure 16 can readily be understood. As the lattice of the PE crystallites expands with increasing temperature, the number of defects in the crystals increases. Apparently, however, the likelihood of defects that are not effective for chain diffusion will also increase. It should be noted that at lower temperatures, the temperature dependence of local and effective jump rates is very similar, indicating that the relative number of effective jumps is almost constant in this regime. Dynamic mechanical measurements of relaxation processes in PEs show a similar temperature dependence of the activation enthalpy.[34] The above findings on a molecular level thus provide a microscopic picture of the macroscopic behaviour of the material.

In the above discussion, the defect-driven mechanisms seem to well explain the temperature dependence of the apparent and effective jump rates. But concerning the nature of the defects in this scenario, at least two mechanisms for chain transport—local conformational defects[5] as opposed to propagating twists[7]—have been proposed and debated in the literature. A recent work from Spiess's group has partially resolved the contention.[11] In that work, a series of well-designed polyolefin samples were synthesised. These samples have a primary structure as PE, but with precisely spaced deuterated methyl branches. The chemical structures of these samples are illustrated in Figure 17, where the number after PE indicates the number of carbon atoms between two adjacent CD_3 branches along the backbone.

The introduction of deuterated methyl branch in the samples provides a possibility to study the mobility of the branches themselves by ^2H NMR. Figure 18 shows the ^2H spectra of the samples at different temperatures. As indicated by the trend of the line shape with temperature, all the three samples show a gradual build-up of molecular dynamics with increasing temperature. The most interesting phenomenon in these spectra, however, is a regular Pake pattern ($\eta = 0$) with half the static line width in the ^2H NMR spectrum observed only for PE15-CD_3 at $T = 303$ K, indicating that the axial rotation in this sample is sufficiently fast to average the ^2H line shape. Because this kind of regular Pake pattern is not observed in linear PE, the axial rotation is attributed to the motions of methyl groups embedded in the crystalline regions. Following this point, the highly asymmetric ^2H line shape observed in all cases is thus probably a result of ill-defined rotations around the local chain axis. Computer simulation shows that such a line shape can be formed by the rotation of the all-*trans* conformer with an amplitude of about 40°.

The dynamic information of chain backbone in these samples is provided by the ^{13}C CSA powder patterns obtained from the 2D CSA recoupling experiment (SUPER).[35] Taking the advantage of the separation of isotropic CS and CSA in the SUPER experiments, the information of the local reorientations of CH_2 groups of

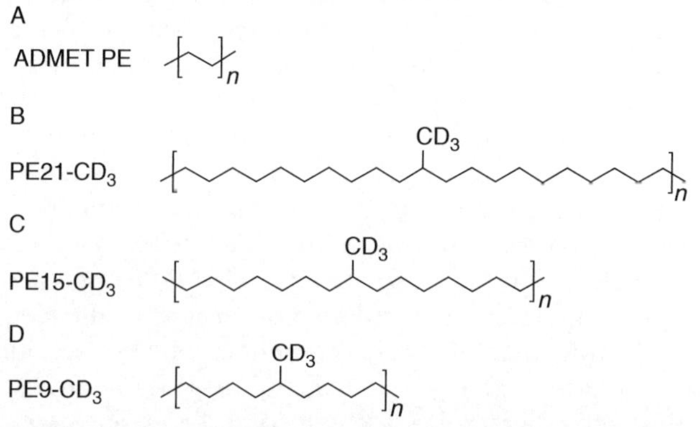

Figure 17 Precisely branched polymers: (A) ADMET PE, with no branching; (B–D) PEn-CD_3, with deuterated methyl groups on each and every nth carbon. From Ref. 11 (Scheme 1).

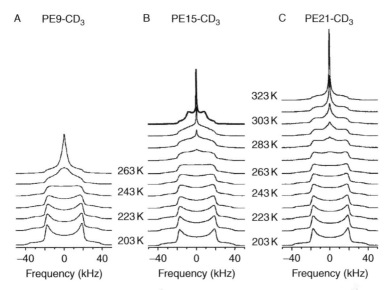

Figure 18   Temperature dependence of ^{2}H NMR spectra for (A) PE9-CD$_3$; (B) PE15-CD$_3$ and (C) PE21-CD$_3$. The special line shape of PE15-CD$_3$ at $T=303$ K (bold) indicates the presence of fast axial rotational dynamics. From Ref. 11 (Figure 1).

the polymer backbone can be site-selectively monitored. Figure 19 shows the isotropic and anisotropic ^{13}C chemical shifts of the samples. A remarkable finding is that the anisotropic ^{13}C chemical shifts of the backbone CH$_2$ groups in PE21-CD$_3$ vary with the distance to the neighbouring branching site. The NMR signal at 33.2 ppm, which is likely from the backbone CH$_2$ groups close to the methyl branch, is associated with the CSA powder pattern with the line shape of an almost axially symmetric CSA tensor, where the principal value of 13 ppm along the crystalline c-axis persists and the two other values are largely averaged, whereas the MAS NMR signal at 33.6 ppm shows a CSA tensor line shape similar to that of PE. This variation on the line shape of the ^{13}C powder pattern indicates the presence of the rotation of the all-*trans* conformer with different amplitudes. Different from PE21-CD$_3$, the CH$_2$ signal in the crystalline regions of PE15-CD$_3$ is observed at 32.9 ppm with a shoulder at about half the signal height at 33.5 ppm. Remarkably, all ^{13}C signals in the NMR spectrum of PE15-CD$_3$ show the powder line shape of well-defined axially symmetric CSA tensors: even at 33.5 ppm, the isotropic chemical shift is characteristic of undistorted all-*trans* units. Thus, in PE21-CD$_3$, the CH$_2$ groups in proximity to the lattice-perturbing methyl branches perform local motions as found in pinned defects, whereas the rotation in PE15-CD$_3$ at $T=303$ K involves all the CH$_2$ groups along a given polymer chain and thus shows collective behaviour as in a rotator phase, which was also deduced from X-ray data.[36]

    Combining the results for PE15-CD$_3$ and PE21-CD$_3$ obtained from ^{2}H and ^{13}C NMR, which probe the branch and the neighbouring chain defect separately, a simple model for the molecular dynamics of regular methyl branched polyethylene in the crystalline regions has been proposed in Figure 20. In the crystalline

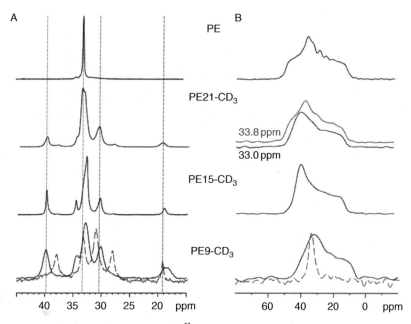

Figure 19 Isotropic (A) and anisotropic (B) ^{13}C NMR chemical shifts for precise polymers. From top to bottom: ADMET PE at 300 K, PE21-CD$_3$ at 300 K, PE15-CD$_3$ at 300 K, PE9-CD$_3$ at 235 K (—) and 245 K (--). From Ref. 11 (Figure 2).

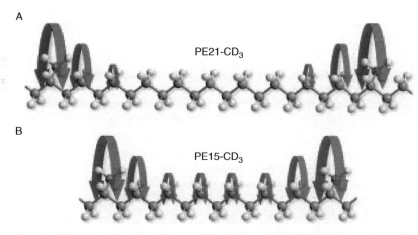

Figure 20 Models for rotational dynamics in (A) PE21-CD$_3$ and (B) PE15-CD$_3$. From Ref. 11 (Figure 3).

regions of PE15-CD$_3$ and PE21-CD$_3$, the lattice-perturbing methyl branches undergo axial oscillations around the polymer backbone. In both samples, this motion of the methyl branches increases in amplitude and rate with increasing temperature. For PE15-CD$_3$, all the sites along the chain show similar rotational dynamics, indicating that the rotation has become a collective behaviour. In the case of PE21-CD$_3$, however, this rotational dynamic mode is restricted to CH$_2$ groups in close proximity to the methyl branch in a twist-like defect, whereas the CH$_2$ groups remote from the methyl branches are not involved in this motion as demonstrated by the regular PE CSA pattern observed at 33.6 ppm, but may undergo 180° flips, as known for crystalline PE. By comparing the intensities of signals at 33.2 and 33.6 ppm, the size of the defect in PE21-CD$_3$ has been estimated, which is 3–4 CH$_2$ units on each side of the branch. The molecular dynamics in PE21-CD$_3$ thus clearly favours the twist propagation model, while the different molecular dynamics in PE15-CD$_3$ and PE21-CD$_3$ exhibits the localised rotational motion in PE21-CD$_3$ turning into collective dynamics owing to the higher density of defect sites in PE15-CD$_3$.

ACKNOWLEDGEMENTS

The authors gratefully acknowledge support from NSFC (Grant No. 20804016) and the National Fundamental Research Programme (Grant No. 2007CB925203).

REFERENCES

1. K. Schmidt-Rohr and H. W. Spiess, Chapter 7, Polymer Dynamics: Multidimensional Exchange Experiments; Chapter 8, Multidimensional Exchange NMR Above the Glass Transition. *Multidimensional Solid-State NMR and Polymers*. Academic Press, London, 1994.
2. A. Hagemeyer, K. Schmidt-Rohr and H. W. Spiess, Two-dimensional nuclear magnetic resonance experiments for studying molecular order and dynamics in static and in rotating solids. *Adv. Magn. Reson.*, 1989, **13**, 85.
3. S. Rastogi, A. B. Spoelstra, J. G. P. Goossens and P. J. Lemstra, Chain Mobility in Polymer Systems: on the Borderline between Solid and Melt. 1. Lamellar Doubling during Annealing of Polyethylene. *Macromolecules*, 1997, **30**(25), 7880–7889.
4. W. G. Hu and K. Schmidt-Rohr, Polymer ultradrawability: the crucial role of α-relaxation chain mobility in the crystallites. *Acta Polym.*, 1999, **50**(8), 271–285.
5. G. Strobl, Chapter 6, Mechanical and Dielectric Response, *The Physics of Polymers*. 2nd edn. Springer, Berlin, 1997.
6. C. R. Ashcraft and R. H. Boyd, A dielectric study of molecular relaxation in oxidized and chlorinated polyethylenes. *J. Polym. Sci. Part B Polym. Phys.*, 1976, **14**(12), 2153–2193.
7. M. Mansfield and R. H. Boyd, Molecular motions, the relaxation, and chain transport in polyethylene crystals. *J. Polym. Sci.: Polym. Phys. Ed.*, 1978, **16**(7), 1227–1252.
8. D. Schaefer, H. W. Spiess, U. W. Suter and W. W. Fleming, Two-dimensional solid-state NMR studies of ultraslow chain motion: glass transition in atactic poly(propylene) versus helical jumps in isotactic poly(propylene). *Macromolecules*, 1990, **23**(14), 3431–3439.
9. L. Shao and J. J. Titman, CAESURA Studies of Helical Jump Motions in Semi-Crystalline Polymers. *Macromol. Chem. Phys.*, 2007, **208**(19-20), 2055–2065.
10. J.-L. Syi and M. L. Mansfield, Soliton models of the crystalline relaxation. *Polymer*, 1988, **29**(6), 987–997.
11. Y. Wei, R. Graf, J. C. Sworen, C.-Y. Cheng, C. R. Bowers, K. B. Wagener and H. W. Spiess, Local and Collective Motions in Precise Polyolefins with Alkyl Branches: A Combination of ^{2}H and ^{13}C Solid-State NMR Spectroscopy. *Angew. Chem. Int. Ed.*, 2009, **48**(25), 4617–4620.
12. E. R. deAzevedo, W. G. Hu, T. J. Bonagamba and K. Schmidt-Rohr, Principles of centerband-only detection of exchange in solid-state nuclear magnetic resonance, and extension to four-time centerband-only detection of exchange. *J. Chem. Phys.*, 2000, **112**(20), 8988–9001.
13. G. C. Rutledge and U. W. Suter, Helix jump mechanisms in crystalline isotactic polypropylene. *Macromolecules*, 1992, **25**(5), 1546–1553.
14. H. W. Spiess, Molecular dynamics of solid polymers as revealed by deuteron NMR. *Colloid Polym. Sci.*, 1983, **261**(3), 193–209.

15. H. W. Spiess, Structure and dynamics of solid polymers from 2D- and 3D-NMR. *Chem. Rev.*, 1991, **91**(7), 1321–1338.
16. K. Zemke, K. Schmidt-Rohr and H. W. Spiess, Polymer conformational structure and dynamics at the glass transition studied by multidimensional ^{13}C NMR. *Acta Polym.*, 1994, **45**(3), 148–159.
17. M. J. Duer, Chapter 1, The Basics of Solid-State NMR, *Introduction to Solid-State NMR Spectroscopy*. Oxford, Blackwell, 2004.
18. Y. F. Yao, R. Graf, H. W. Spiess and S. Rastogi, Restricted Segmental Mobility Can Facilitate Medium-Range Chain Diffusion: A NMR Study of Morphological Influence on Chain Dynamics of Polyethylene. *Macromolecules*, 2008, **41**(7), 2514–2519.
19. K. Saalwachter and I. Schnell, REDOR-Based Heteronuclear Dipolar Correlation Experiments in Multi-Spin Systems: Rotor-Encoding, Directing, and Multiple Distance and Angle Determination. *Solid State Nucl. Magn. Reson.*, 2002, **22**(2–3), 154–187.
20. Y. Yao, R. Graf, H. W. Spiess and S. Rastogi, Influence of Crystal Thickness and Topological Constraints on Chain Diffusion in Linear Polyethylene. *Macromol. Rapid Commun.*, 2009, **30**(13), 1123–1127.
21. A. P. M. Kentgens, E. de Boer and W. S. Veeman, Ultraslow molecular motions in crystalline polyoxymethylene. A complete elucidation using two-dimensional solid state NMR. *J. Chem. Phys.*, 1987, **87**(12), 6859–6866.
22. W. G. Hu, C. Boeffel and K. Schmidt-Rohr, Chain Flips in Polyethylene Crystallites and Fibers Characterized by Dipolar ^{13}C NMR. *Macromolecules*, 1999, **32**(5), 1611–1619.
23. I. Schnell, A. Watts and H. W. Spiess, Double-Quantum Double-Quantum MAS Exchange NMR Spectroscopy: Dipolar-Coupled Spin Pairs as Probes for Slow Molecular Dynamics. *J. Magn. Reson.*, 2001, **149**(1), 90–102.
24. K. Schmidt-Rohr and H. W. Spiess, Chain diffusion between crystalline and amorphous regions in polyethylene detected by 2D exchange carbon-13 NMR. *Macromolecules*, 1991, **24**(19), 5288–5293.
25. D. A. Torchia, The measurement of proton-enhanced carbon-13 T_1 values by a method which suppresses artifacts. *J. Magn. Reson.*, 1978, **30**(3), 613–616.
26. M. Doi and S. F. Edwards, Chapter 4, Dynamics of flexible polymers in dilute solution; Chapter 6, dynamics of a polymer in a fixed network, *The Theory of Polymer Dynamics*. Oxford University Press, Oxford, 1986.
27. N. A. J. M. Van Aerle and A. W. M. Braam, A structural study on solid state drawing of solution-crystallized ultra-high molecular weight polyethylene. *J. Mater. Sci.*, 1988, **23**(12), 4429–4436.
28. A. Peterlin, Molecular model of drawing polyethylene and polypropylene. *J. Mater. Sci.*, 1971, **V6**(6), 490–508.
29. S. Rastogi and A. E. Terry, Morphological implications of the interphase bridging crystalline and amorphous regions in semi-crystalline polymers: Interphases and Mesophases in Polymer Crystallization I. *Adv. Polym. Sci.*, 2005, **180**, 161–194.
30. H. Luo, Q. Chen and G. Yang, Studies on the minimum crystallizable sequence length of semicrystalline copolymers. *Polymer*, 2001, **42**(19), 8285–8288.
31. W. X. Lin, Q. J. Zhang, G. Yang and Q. Chen, Biexponential ^{13}C Spin-Lattice Decay Behavior of the Crystalline Region of Ethylene Copolymers and Its Origin. *Mol. Struct.*, 2003, **655**(1), 37–45.
32. L. Zhang, Q. Chen and E. W. Hansen, Morphology and Phase Characteristics of High-Density Polyethylene Probed by NMR Spin Diffusion and Second Moment Analysis. *Macromol. Chem. Phys.*, 2005, **206**, 246–257.
33. Y. F. Yao, R. Graf, H. W. Spiess, D. R. Lippits and S. Rastogi, Morphological differences in semi crystalline polymers: Implications for local dynamics and chain diffusion. *Phys. Rev. E*, 2007, **76**(6), 060801–060804.
34. M. Matsuo, Y. Bin, C. Xu, L. Ma, T. Nakaoki and T. Suzuki, Relaxation mechanism in several kinds of polyethylene estimated by dynamic mechanical measurements, positron annihilation, X-ray and ^{13}C solid-state NMR. *Polymer*, 2003, **44**(15), 4325–4340.
35. S. F. Liu, J. D. Mao and K. Schmidt-Rohr, A Robust Technique for Two-Dimensional Separation of Undistorted Chemical-Shift Anisotropy Powder Patterns in Magic-Angle-Spinning NMR. *J. Magn. Reson.*, 2002, **155**(1), 15–28.
36. G. Lieser, G. Wegner, J. A. Smith and K. B. Wagener, Morphology and packing behavior of model ethylene/propylene copolymers with precise methyl branch placement. *Colloid Polym. Sci.*, 2004, **282**(8), 773–781.